Engineering of Power Systems Economics

Rui Castro

Engineering of Power Systems Economics

Springer

Rui Castro
Department of Electrical and Computer Engineering, Técnico Lisboa
University of Lisbon
Lisbon, Portugal

ISBN 978-3-031-55250-2 ISBN 978-3-031-55251-9 (eBook)
https://doi.org/10.1007/978-3-031-55251-9

This Springer imprint is published by the registered company Springer Nature Switzerland AG
The registered company address is: Gewerbestrasse 11, 6330 Cham, Switzerland

Paper in this product is recyclable.

To my wife Lina and my sons João Pedro and Guilherme.

Foreword by João Tomé Saraiva

My good friend Rui Castro invited me to write the Preface of his new book entitled "Engineering of Power Systems Economics". To start with, I have to congratulate Rui Castro for devoting an important part of his time to writing, which is, in my opinion, one of the most noble activities that humans can develop. In fact, writing has a double function in terms of externalizing and sharing knowledge, views, opinions or the emotions of the writers with others, whoever they are, and of fixing in a stable and non-volatile way this knowledge for the future.

In this case, we are talking about a book that addresses a timely matter—the operation of modern power systems in view of the introduction of competitive mechanisms in some of its activities together with a set of regulatory mechanisms and controls. The book is written in a very didactic way and, at the end of each chapter, it includes exercises that are proposed to the reader. This means that Rui Castro certainly had his students at Instituto Superior Técnico in his mind as a way of providing them with more material, quality material, for them to study and to improve their performances. For all these reasons, congratulations again for having written and published this book.

Once I received the mail from Rui Castro inviting me to write this Preface, I immediately accepted it. I have just checked my mailbox and I realize that I answered him less than one hour after having received his initial mail, in fact I think as soon as I read it. It is true that I always try to answer quickly but, in this case, I had no doubt in accepting this invitation for a few reasons that I will now briefly enumerate.

In the first place, I have known Rui Castro for many years. We met in some conferences in the 90s of the last century and then I remember I invited him for a MSc Jury at Faculdade de Engenharia da Universidade do Porto back in 1997. Since then, I have been admiring how rigorous Rui Castro is in evaluating academic works, always adopting a gentle but simultaneously assertive approach. Since then, we participated in numerous academic juries and other activities both in Porto and in Lisbon and this perspective was always reinforced.

Secondly, the topics addressed in the book are very close to the ones I have devoted the last 20 years of my academic path at Faculdade de Engenharia da Universidade do Porto, namely topics related with the restructuring and unbundling of traditionally vertically integrated power systems, the introduction of competition and market mechanisms in some of its activities, the development of regulation and the application of regulatory procedures in some others. In recent years and certainly till the end of the current decade, the sharp increase of PV and wind generation, the electrification of several activities of our societies (as mobility), the increasing maturity of several storage technologies, the development of local markets, of flexibility mechanisms and of energy communities present new challenges, and also certainly new opportunities, that are addressed by Rui Castro in the final chapter of his book. How could I refuse writing this Preface given these are the topics addressed in this book?

Finally, the title selected by Rui Castro for his book caught my immediate attention, "Engineering of Power Systems Economics". For many years economics were not so present in power systems, we certainly aimed at supplying the demand at the least possible cost, but we were eventually more concerned with the security of supply, with the reinforcement of generation and transmission systems and with the electrification of all households. Then, economy and finance started to emerge in power systems with the unbundling of traditional utilities, with the privatization programs implemented in many countries and with the increasing presence of financial mechanisms in the sector. Economists prevailed and eventually engineers had to follow economists in the sense that technical solutions had to be created or developed to implement economic "mythical" models. In fact, we all—engineers and economists-should accept that power systems do have a double and complementary nature. Certainly, we have to be concerned with economic issues, but we are facing a system that has some peculiar technical characteristics that simply don't allow us to do everything we want, and that sometimes prevent us from adopting the pure economic optimal solutions. As I once heard a colleague commenting on this issue, it is "unfortunate" that power systems have to comply with the Kirchoff Laws that, from time to time, remind us of their presence and force us to introduce some constraints in optimization problems. As we know, this can determine that the mentioned pure economic optimal solutions are technically unfeasible and thus cannot be implemented.

In this sense the timely title of the book is very well chosen. It reminds us all that in fact "some" Engineering does exist and is required in power systems even if economics eventually dominates the arena. This is another and a final reason to congratulate Rui Castro again for having written this book and to wish that this publication has an editorial success as large as possible.

October 2023

João Tomé Saraiva
Professor at the Faculty
of Engineering
University of Porto
Porto, Portugal

Foreword by Jorge Sousa

Energy systems are facing unprecedented challenges to achieve the carbon neutrality established by the Paris agreement from which we are still very far away as about 80% of worldwide final energy consumption is still coming from fossil fuels.

The energy crisis of 2022 has increased the urgency of the energy transition to a sustainable energy world fostering the energy consumption electrification, both directly and indirectly for the hard to abate sectors, and the decarbonization of the electricity generation. Harmonizing the energy trilemma is key in this decarbonization path to ensure that energy is secure and competitive for both domestic and industry consumers.

The competitiveness of the European economy very much depends on the way we will be able to reduce external energy dependency and decarbonize energy systems, while keeping energy affordable.

In this context, energy markets are critical to create the right framework for the needed investment in the sector and to adequately operate power systems. Therefore, it is of utmost relevance this textbook that Rui Castro brings to light as a very timely and useful contribution to understand electricity markets.

Rui is a talented and highly dedicated professor that has inspired generations of students in the topics of renewable energy and power systems economics. This textbook represents another remarkable milestone in the interdisciplinary topic of energy markets, merging the perspectives of both engineering and economics.

With this textbook Rui shed light into the complex world of electricity markets, guiding the reader through the economics principles behind producers (Chap. 1) and consumers (Chap. 2), which supports the market equilibrium outcome (Chap. 3), and market structures (Chap. 4). In addition, the regulation of natural monopolies is developed (Chap. 6) and the specific features of energy markets are broadly explained in terms of cost structures (Chap. 5), the functioning of electricity markets (Chap. 7) and its future perspectives (Chap. 8).

It is a comprehensive approach, yet very didactic, which delivers the right tools needed to address the challenges that electricity markets face today. The concern with significative learning is clear throughout the book by the systematic presentation of numerical examples and proposed exercises.

This textbook is an opportunity not only for engineering and economics undergraduate and post-graduate students but also for a wide range of professionals willing to dive into the world of energy economics. Its underlying concepts foster a deeper understanding of the functioning of electricity markets at present and the future challenges faced under the complex energy transition process we are in.

As it takes two to tango it also takes engineering and economics to energy markets. Rui has delivered a great example of this in this book and the energy academic and professional community will surely be very thankful for his contribution.

Lisbon, Portugal
November 2023

Jorge Sousa
President of the Portuguese
Association for Energy Economics
(APEEN/IAEE)
and Professor at the Polytechnics
of Lisbon (ISEL)

Preface

In today's rapidly evolving world, the power industry stands at the intersection of technological innovation, environmental sustainability, and economic stability. The Engineering of Power Systems Economics book is a comprehensive exploration of the intricate web of factors that shape the energy landscape. From the generation of electricity to its consumption, from the role of market structures to the complexities of regulation, this book offers a profound understanding of the economic and engineering dimensions underpinning power systems.

This book is mainly directed at university master's students attending economics of power systems courses. It approaches the economics of power systems from the perspective of engineering-related aspects. It may serve as a textbook for these courses as all the fundamentals of power systems economics are addressed in the book.

The book is also adequate for Ph.D. students who wish to deepen their knowledge of power systems economics from an engineering point of view. The book is helpful because some of the presented models are reported with high-level depth and detailed specifics of economic engineering aspects.

Finally, engineers of all specialities may find the book interesting because it answers problems found in their professional life, namely the worked examples and solved problems based on real-world engineering issues, with real data taken from existing installations. From a broader perspective, the book will certainly be useful to all those, and more and more of them are interested, by duty or self-education, in the topic of the economics of power systems.

The book differs from existing literature precisely because it establishes a link between economics and engineering. Traditional literature usually addresses the topic from an economic theory approach with weak links to engineering problems. This book fills the gap by building a bridge between economic theory and engineering applications. In the first part of the book, classic economic theory is exposed. Traditional subjects, such as the theory of producers and consumers, the behaviour of typical markets, and market general structures, are introduced always with a view to the power system application. The examples are often taken from the power system industry. In the second part of the book, the theory exposed in the first part is applied to power system-specific problems, such as the discussion of the different costs of power systems, regulation of power systems, and current and future electricity markets specifics.

In Chap. 1—Theory of Producers, we delve into the fundamental theories that govern the behaviour of power producers. We examine the strategies, decision-making processes, and economic incentives that drive these vital players in the power industry. Topics such as production function, returns to scale, formulation of the optimisation problem, production costs, and the supply curve are addressed. By understanding the motivations of producers, we lay the foundation for comprehending the broader economic dynamics at play.

Chapter 2—Theory of Consumers shifts the focus to the consumers of electricity. We explore how individuals, businesses, and industries make choices regarding their energy consumption, considering factors such as utility function, budget line, formulation of the optimisation problem, and demand functions. Understanding consumer behaviour is essential for designing efficient and sustainable energy systems.

In Chap. 3—The Single Market Behaviour, we examine the dynamics of a single, interconnected power market. We investigate how producers and consumers interact within this market, exploring concepts like market equilibrium, market adjustment, and dynamic equilibrium from the perspective of both Marshall and Walras.

Chapter 4—Market Structures introduces various market structures and their impact on power systems. We explore the differences among competitive markets, classical and regulated monopolies, and the market structures that lay between these two. By grasping the intricacies of market structures, readers gain insights into the economic forces shaping the industry.

Understanding the costs associated with power generation, transmission, and distribution is crucial for economic decision-making. In Chap. 5—Costs in Power Systems, we delve into cost analysis, examining fixed and variable costs, the levelized cost of energy, economies of scale, and the role of technology in cost reduction.

Chapter 6—Theory of Regulation delves into the theory and practice of regulation in the power sector, addressing the different models available in the literature. We explore the various regulatory frameworks that govern electricity markets, the objectives of regulation, and the challenges regulators face in balancing economic efficiency with social and environmental goals. We also detail the structure of selling prices of electricity.

Chapter 7—Energy Markets takes a deep dive into energy markets, including spot markets, forward markets, and balancing markets. We analyse market design, trading strategies, price mechanisms, and risk management techniques used by market participants to navigate the complexities of energy trading.

The final chapter, Chap. 8—Emerging Energy Markets, explores the exciting world of emerging energy markets. We investigate new technologies, trends, and innovations that are reshaping the energy landscape, such as Peer-to-Peer markets, local flexibility markets, and new renewable models to face the increasing penetration of renewables in energy markets. From renewable energy integration to smart grids and decentralised energy systems, this chapter offers a glimpse into the future of power systems markets.

The Engineering of Power Systems Economics book is designed to be a valuable resource for students, professionals, and policymakers seeking to navigate the intricate economic terrain of the power industry. We hope that the insights presented in this book will empower readers to make informed decisions that promote economic efficiency, environmental sustainability, and the reliable supply of electricity in a rapidly changing world.

Lisbon, Portugal
October 2023

Rui Castro

Acknowledgements

This book is the culmination of years of research, analysis, and collaboration within the field of the link between power systems engineering and economics. I recognise that my journey would have been significantly more challenging without the support of former professors, current colleagues, ex- and present students, institutions, and my family, who have fostered an environment of exploration and learning. I am going to thank some of the minds and institutions that somehow paved the way for my understanding. I apologise if I forgot to mention anyone that deserves acknowledgment.

In what regards the institutions, I wish to thank Técnico Lisboa, the science, technology, and engineering school of my home University of Lisbon, and INESC-ID, my research institution, for making available all the necessary material resources, including computers and dedicated software, indispensable for the writing of the book, and for all the facilities and support granted to the project. This project was funded by national funds through Fundação para a Ciência e a Tecnologia (FCT) with reference UIDB/50021/2020.

The first individual that I wish to thank is Retired Full Prof. João Santana, from the University of Lisbon. Some years ago, he trusted me to pursue his legacy in the course "Economics and Energy Markets" by inviting me to lecture with him. The groundings of my current knowledge on these subjects were taught by him, with his endless passion for economics and energy markets. Furthermore, he allowed me to fully use his book, written in Portuguese, to extend and develop mine. For all of this, and for his friendship, I owe him a big thank you.

A special thanks is due to my colleagues Jorge Sousa, a professor at the Polytechnics of Lisbon (ISEL), and Ricardo Francisco Pacheco, a senior advisor of energy markets at the Portuguese Energy Regulator (ERSE). Their expertise was valuable in revising some chapters of the book, namely Chap. 7—Energy Markets. I must say that I learned a lot from them and the book would have been worse without their useful contribution.

I am also in debt to my former students Sanha Almeida, from the Masters in Engineering and Management of Innovation and Entrepreneurship (MEGIE), and Tiago Felizardo, from the Masters in Electrical and Computer Engineering (MEEC), who provided the raw material to write Chap. 8—Emerging Energy Markets. Their help was unspeakably valuable and a source of inspiration.

During my research for literature on the book's topics, I came across some relevant papers that shaped my learning. On this matter, I would like to mention, Mikael Amelin, from the Royal Institute of Technology (KTH), Sweden, Tom Brown and Mirko Schäfer, from the Frankfurt Institute for Advanced Studies (FIAS), Germany, Athanasios Papakonstantinou, from the Technical University of Denmark (DTU), Denmark, and Pierre Pinson, from the Imperial College of London, UK.

I have to mention that, for the first time, I used ChatGPT to help me write the abstracts that come before each chapter. I can share that this is a powerful tool but I had to make some corrections and adapt the output to my own writing style. ChatGPT is not human, and, nevertheless, it struggles to replace humans!

I would like to express my deepest appreciation to my beloved family, my wife Lina, and our two wonderful sons, João Pedro and Guilherme. Your unwavering support, patience, and understanding have been the cornerstone of my journey in writing this book on "Engineering of Power Systems Economics." Throughout the countless hours of research, writing, and revisions, you stood by me with boundless love and encouragement. Your sacrifices and understanding during the moments when I needed to immerse myself in my work were truly remarkable.

Lisbon, Portugal
October 2023

Rui Castro

Contents

About the Author

Rui Castro received a Ph.D. degree from Técnico Lisboa (IST), University of Lisbon, and the Habilitation title from the same University. Currently, he is a Full Professor at the Power Systems Section, Electrical and Computer Engineering Department at Técnico Lisboa, and a researcher at INESC- ID. He lectures the IST Master's Courses on "Renewable Energy" and "Economics and Energy Markets" and the Ph.D. Course on "Renewable Energy Resources and Energy Conversion". He formerly published a book with Springer on "Electricity Production from Renewables". He has participated in several projects with the industry, namely with EDP (Portuguese utility), REN (Portuguese Transmission System Operator), and ERSE (Portuguese Energy Regulator). Rui published more than 100 papers in top international journals, covering topics on renewable energy, demand-side management, offshore wind farms, wind power forecasting, energy resource scheduling on smart grids, battery energy storage systems, electrical vehicles, hydrogen systems, energy transition, energy economics, etc. Rui was the recipient of the University of Lisbon Scientific Award in 2021 and is on the list of the World Ranking Top 2% Scientists published by the University of Stanford.

Personnel: https://sites.google.com/site/ruigameirocastro

Google Scholar: https://tinyurl.com/oIRnXSkAAAAJ

ORCID: https://orcid.org/0000-0002-3108-8880

SCOPUS: https://tinyurl.com/55937371000

Acronyms

AC	Average Cost
ACER	Agency for the Cooperation of Energy Regulators
ADMM	Alternating Direction Method of Multipliers (decomposition method)
AFC	Average Fixed Cost
aFRR	Frequency Restoration Reserve with automatic activation
ANM	Active Network Management
AP	Average Productivity
APP	Auxiliary Problem Principle (decomposition method)
AR	Average Revenue
ATC	Analytical Target Cascading (decomposition method)
ATC	Available Transfer Capacity
ATC	Average Total Cost
AVC	Average Variable Cost
BL	Budget Line
BRP	Balance Responsible Party
BSP	Balance Service Provider
CAPEX	Capital Expenditure
CCGT	Combined Cycle Gas Turbine
CCS	Carbon Capture and Storage
CES	Constant Elasticity of Substitution
CfD	Contract for Differences
CHP	Combined Heat and Power
CO2	Carbon dioxide
CRS	Constant Returns to Scale
CS	Consumer Surplus
CSP	Concentrating Solar Power
DER	Distributed Energy Resources
DRS	Decreasing Returns to Scale
DSO	Distribution System Operator
DWL	Dead Weight Loss
EHV	Extremely High Voltage
ENTSO-E	European Network Transmission System Operator for Electricity
ESS	Energy Storage Systems

ETS	Emissions Trading System
EV	Electrical Vehicles
FACTS	Flexible AC Transmission Systems
FC	Fixed Cost
FCR	Frequency Containment Reserve
FiT	Feed-in-Tariff
GENCO	Generating Company
HV	High Voltage
HVAC	Heat, Ventilation and Air Conditioning
IC	Indifference Curve
IRR	Internal Rate of Return
IRS	Increasing Returns to Scale
KKT	Karush-Kuhn-Tucker conditions (decomposition method)
LCOE	Levelized Cost of Energy
LEM	Local Energy Markets
LFM	Local Flexibility Markets
LFMO	Local Flexibility Markets Operator
LP	Linear Programming
LRT	Last Resort Trader
LV	Low Voltage
MC	Marginal Cost
MFC	Marginal Factor Cost
mFRR	Frequency Restoration Reserve with manual activation
MIBEL	Iberian Electricity Market
MILP	Mixed Integer Linear Programming
MINLP	Mixed Integer Non-Linear Programming
MIQP	Mixed Integer Quadratic Programming
MP	Marginal Productivity
MR	Marginal Revenue
MRP	Marginal Revenue Product
MRS	Marginal Rate of Substitution
MRTS	Marginal Rate of Technical Substitution
MU	Marginal Utility
MV	Medium Voltage
NOx	Nitrogen oxides
NPV	Net Present Value
NTC	Net Transfer Capability
O&M	Operation and Maintenance
OMC	Operation and Maintenance Costs
OMIE	Iberian Market Operator—Spain
OPEX	Operational Expenditure
OTC	Over the Counter
P2P	Peer-to-Peer (markets)
PAB	Pay-As-Bid
PAC	Pay-As-Clear

PLC	Public Limited Company
PMP	Proximal Message Passing (decomposition method)
PPA	Power Purchase Agreement
PPPA	Physical Power Purchase Agreement
PS	Producer Surplus
PtX	Power-to-X
PV	Photovoltaic
QP	Quadratic Programming
RAV	Regulatory Asset Value
RBM	Renewable Business Models
RES	Renewable Energy Sources
RPI	Retail Price Index
RR	Replacement Reserve
SME	Small and Medium Enterprises
SO2	Sulphur dioxide
SOE	State-Owned Enterprise
STATCOM	STATic synchronous COMpensator
TC	Total Cost
ToU	Time of Use (tariffs)
TR	Total Revenue
TSO	Transmission System Operator
TU	Total Utility
VC	Variable Cost
VPP	Virtual Power Plant
VPPA	Virtual Power Purchase Agreement
vRES	Variable Renewable Energy Sources

List of Figures

List of Tables

Theory of Producers 1

Abstract

In this chapter, the theory of producers is approached with a focus on their relevance to power systems. It encompasses key concepts that elucidate firms' behaviour and decision-making in the energy production process. Returns to scale are key in power systems economics as understanding the relationship between input adjustments and output changes enables power system engineers to optimise the capacity and size of power generation facilities. The production function, a mathematical representation of the relationship between inputs and outputs, allows engineers to assess the productivity of different combinations of resources in power generation. The Cobb-Douglas function is particularly relevant in power systems economics due to its flexibility in capturing the relationships between multiple inputs. The optimisation problem is at the core of power systems economics, as engineers aim to maximise power generation efficiency while minimising costs and meeting regulatory requirements. Engineers must carefully manage production costs, considering factors such as fuel expenses, maintenance and operation costs, and capital investments, to ensure the economic sustainability of power systems. The supply function denotes the relationship between prices and quantities supplied. Engineers utilise the supply function to determine the optimal level of electricity generation to meet demand at different price levels. By leveraging all these concepts, power system engineers can optimise resource allocation, improve power generation efficiency, manage costs effectively, and ensure a reliable supply of electricity to meet the demands of a dynamic and evolving energy market.

1.1 Introduction

The theory of producers, also known as the theory of the firm, forms a core component of economic analysis. It focuses on how firms make decisions regarding production, cost minimisation, and profit maximisation.

R. Castro, *Engineering of Power Systems Economics*,
https://doi.org/10.1007/978-3-031-55251-9_1

Returns to scale refer to the relationship between the inputs used in production and the resulting output. In the case of the power system, the output is the produced energy. It examines how an increase or decrease in inputs affects the scale of production and output levels. There are three possibilities: increasing returns to scale, constant returns to scale, and decreasing returns to scale.

The production function represents the relationship between the quantities of inputs, such as labour and capital, and the resulting output of goods or services, energy and related services, in the case of the power system. It showcases how inputs are transformed into outputs using technology and the firm's production methods. The production function can take various forms, including linear, quadratic, and Cobb-Douglas functions, among others. It enables engineers and economists to analyse the efficiency of production and the impact of changes in input quantities on output levels.

Power production firms face an optimisation problem, seeking to maximise their profits by selecting the optimal combination of inputs to produce a given level of output. This problem involves determining the input quantities that minimise costs while achieving the desired output level. In this chapter, mathematical optimisation methods are presented to deal with this problem and enable producers to come up with informed optimal decisions.

Production costs are the expenses incurred by firms in the process of producing goods or services. They consist of both explicit costs, such as wages, raw materials, and rent, and implicit costs, such as the opportunity cost of using the firm's resources. Production costs can be divided into fixed costs, which do not vary with the level of output, and variable costs, which change as output levels change. Understanding production costs is essential for firms to determine the profitability of their operations and make informed decisions regarding production levels.

The supply function represents the relationship between the price of a good and the quantity supplied by producers. It shows how producers respond to changes in market prices while aiming to maximise their profits. The supply function captures the behaviour of firms in the market and helps determine the equilibrium price and quantity.

These concepts provide a framework for firms to make rational decisions regarding production levels, input allocation, cost management, and pricing strategies.

Firms produce products. A product is an economic good since it has a cost and a value. As such, a product may be a tangible good, service, or information. Each product is characterised by its physical and intangible properties. To quantify the cost and value of a product, it is necessary to establish a unit that is used as an economic measure of the product. The currency, because it facilitates the functioning of the market, is the standard choice. The real numbers obtained in the exchange relations are designated as prices.

One may distinguish between the cost price, which is related to the producer's view, and the valuation price, i.e., the price expressing the valuation of the product by the consumer. It follows that economic agents (producers and consumers) may show or expect different prices for the same product.

In this chapter, we approach the groundings of the theory of the producers, one of the fundamental agents that operate in a market. We begin by introducing the basics of the *firm*, the producers' organisational unit. We then delve into the related important concepts of *returns to scale* and *production function*. In regards to the latter, the concepts of *average product* and *marginal product*, *diminishing marginal returns* and *marginal rate of technical substitution* are dealt with. The *Cobb-Douglas production function*, the production function typically used as an example in Economics textbooks, is developed next. An *optimisation problem* can be formulated by the producers as follows *what is the minimum cost at which a certain quantity can be produced?* The problem is solved using the *Lagrangian method*, where the concept of *Lagrange multiplier* is introduced. Special cases of *corner solution* and *kink solution* are also addressed. The *production costs* are next approached, where the *average cost* and the *marginal cost* are discussed. The chapter ends by teaching how the *supply curve* can be built. At the very end, some exercises are proposed, alongside the respective solution.

Before we begin, let us recall some maths that are useful in the following chapters.

$$\frac{d}{dx}\left[kx^j\right] = kjx^{j-1} \tag{1.1}$$

$$x^j x^i = x^{j+i} \tag{1.2}$$

$$\frac{1}{x^j} = x^{-j} \tag{1.3}$$

$$y = x^j \overset{\text{yields}}{\rightarrow} x = y^{\frac{1}{j}} \tag{1.4}$$

1.2 The Firm

When addressing the subject of the class of the producers in an economic market, perhaps the first question that comes to mind is: "What is a firm?" One straightforward answer would be: A firm is an economic agent that produces some goods (outputs) using other goods (inputs) with the aim of selling them and, therefore, making a profit. Company or business are commonly synonymous with the firm. One could list several types of firms, for instance (more information available in https://www.onlineschoolsreport.com):

- Entrepreneurs working for themselves: Self-employed individuals who own and run their own companies.
- Private company: Companies that are small to medium in size (SME – Small and Medium Enterprises) and are usually owned by a small group of individuals.

- Public limited company (PLC): Large companies that are on the stock market and where members of the public may buy individual shares.
- Co-operatives or social ventures: Companies whose objective is not maximising their profits, instead, they seek to achieve goals for social or economic benefit.
- State-owned enterprise (SOE): A company owned by the government for the purpose of engaging in commercial activities.

Companies undertake production plans, where actions to efficiently design and produce a given product or service are established. This involves, for instance, setting the optimal quantities of the inputs and outputs, considering the limitations imposed by the production system.

The production system is the system that allows for the transformation of the input products into output products. The production system uses common resources (or inputs, for instance, capital, labour, land, and technical knowledge) to perform this change. Production managers have referred to them as the "five M's": men, machines, methods, materials, and money (https://www.britannica.com/). Production equipment is the physical equipment that allows said transformation to effectively take place.

A production set represents the possible inputs and outputs of a production process. Inputs may comprehend, raw materials, semi-manufactured goods, buildings, labour, etc. The outputs are the resulting products of the production system.

Operating cost minimisation is the objective of the producer. This criterion is equivalent to profit maximisation. It should be mentioned that other objectives, such as gaining market share, may be established.

1.3 Returns to Scale and Production Function

Two basic concepts of the theory of producers are returns to scale and production function. With regards to the concept of returns to scale, let us consider a production system in which n inputs are used to produce one single output. $v_1, v_2, \ldots, v_n$ are the inputs quantities and q is the output quantity. We note that we are assuming a single output for simplification purposes, but, in general, more than one output product can be produced. Also, it should be highlighted that another name for the inputs is production factors.

A simple question can be asked. If the input quantities change, how does the output quantity change? The concept of returns to scale intends to provide an answer to this question. Returns to scale are the rate at which output increases in response to proportional increases in all the inputs.

We may devise three possibilities:

- Constant Returns to Scale (CRS): Doubling the inputs results in a precise doubling of outputs.

- Increasing Returns to Scale (IRS): Doubling the inputs results in more than doubling of outputs.
- Decreasing Returns to Scale (DRS): Doubling the inputs yields less than doubling of outputs.

For instance, if passing from the initial input quantities, $v_1, v_2, \ldots, v_n$, to $2v_1, 2v_2, \ldots, 2v_n$ implies that the output quantity changes from q to $2q$, we have CRS. However, if the same change in the input quantities results in an increase in the output quantity to $3q$, it is said IRS are present. Obviously, if the increase in the output quantity is to $1.5q$, we have DRS. An example of DRS is when a firm expands beyond its ability to be effectively managed.

It should be highlighted that returns to scale should not be confused with economies of scale. Economies of scale are the cost benefits resulting from increasing the quantity produced (scale of operation). When average costs start falling as output increases, then economies of scale occur. Economies of scale show the effect of an increased output level on unit costs (average costs), whereas returns to scale focus only on the relation between input and output quantities.

Let us now look at the production function concept. The production function is a function that describes the output quantity that may be obtained from given input quantities. Using the same nomenclature that we have been using, we can write that the production function is:

$$q = f(v) \tag{1.5}$$

where $v = [v_1\ v_2 \ldots v_n]$ is the input vector. We note that often we write simply input (production factor) and output when we really mean input quantity and output quantity. The reader is advised that this is a language simplification. Let us look at an example of a production function (Table 1.1). We note that this is an example using a discrete function. Continuous functions can be and are used, as we will see further in this section.

In this table, we have a production function whose production factors are Labour (L) and Capital (K) (quantities). For each pair (L, K), the table shows the output (quantity) that can be produced with the said combination of inputs (quantities). The combinations of inputs that lead to the same output level (quantity) are marked. For instance, the output of 75 units can be obtained using the following combinations of inputs (quantities) $(L, K): (1, 5), (2, 3), (3, 2)$ *and* $(5, 1)$.

Table 1.1 Example of a production function

Capital ↓ Labour →	1	2	3	4	5
1	20	40	55	65	**75**
2	40	60	**75**	85	90
3	55	**75**	90	100	105
4	65	85	100	110	115
5	**75**	90	105	115	120

Taking the same example, we can represent the combinations of labour and capital which generate 75 units of output in the plan (L, K), as in Fig. 1.1.

This leads us to the definition of isoquant. An isoquant is a curve that contains all combinations of the inputs that generate the same level of output. Isoquants are represented in an isoquant map as in Fig. 1.2.

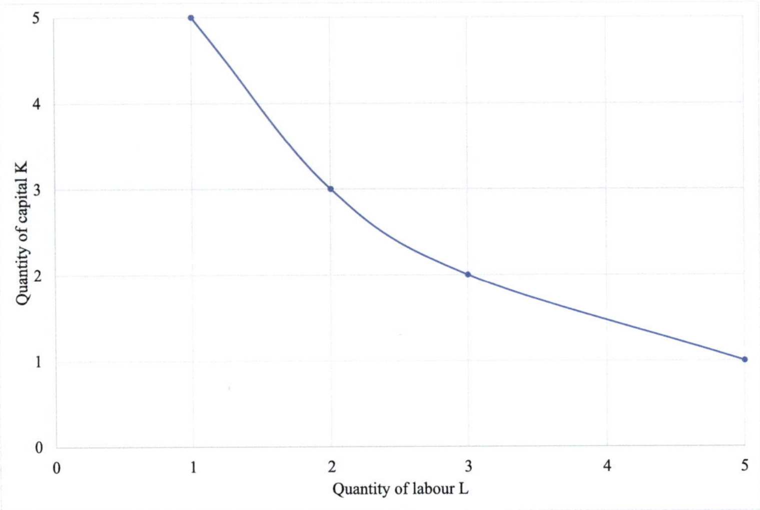

Fig. 1.1 Combinations of labour and capital which generate 75 units of output

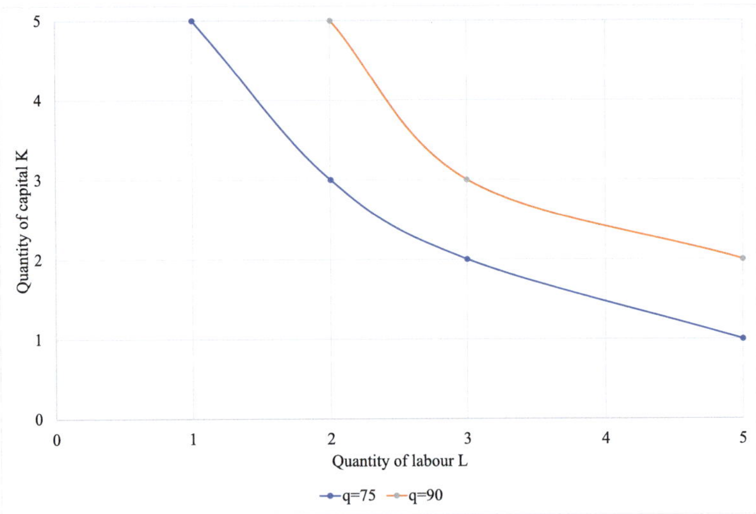

Fig. 1.2 Isoquants describing the combinations of capital and labour which generate output levels of 75 and 90

Isoquants show the firm's possibilities for substituting inputs (quantities) without changing the level of output. For instance, in the example we are following, the same level of output ($q = 75$) is obtained with 3 units of labour and 2 units of capital or 2 units of labour and 3 units of capital. These possibilities allow the producer to react to changes in the prices of inputs.

The relative position of the isoquants in the isoquant map is determined by the type of returns to scale the production function holds. If the production function holds CRS, the isoquants are equidistant as can be seen in Fig. 1.3.

As can be seen in Fig. 1.3, if we double the labour from $L_1 = 2$ to $L_2 = 4$, and the capital from $K_1 = 50$ to $K_2 = 100$, the output doubles from $q_1 = 10$ to $q_2 = 20$.

What if the production function is IRS? In this case, the isoquants get closer when the output increases, as depicted in Fig. 1.4. When the inputs double (from $L_1 = 2$ to $L_2 = 4$ and from $K_1 = 5$ to $K_2 = 10$), the output passes from $q_1 = 10$ to $q_2 = 40$, i.e., more than doubles.

It should be now apparent that when the production function is DRS, the isoquants get further apart when the output increases (see Fig. 1.5).

As can be seen in Fig. 1.5, when the two inputs are multiplied by 2, the output gets only multiplied by 1.415.

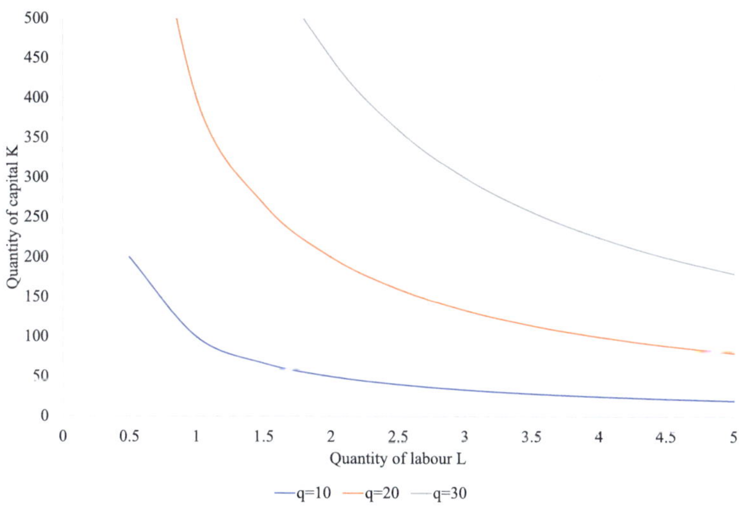

Fig. 1.3 Isoquant map for a production function with Constant Returns to Scale (CRS); equidistant isoquants

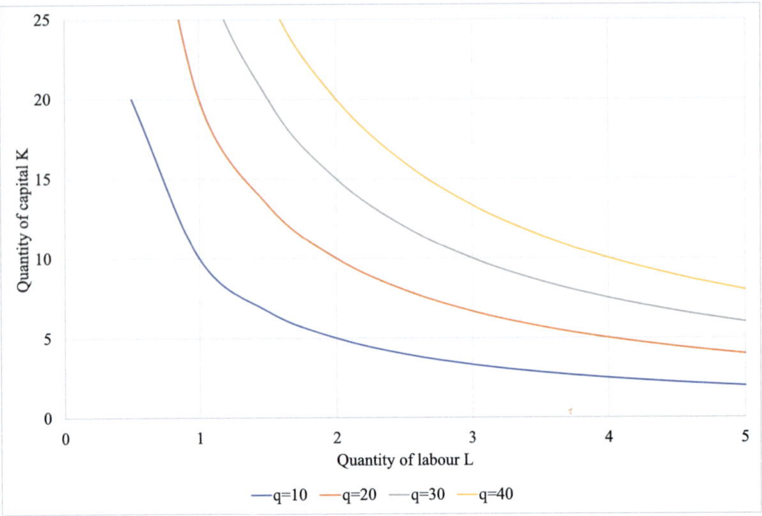

Fig. 1.4 Isoquant map for a production function with Increasing Returns to Scale (IRS); isoquants get closer when output increases

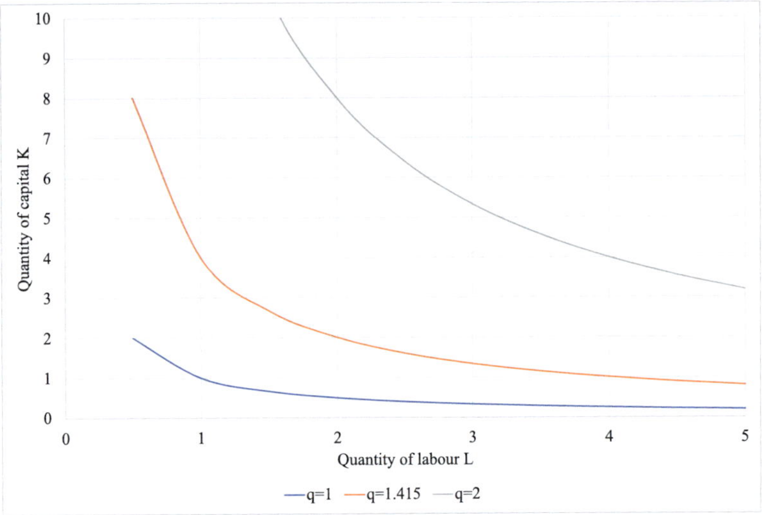

Fig. 1.5 Isoquant map for a production function with Decreasing Returns (DRS); isoquants get further apart when output increases

Example 1.1

Find the type of returns to scale featured in the following production functions:

- $q = f(L, K) = L^{\frac{1}{5}} K^{\frac{4}{5}}$
- $q = f(L, K) = LK$
- $q = f(K) = 4K^{\frac{1}{2}}$
- $q = f(L, K) = L + K$

Draw each production function in (L,K) plan for $q_1 = 10$, $q_2 = 20$, $q_3 = 30$ and verify that the distance between curves complies with what we have learned so far regarding production function and returns to scale. ◀

Solution

A good strategy to find the type of returns to scale is to multiply the production factors by a scalar greater than 1 and look at what happens to the output. Let $\rho > 1$.

$$f(\rho L, \rho K) = (\rho L)^{\frac{1}{5}} (\rho K)^{\frac{4}{5}} = \rho L^{\frac{1}{5}} K^{\frac{4}{5}} = \rho f(L, K) = \rho q$$
$$f(\rho L, \rho K) = \rho L \rho K = \rho^2 LK = \rho^2 f(L, K) = \rho^2 q > \rho q$$
$$f(\rho K) = 4(\rho K)^{\frac{1}{2}} = \rho^{\frac{1}{2}} 4K^{\frac{1}{2}} = \rho^{\frac{1}{2}} f(K) = \rho^{\frac{1}{2}} q < \rho q$$
$$f(\rho L, \rho K) = \rho L + \rho K = \rho (L + K) = \rho f(L, K) = \rho q$$

We can now easily conclude that the returns to scale are CRS, IRS, DRS, and CRS respectively.

The plotting of the four production functions is in Figs. 1.6, 1.7, 1.8 and 1.9, respectively. Looking, for instance, at Fig. 1.7, we can see that for $(L, K) = (1, 10)$ the output is 10. If we double both inputs, to $(L, K) = (2, 20)$, we will go to an isoquant corresponding to an output quantity higher than 30. This means the production function is IRS. Looking now at Fig. 1.8 for $(L, K) = (1, 6.25)$, the output is 10. Doubling the two inputs, the output will be lower than 20, therefore, meaning the production function is DRS. ◀

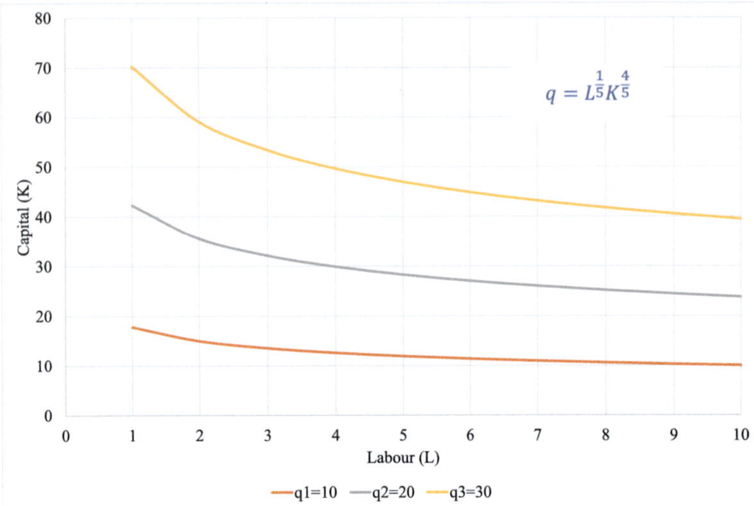

Fig. 1.6 Isoquant map for the production function $q = L^{\frac{1}{5}}K^{\frac{4}{5}}$ with Constant Returns to Scale (CRS)

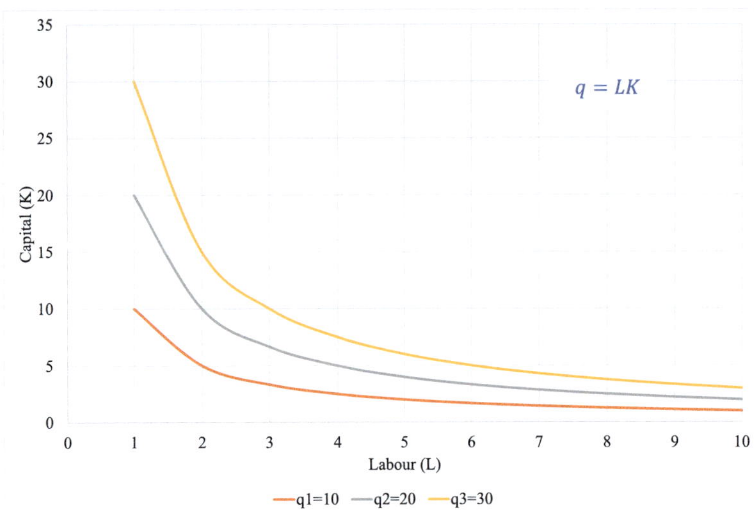

Fig. 1.7 Isoquant map for the production function $q = LK$ with Increasing Returns to Scale (IRS)

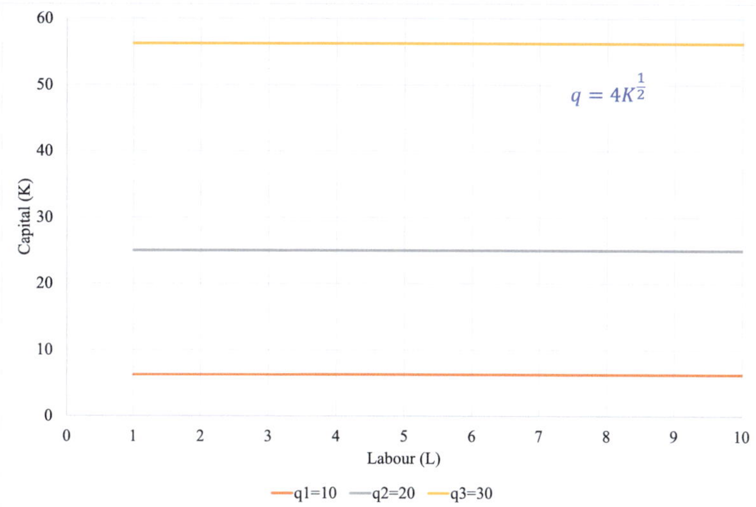

Fig. 1.8 Isoquant map for the production function $q = 4K^{\frac{1}{2}}$ with Decreasing Returns to Scale (DRS)

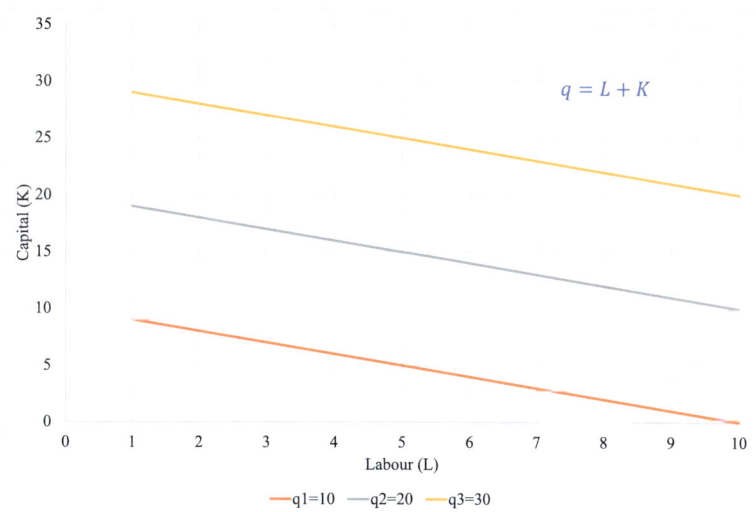

Fig. 1.9 Isoquant map for the production function $q = L + K$ with Constant Returns to Scale (CRS)

1.4 Features of the Production Function

1.4.1 Average Productivity and Marginal Productivity

With the production function in the background, there are some concepts that are usually defined. Two fundamental concepts are average productivity (or average product) and marginal productivity (or marginal product).

Average Productivity (AP)

$$AP = \frac{q}{v_k} \tag{1.6}$$

Marginal Productivity (MP)

$$MP = \frac{\partial q}{\partial v_k} \approx \frac{\Delta q}{\Delta v_k} \tag{1.7}$$

q is the output quantity and v_k is the quantity of input k. Looking at the two equations the definition of average productivity and marginal productivity is apparent. The former is the produced output per unit of the considered input, the latter is the additional output obtained by increasing the considered input in one unity. It follows that there is a AP and a MP for each input. In the MP computation, we are assuming that for small increments, the differential equation may be approximated by the differences equation. In this context, the output quantity is also known as the total product.

For example, over one week, Charles produces 100 pairs of jeans. When he works with a partner, Max, the output of the pair increases to 400 pairs of jeans. It follows that Charles's average productivity is 100 (100 units of total product divided by one unit of labour), and the marginal productivity of adding a second worker is 300 ($\Delta q = 400 - 100$; $\Delta L = 2 - 1$).

Table 1.2 shows an example of the practical computation of both average and marginal products, for a production function with two production factors: labour and capital (it is assumed the capital is constant, i.e., it does not affect the output).

1.4.2 Law of Diminishing Marginal Returns

There is an economic principle that states that while increasing one input and keeping other inputs at the same level may initially increase output, further increases in that input will have a limited effect, and eventually no effect or a negative effect, on output. This is known as the *Law of Diminishing Marginal Productivity* or *Law of Diminishing Marginal Returns*. In other words, this means that, as we increase the quantity of one input, leaving the other unchanged, the marginal product initially increases, reaches a maximum, and eventually decreases, possibly attaining negative values. This law can be seen graphically in Fig. 1.10.

An example of the law of diminishing marginal returns involves land and agricultural labour. When there is plenty of land and few people to work on it, the addition of one more labourer significantly increases the harvest. Adding more land is relatively unproductive because there are no people to work on it. As we move along the

Table 1.2 Example of a production function and computation of both the average product and the marginal product of labour

Labour	Capital	Output	AP	MP
0	10	0	0	
1	10	10	10	10
2	10	30	15	20
3	10	60	20	30
4	10	80	20	20
5	10	95	19	15
6	10	108	18	13
7	10	112	16	4
8	10	112	14	0
9	10	108	12	−4
10	10	100	10	−8

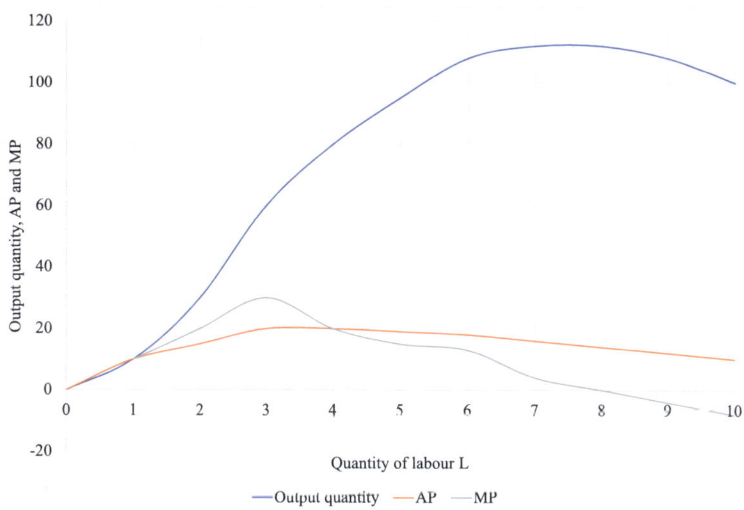

Fig. 1.10 Total product, average product, and marginal productivity. Law of diminishing marginal returns

isoquant, i.e., adding more workers, the land becomes scarce and there are plenty of people, so adding one more labourer is relatively unproductive, whereas the existence of more land is productive. The marginal product of a given factor decreases as the ratio between the considered factor and the other factors increases. This is the law of diminishing marginal returns.

Example 1.2

Consider the production function given by $q = f(L, K) = L^2K - L^3$ Suppose the capital is fixed in $K = 18$.
Compute the total product, average product, and marginal product of labour. Plot the quantities in the (L, K) plan. Show that the law of marginal diminishing returns is verified. ◀

Solution

$$q = 18L^2 - L^3$$

$$AP = \frac{q}{L} = 18L - L^2$$

$$MP = \frac{\partial q}{\partial L} = 36L - 3L^2$$

The requested plotting is in Fig. 1.11. The maximum total output occurs for $L = 12$, the maximum average productivity occurs for $L = 9$, and the maximum marginal productivity occurs for $L = 6$. Of course, this can be analytically confirmed. For example:

$$\max q \overset{\text{yields}}{\rightarrow} \frac{\partial q}{\partial L} = 0 \overset{\text{yields}}{\rightarrow} L = 12$$

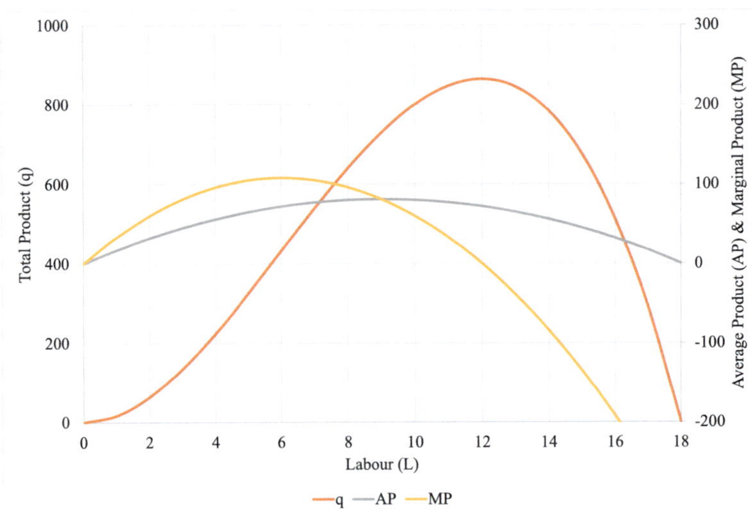

Fig. 1.11 Total product, average product, and marginal product for Example 1.2

Note that there is an inflection point of the total product at $L = 6$. This is why the marginal product starts to decrease at this point.
The marginal productivity starts by increasing, reaches a maximum, and then decreases, eventually becoming negative. This complies with the law of diminishing marginal returns. ◀

1.4.3 Marginal Rate of Technical Substitution

Another important concept associated with the production function is the *Marginal Rate of Technical Substitution* (MRTS). Let us suppose we have a production function $q = f(v)$ and we are interested in observing the influence of two production factors, v_k and v_i, while keep the output constant ($dq = df = 0$). Changes in output (in the long run) is measured as the sum of changes in input v_k (via the marginal productivity) and/or changes in input v_i (via the marginal productivity). Therefore, it is possible to write:

$$df = \frac{\partial f}{\partial v_k} dv_k + \frac{\partial f}{\partial v_i} dv_i = 0 \tag{1.8}$$

This leads to the definition of MRTS as:

$$MRTS_{k,i} = \frac{\frac{\partial f}{\partial v_k}}{\frac{\partial f}{\partial v_i}} = -\frac{dv_i}{dv_k} \approx -\frac{\Delta v_i}{\Delta v_k} \tag{1.9}$$

MRTS is the rate at which we may substitute input k by input i so that the output remains constant. It is computed by dividing the marginal productivities of the two production factors. The slope of the isoquant establishes the ratio at which we may substitute the input k by input i, while keeping the same output level.

Example 1.3

Suppose a firm with an isoquant for a certain output quantity as depicted in Fig. 1.12.
Compute the MRTS of replacing labour by capital when the firm moves up the isoquant from point B to point A. ◀

Solution

$$MRTS_{(L,K)} = -\frac{\Delta K}{\Delta L} = -\frac{(5-3)}{(1-2)} = 2$$

We conclude that two units of capital are needed to replace one unit of labour if one wishes to maintain the level of output. The dual equivalent problem is that when labour is increased in one unit, capital may decrease by two units without changing the output. We say the MRTS is equal to 2.

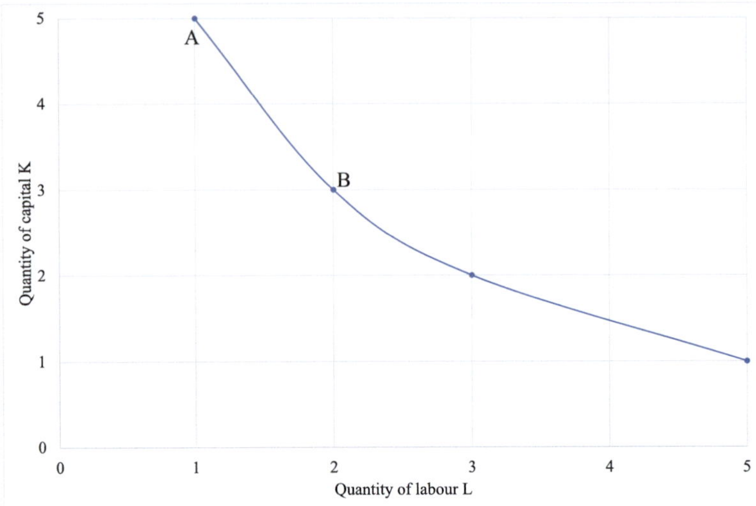

Fig. 1.12 Isoquant for Example 1.3

We note that the MRTS changes along the isoquant, according to its slope. For instance, if operating at point B, the firm decides to hire another employee, the capital can be reduced in one unit without compromising the output level. ◀

1.5 The Cobb-Douglas Production Function

The most used example in Economics textbooks of a production function is the Cobb-Douglas function given by:

$$f(L, K) = AK^{\alpha}L^{\beta} \tag{1.10}$$

where L is Labour, K is Capital and A, α, β are constant parameters. α and β are the output elasticities of capital and labour, respectively. Output elasticity measures the responsiveness of output to a change in levels of either labour or capital used in production. For example, if $\alpha = 0.45$, a 1% increase in capital usage would lead to approximately a 0.45% increase in output, assuming that the labour is constant.

1.5.1 Returns to Scale

It is worth verifying some interesting properties of the Cobb-Douglas production function. We will begin by looking at the type of returns to scale the Cobb-Douglas function show. Let us check, using the method seen before in this chapter.

$$f(\rho L, \rho K) = A(\rho K)^{\alpha}(\rho L)^{\beta} = \rho^{\alpha+\beta} A K^{\alpha} L^{\beta} = \rho^{\alpha+\beta} f(L, K) \tag{1.11}$$

From here, we can conclude that (remember that $\rho > 1$):

- $\alpha + \beta = 1$ leads to Constant Returns to Scale (CRS).
- $\alpha + \beta > 1$ leads to Increasing Returns to Scale (IRS).
- $\alpha + \beta < 1$ leads to Decreasing Returns to Scale (DRS).

1.5.2 Marginal Product

The two marginal productivities are:

$$MP_K = \frac{\partial f}{\partial K} = \alpha A K^{\alpha-1} L^{\beta} \tag{1.12}$$

$$MP_L = \frac{\partial f}{\partial L} = \beta A K^{\alpha} L^{\beta-1} \tag{1.13}$$

1.5.3 Marginal Rate of Technical Substitution

The marginal rate of technical substitution of capital by labour is:

$$MRTS_{K,L} = \frac{\frac{\partial f}{\partial K}}{\frac{\partial f}{\partial L}} = \frac{\alpha A K^{\alpha-1} L^{\beta}}{\beta A K^{\alpha} L^{\beta-1}} = \frac{\alpha L}{\beta K} \tag{1.14}$$

For a Cobb-Douglas production function with CRS, it is easy to verify that:

$$\lim_{K\to\infty} MP_K = \lim_{K\to\infty} \alpha A K^{\alpha-1} L^{\beta} = 0, \qquad \alpha < 1 \tag{1.15}$$

$$\lim_{L\to\infty} MP_L = \lim_{L\to\infty} \beta A K^{\alpha} L^{\beta-1} = 0, \qquad \beta < 1 \tag{1.16}$$

We recall that in the law of diminishing marginal returns, the marginal productivity initially may increase when more of an input (say labour) is employed, keeping the other input (say capital) constant. At some point, as more and more variable input (labour) is employed, marginal productivity starts to fall and eventually tends to zero.

1.6 The Optimisation Problem

Bearing in mind the knowledge we have learned so far, a relevant question may be posed. What is the minimum cost at which a certain quantity can be produced? To answer this, an optimisation problem must be formulated and solved. The solution to the optimisation problem will output the optimal input quantities that should be used to obtain the minimum production cost for a given quantity.

1.6.1 The Lagrangian Method

Optimisation problems and their solution is a well-known topic widely discussed in the literature. There are several methods that can be used to solve optimisation problems. In this textbook, we will use a calculus-based method, the Lagrangian method. Let us begin by presenting the general form of the Lagrangian method.

The objective of the optimisation is to maximise or minimise a mathematical function called the objective function. The solution must be found considering some auxiliary condition that needs to be satisfied, called the restriction function. So, we can mathematically formulate the problem as follows ($i = 1, 2, \ldots, n$):

$$\min\left[f\left(x_i\right)\right] \tag{1.17}$$

subject to s.t.:

$$g\left(x_i\right) = 0 \tag{1.18}$$

where $f(x_i)$ is the objective function and $g(x_i)$ is the restriction function. This is called a constrained optimisation problem.

To solve this optimisation problem, we first begin by building the Lagrangian function, Lag, as follows:

$$Lag\left(x_i, \lambda\right) = f\left(x_i\right) - \lambda g(x_i) \tag{1.19}$$

λ is a parameter, called the *Lagrange Multiplier*, whose physical meaning will be given later. It is possible to demonstrate that at the optimal point the partial derivatives of the Lagrangian function are null, i.e., at the optimal point:

$$\frac{\partial Lag}{\partial x_i} = \frac{\partial Lag}{\partial \lambda} = 0 \tag{1.20}$$

We have $n + 1$ equations and $n + 1$ unknowns: x_i, $i = 1, \ldots, n$ and λ. Therefore, solving the system of $n + 1$ equations, we can find the optimal variables (x_i^* and λ) that minimise the objective function subject to the restriction function. We note that the exact same formulation and solution method can be applied to a maximisation problem.

We will now apply the Lagrangian method to solve our cost minimisation problem. The objective is to determine the right (optimal) combination of inputs to minimise the production cost for a certain output quantity (restriction). To simplify, but without losing generality, we assume only two inputs. In mathematical terms, the problem is formulated as follows:

$$\min\left[\omega_1 v_1 + \omega_2 v_2\right] \tag{1.21}$$

s.t.

$$f\left(v_1, v_2\right) - q = 0 \tag{1.22}$$

where:

- ω_1, ω_2 are the unit costs of input 1 and 2, respectively.
- v_1, v_2 are the quantities of input 1 and 2, respectively.
- $\omega_1 v_1 + \omega_2 v_2$ is the cost function we want to minimise, i.e., the objective function.
- $f(v_1, v_2)$ is the production function.
- q is the desired output quantity that must be specified.
- $f\left(v_1, v_2\right) - q = 0$ is the restriction function.

To solve this problem, we build the Lagrangian function:

$$Lag\left(v_1, v_2, \lambda\right) = \omega_1 v_1 + \omega_2 v_2 - \lambda\left[f\left(v_1, v_2\right) - q\right] \tag{1.23}$$

Now, we apply the optimality conditions:

$$\frac{\partial Lag}{\partial v_1} = \omega_1 - \lambda \frac{\partial f}{\partial v_1} = 0 \tag{1.24}$$

$$\frac{\partial Lag}{\partial v_2} = \omega_2 - \lambda \frac{\partial f}{\partial v_2} = 0 \tag{1.25}$$

$$\frac{\partial Lag}{\partial \lambda} = q - f(v_1, v_2) = 0 \tag{1.26}$$

The unknowns are (v_1, v_2, λ), which can be obtained by solving the system of equations given by the previous three equations. The optimal quantities of the inputs (the ones that lead to the minimum cost) are denoted by $(v_1^*,\ v_2^*)$.

The result that we have just discussed has a graphical interpretation, which is depicted in Fig. 1.13.

In the input's plan (v_1, v_2), we plot the production function's isoquant for the specified quantity, $f\left(v_1, v_2\right) = q$. Different cost functions, $TC = \omega_1 v_1 + \omega_2 v_2 \overset{\text{yields}}{\rightarrow} v_2 = \frac{TC}{\omega_2} - \frac{\omega_1}{\omega_2} v_1$, (TC is the Total Cost) may be plotted in the same plane, for varying specified cost values. We note that, for fixed unit costs (ω_1, ω_2), the cost functions are straight lines with the same slope, $-\frac{\omega_1}{\omega_2}$. Three situations may happen:

- The cost function does not intercept the isoquant, meaning that it is impossible to produce the specified quantity at the given cost value.
- The cost function intercepts the isoquant in more than one point, meaning that it is possible to produce the specified quantity at a lower cost.
- The cost function is tangent to the isoquant (tangency point P in Fig. 1.13), meaning that the given cost value is the minimum cost at which the specified quantity can be produced.

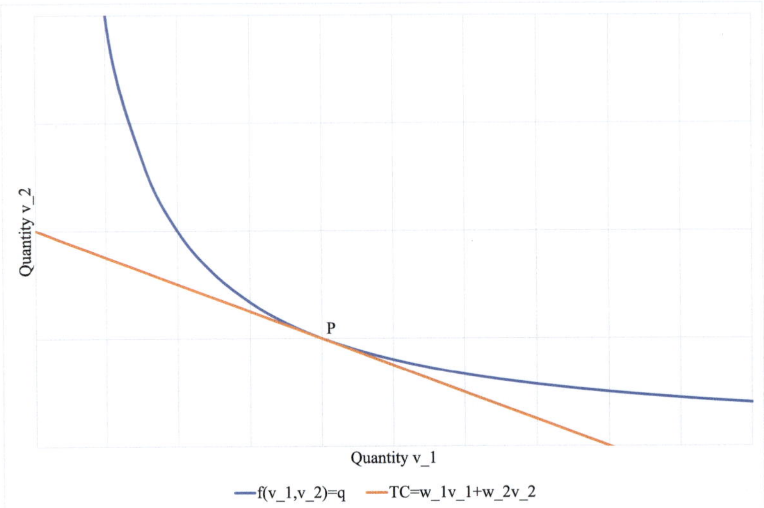

Fig. 1.13 Graphical interpretation of cost minimisation problem

The cost lines are called isocosts. An isocost shows the combination of inputs quantities that deliver the same total cost.

1.6.2 Corollary of the Lagrangian Method

Recalling the concept of marginal product, which we have seen before, from Eqs. 1.24 and 1.25, one can write:

$$\frac{MP_1}{MP_2} = \frac{\omega_1}{\omega_2} \tag{1.27}$$

meaning that, at the optimal point, the slope of the isoquant must be equal to the slope of the isocost.

We can rewrite Eq. 1.27 as:

$$\frac{MP_1}{\omega_1} = \frac{MP_2}{\omega_2} \tag{1.28}$$

This is equivalent to saying that, at the optimal point, the marginal product of input v_1 per monetary unit spent on it must be equal to the marginal product of input v_2 per monetary unit spent on input v_2. In general, the optimal point is called an interior optimum, i.e., the optimum occurs for positive quantities of the inputs, as it is needed, and at the tangency point of the isoquant and the isocost line.

1.6.3 The Lagrange Multiplier

It is now time to take a deeper look at the Lagrange multiplier. Based on Eq. 1.23, we can write:

$$\lambda = \frac{\partial Lag}{\partial q} \tag{1.29}$$

We note that, at the optimal point, the Lagrangian is the cost function because:

$$Lag\left(v_1^*, v_2^*, \lambda\right) = \omega_1 v_1^* + \omega_2 v_2^* - \lambda\left[f\left(v_1^*, v_2^*\right) - q\right] = \omega_1 v_1^* + \omega_2 v_2^* - 0 \tag{1.30}$$

Therefore, it is possible to conclude that, generally speaking, the Lagrange multiplier, λ, measures how much the optimal value of the objective function would change if the constraint were relaxed (changed) marginally, say by one infinitesimal unit.

In the specific case of the cost minimisation problem, λ is the cost of producing one more unit of output. This is called the marginal cost. We shall return to this later in this chapter.

1.6.4 The Case of the Cobb-Douglas Production Function

Looking at Eqs. 1.24 and 1.25, one can write:

$$\frac{\frac{\partial f}{\partial v_2}}{\frac{\partial f}{\partial v_1}} = \frac{\omega_2}{\omega_1} = MRTS \tag{1.31}$$

The conclusion is that the marginal rate of technical substitution between any two production factors equals the ratio of its unitary costs.

We can now use what we have learned to derive the general equations for the optimal inputs' quantities and the Lagrange multiplier that led to the minimum production cost to produce a given output quantity, q, of a firm showing a Cobb-Douglas production function with given parameters.

The Cobb-Douglas production function is $q = f(L, K) = AK^{\alpha}L^{\beta}$, where K and L are capital and labour (inputs) quantities, respectively, and q is output quantity.

Considering ω_L and ω_K as the given prices per unit of labour and capital, respectively, the optimisation problem that we face is:

$$\min\left[\omega_L L + \omega_K K\right] \tag{1.32}$$

s.t.

$$AK^{\alpha}L^{\beta} - q = 0 \tag{1.33}$$

To solve this optimisation problem, the Lagrange method implies that:

$$Lag = \omega_L L + \omega_K K - \lambda\left(AK^{\alpha}L^{\beta} - q\right) \tag{1.34}$$

and the optimal conditions are:

$$\frac{\partial Lag}{\partial L} = \omega_L - \lambda A\beta K^{\alpha} L^{\beta-1} = 0 \tag{1.35}$$

$$\frac{\partial Lag}{\partial K} = \omega_K - \lambda A\alpha K^{\alpha-1} L^{\beta} = 0 \tag{1.36}$$

$$\frac{\partial Lag}{\partial \lambda} = q - AK^{\alpha} L^{\beta} = 0 \tag{1.37}$$

Dividing the second equation by the first, one obtains:

$$\frac{\omega_K}{\omega_L} = \frac{\alpha L}{\beta K} \overset{\text{yields}}{\to} L = \frac{\omega_K \beta}{\omega_L \alpha} K \tag{1.38}$$

Replacing Eq. 1.38 in Eq. 1.37 leads to the optimal capital quantity, K^*, as:

$$K^* = \left(\frac{w_L \alpha}{w_K \beta}\right)^{\frac{\beta}{\alpha+\beta}} \left(\frac{q}{A}\right)^{\frac{1}{\alpha+\beta}} \tag{1.39}$$

The optimal labour quantity, L^*, is (see Eq. 1.38):

$$L^* = \frac{\omega_K \beta}{\omega_L \alpha} K^* = \left(\frac{w_K \beta}{w_L \alpha}\right)^{\frac{\alpha}{\alpha+\beta}} \left(\frac{q}{A}\right)^{\frac{1}{\alpha+\beta}} \tag{1.40}$$

The Lagrange multiplier is obtained from Eq. 1.36, for instance (the same result is obtained if we use Eq. 1.35):

$$\lambda = \frac{\omega_K}{A\alpha K^{*\alpha-1} L^{*\beta}} \tag{1.41}$$

The minimum production cost to produce the given output quantity is:

$$TC^* = \omega_L L^* + \omega_K K^* \tag{1.42}$$

Example 1.4

Consider a firm with a Cobb-Douglas production function given by $q = f(L, K) = 10K^{0.5}L^{0.5}$. The unit prices of labour and capital are 4 and 1, respectively.

1. Compute the minimum production cost to produce 80 units of output.
2. Solve the dual optimisation problem.
3. What happens if the cost of capital doubles?

◀

Solution

1.

The problem is formulated as follows:

$$\min[4L + K]$$

s.t.

$$10K^{0.5}L^{0.5} - 80 = 0$$

The problem is solved using the Lagrangian method. We can develop the application of the Lagrangian method as we learned, or simply applying the appropriate equations that we have just derived:

$$K^* = \left(\frac{w_L\alpha}{w_K\beta}\right)^{\frac{\beta}{\alpha+\beta}}\left(\frac{q}{A}\right)^{\frac{1}{\alpha+\beta}} = \left(\frac{4\times 0.5}{1\times 0.5}\right)^{\frac{0.5}{0.5+0.5}}\left(\frac{80}{10}\right)^{\frac{1}{0.5+0.5}} = 16$$

$$L^* = \frac{\omega_K\beta}{\omega_L\alpha}K^* = \frac{1\times 0.5}{4\times 0.5}\times 16 = 4$$

$$\lambda = \frac{\omega_K}{A\alpha K^{*\alpha-1}L^{*\beta}} = \frac{1}{10\times 0.5\times 16^{0.5-1}\times 4^{0.5}} = \frac{1}{2.5} = 0.4$$

$$TC^* = \omega_L L^* + \omega_K K^* = 4\times 4 + 1\times 16 = 32$$

The following conclusions can be drawn:

- To produce 80 units of output at minimum cost, 16 units of capital and 4 units of labour should be employed.
- The minimum production cost of producing 80 units of output is 32.
- The marginal cost is 0.4, i.e., to produce an additional unit of output will cost 0.4.

Let us look in more detail at this last conclusion and compute the minimum cost of producing 81 units of output.

$$K^* = \left(\frac{4\times 0.5}{1\times 0.5}\right)^{\frac{0.5}{0.5+0.5}}\left(\frac{81}{10}\right)^{\frac{1}{0.5+0.5}} = 16.2$$

$$L^* = \frac{1\times 0.5}{4\times 0.5}\times 16.2 = 4.05$$

$$TC^* = 4\times 4.05 + 1\times 16.2 = 32.4$$

which is the expected result.

The obtained results have a graphical interpretation. In Fig. 1.14, we plotted the isoquant for the output quantity of 80—$K = \left(\frac{80}{10L^{0.5}}\right)^2$—and three cost lines—$K = TC - 4L$, for $TC = 24$, $TC = 32$, $TC = 40$.

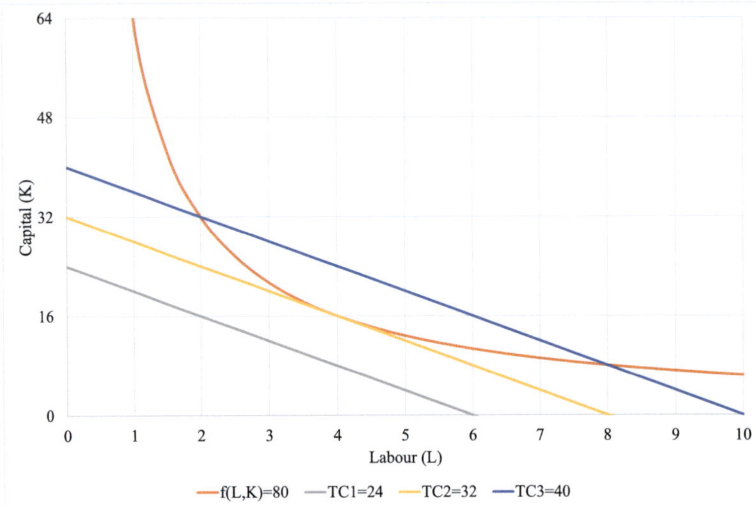

Fig. 1.14 Graphical interpretation of Example 1.4 1

Conclusions that can be drawn are as follows:

- It is impossible to produce 80 units of output at a cost of 24.
- It is possible to produce 80 units of output at a cost lower than 40.
- 32 is the minimum cost of producing 80 units of output (tangency point). The optimal quantities are $L^* = 4, \ K^* = 16$

2.

The dual optimisation problem can be formulated as:

- What is the maximum output that can be produced at a cost of 32?

The mathematical formulation is:

$$\max \left[10K^{0.5}L^{0.5} \right]$$

s.t.

$$4L + K - 32 = 0$$

The Lagrangian is:

$$Lag = 10K^{0.5}L^{0.5} - \lambda\,(4L + K - 32)$$

The optimal conditions are:

$$\frac{\partial Lag}{\partial K} = 10 \times 0.5K^{-0.5}L^{0.5} - \lambda = 0$$

$$\frac{\partial Lag}{\partial L} = 10 \times 0.5L^{-0.5}K^{0.5} - 4\lambda = 0$$

$$\frac{\partial Lag}{\partial \lambda} = 32 - 4L - K = 0$$

Dividing the first equation by the second, we obtain:

$$L = \frac{1}{4}K$$

and plugging in the third equation, the expected result is obtained: $K^* = 16$, $L^* = 4,\ \lambda = 2.5,\ q^* = 80$.

The Lagrange multiplier is now the increase in the output quantity when we increase the cost by one unit. It is the inverse of the marginal cost.

We note that:

$$MRTS_{K,L} = \frac{\alpha L^*}{\beta K^*} = \frac{1}{4} = \frac{\omega_K}{\omega_L} = \frac{1}{4}$$

This means that 4 units of capital are needed to replace 1 unit of labour (or 0.25 units of labour replace 1 unit of capital) while keeping the same output level.

3.

If the cost of capital goes up to double, $\omega_K = 2$ (all the rest remains unchanged), the application of the Lagrange method leads to the following results (the reader is invited to verify):

$$L^* = 5.6569$$

$$K^* = 11.3136$$

$$TC^* = 45.2548$$

To produce 80 units of output, 45.2548 monetary units are now needed instead of the previous 32. This is due to the increase in capital cost. The graphical representation of Fig. 1.15 helps in the interpretation of the results.

Before the capital cost increase, $\omega_L = 4,\ \omega_K = 1$, the operating point was point A, corresponding to a total cost of 32. We note that the cost line (isocost) equation is:

$$K = \frac{\text{TC}}{\omega_K} - \frac{\omega_L}{\omega_K}L$$

The x-axis and y-axis intercepts are, respectively:

$$L_0\,(K = 0) = \frac{\text{TC}}{\omega_L}$$

$$K_0\,(L = 0) = \frac{\text{TC}}{\omega_K}$$

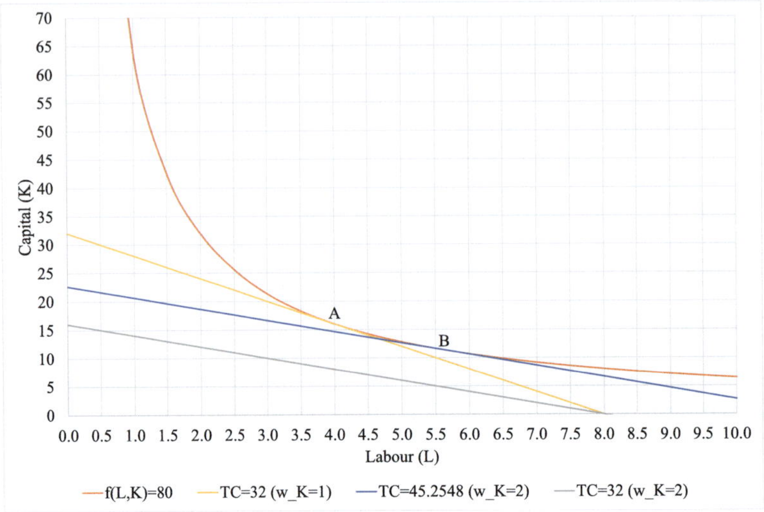

Fig. 1.15 Graphical interpretation of Example 1.4 3

After the capital cost increase, we have $\omega_L = 4,\ \omega_K = 2$. The new operating point is point B. It is still possible to produce 80 units of output, with a different combination of inputs, but at a higher cost: 45.2548 instead of 32. We can see this because the new x-intercept is higher, indicating that the total cost has increased (the labour cost is kept constant). In Fig. 1.15, we have also plotted the isocost line for a total cost of 32 with the new capital cost of 2. It can be seen that, with the new capital cost of 2, it is impossible to produce 80 units of output with a total cost of 32. Also, we can see that the slope, $-\frac{\omega_L}{\omega_K}$, of the new isocost line is lower (in absolute value)—more flattened—than the initial one because the capital cost has increased.

And what if the two inputs suffer the same proportionate increase in the cost, let us say, that both inputs double the respective unit cost, $\omega_L = 2 \times 4 = 8,\ \omega_K = 2 \times 1 = 2$? The reader is invited to verify that for this case the solution is:

$$L^* = 4$$

$$K^* = 16$$

$$TC^* = 64$$

The combination of inputs is the same as initially. What changes is the total cost which now has doubled. The isocost line remains the same: the slope is the same because the ratio of the two costs keeps unchanged; also, both the x-intercept and the y-intercept do not change, because the total cost has also doubled.

We conclude that it is possible to produce 80 units of output using the same combination of input quantities, but at a cost that is the double than before. ◀

1.7 The Optimisation for Special Cases

However, there are special cases that do not follow the methodology outlined above. Let us look at a couple of those cases. A good lecture on this subject can be found in https://kmlv.github.io/Econ100A_F17/S9_CostMinimization_Ch21/

1.7.1 Corner Solution, Perfect Substitutes

The production function:

$$f(\nu_1, \nu_2) = a\nu_1 + b\nu_2 = q \tag{1.43}$$

is called a *perfect substitutes* production function. In this case, we say the technology is linear. The Lagrange method is not applicable because the marginal product of input ν_1 and the marginal product of input ν_2 are constant, they do not depend on the Inputs, $MP_1 = a;\ MP_2 = b$. So, we would have a problem applying Eq. 1.28. We say we have a *corner solution* whenever solving for optimal quantities, we face a solution that involves one of the quantities being equal to zero, meaning that we are either at the y-intercept or x-intercept of the cost line. There are three options, corresponding to corner solutions:

- If $\frac{MP_1}{\omega_1} > \frac{MP_2}{\omega_2}$, it means the marginal product of input ν_1 per unit of cost is greater than the marginal product of input ν_2 per unit of cost. This implies there is no reason to use input ν_2 and all the output quantity should be satisfied by using only the input ν_1, the respective quantity being computed through $a\nu_1$+0= $q \overset{\text{yields}}{\rightarrow} \nu_1 = \frac{q}{a}$.
- If $\frac{MP_1}{\omega_1} < \frac{MP_2}{\omega_2}$, it means there is no reason to use input ν_1 and all the output quantity should be satisfied by using only input ν_2, the respective quantity being computed through 0+b$\nu_2 = q \overset{\text{yields}}{\rightarrow} \nu_2 = \frac{q}{b}$.
- If $\frac{MP_1}{\omega_1} = \frac{MP_2}{\omega_2}$, it means any point along the isoquant minimises the cost, so any quantity of inputs that obey the production function could be used.

To summarise, we can write:

$$TC = \begin{cases} q\frac{\omega_1}{a} \text{ if } \frac{a}{\omega_1} \geq \frac{b}{\omega_2} \\ q\frac{\omega_2}{b} \text{ if } \frac{a}{\omega_1} < \frac{b}{\omega_2} \end{cases} \tag{1.44}$$

or

$$TC = q \min\left\{\frac{\omega_1}{MP_1}, \frac{\omega_2}{MP_2}\right\} \tag{1.45}$$

1.7.2 Kink Solution, Perfect Complements

The Leontief production function or fixed proportions production function implies the production factors (inputs) will be used in fixed proportions, i.e., pre-determined by the technology, because there is no substitutability between factors. We say these production factors are *perfect complements*.

$$f(v_1, v_2) = \min\left\{\frac{v_1}{a}, \frac{v_2}{b}\right\} = q \tag{1.46}$$

The cost is minimized if:

$$\frac{v_1}{a} = q;\ \frac{v_2}{b} = q \tag{1.47}$$

which results in:

$$TC = (\omega_1 a + \omega_2 b)\, q \tag{1.48}$$

Example 1.5

Consider a firm whose market wage is $\omega_L = 5$ and market rental rate is $\omega_K = 10$. Compute the minimum cost of producing 20 units of output and the optimal input quantities for the following production functions:

1. $f(L, K) = 2L + 5K$
2. $f(L, K) = \min\{2L, 5K\}$
3. $f(L, K) = L^{\frac{1}{3}} K^{\frac{2}{3}}$

◀

Solution

1.
This is a perfect substitutes production function with a corner solution. Therefore, recalling what we have learned, we can write:

$$\frac{MP_L}{\omega_L} = \frac{2}{5} < \frac{MP_K}{\omega_K} = \frac{5}{10}$$

$$L^* = 0;\ K^* = 4$$

$$TC = 10 \times 4 = 20 \times \frac{10}{5} = 40$$

2.
This is a perfect complements production function with a kink solution.

$$L^* = \frac{20}{2} = 10;\ K^* = \frac{20}{5} = 4$$

$$TC = 5 \times 10 + 10 \times 4 = \left(5 \times \frac{1}{2} + 10 \times \frac{1}{5}\right) \times 20 = 90$$

3.
This is a Cobb-Douglas production function with an interior point solution.

$$MP_L = \frac{1}{3}L^{-\frac{2}{3}}K^{\frac{2}{3}};\ MP_K = \frac{2}{3}L^{\frac{1}{3}}K^{-\frac{2}{3}}$$

$$\frac{MP_L}{MP_K} = \frac{\omega_L}{\omega_K} \overset{\text{yields}}{\to} \frac{K}{2L} = \frac{5}{10} \overset{\text{yields}}{\to} K = L$$

Replacing this result in the production function, we obtain:

$$L^* = K^* = 20$$

$$TC = 5 \times 20 + 10 \times 20 = 300$$

◀

1.8 Production Costs

In general, the total production cost (TC) is composed of two parcels: the fixed cost, FC, which is independent of the output quantity, q, and the production factors' unitary costs, ω, and the variable cost, VC, which depends on these both. Mathematically, we can write:

$$TC\,(q, \omega) = FC + VC(q, \omega) \tag{1.49}$$

1.8.1 Average Cost and Marginal Cost

We define the average cost as the total cost per unit of output quantity. In the examples of Economy textbooks, constant production factors' unit prices are normally considered.

Average cost

We would like to say that the literature uses the acronyms ATC and AC to express the Average Total Cost (or simply Average Cost).

$$ATC\,(q) = \frac{TC}{q} = \frac{FC + VC(q)}{q} = AFC\,(q) + AVC(q) \tag{1.50}$$

where $AFC(q)$ is the average fixed cost and $AVC(q)$ is the average variable cost. Note that FC does not depend on quantity but $AFC(q)$ does.

Considering that the selling price of the product is p_S, we will see later that if $p_S < ATC$, the producer will not survive in the long-term, whereas if $p_S \geq ATC$, the long-term economic viability is assured. This is pretty much straightforward.

We now define marginal cost as the cost of producing one additional unit of output.

Marginal cost

$$MC\,(q) = \frac{dTC(q)}{dq} = \frac{dVC\,(q)}{dq} \tag{1.51}$$

The marginal cost is a dynamic economic concept that is very important in short-term analysis. If $p_S < MC$, the producer is not economically viable in the short term. The opposite happens if $p_S \geq MC$.

If there is a minimum average cost, it occurs when:

$$\frac{dATC}{dq} = \frac{d}{dq}\left[\frac{FC + VC}{q}\right] = -\frac{FC}{q^2} + \frac{1}{q}\frac{dVC}{dq} - \frac{VC}{q^2} = 0 \tag{1.52}$$

This leads to:

$$\frac{dVC}{dq} = \frac{FC + VC}{q} \overset{\text{yields}}{\rightarrow} MC = AC \tag{1.53}$$

The minimum average cost, if it exists, occurs when the average cost is equal to the marginal cost.

Figure 1.16 shows a general representation of both the ATC and the MC.

The following conclusions are apparent from the figure.

- The minimum ATC occurs when $ATC = MC$.
- When ATC is decreasing, it follows that $MC < ATC$.
- When ATC is increasing, then $MC > ATC$.

For example:

$$TC(q) = 1 + q^2$$

$$FC = 1$$

$$VC(q) = q^2$$

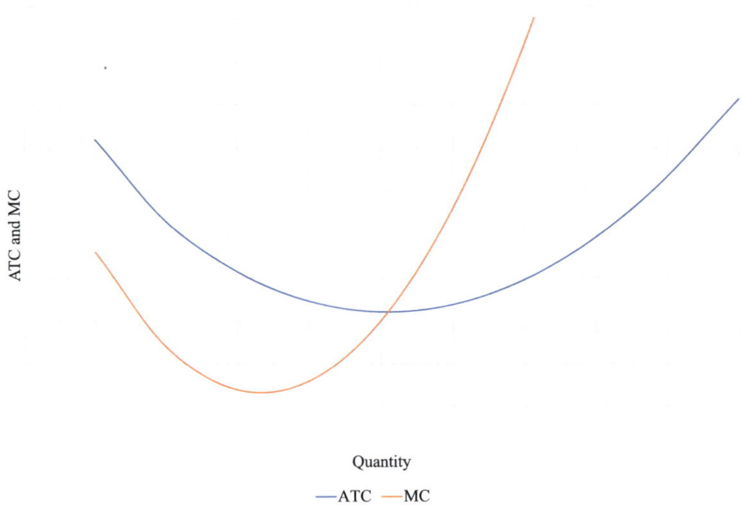

Fig. 1.16 Average Total Cost (ATC) and Marginal Cost (MC)

$$AFC\left(q\right)=\frac{1}{q}$$

$$AVC\left(q\right)=q$$

$$ATC\left(q\right)=\frac{1}{q}+q$$

$$MC\left(q\right)=2q$$

1.8.2 Explicit and Implicit Costs

In economics, explicit and implicit costs are two types of costs that play a crucial role in decision-making and resource allocation. Explicit costs are tangible, out-of-pocket expenses that a firm incurs in its production process. These costs can include payments for labour, raw materials, rent, utilities, and other expenses that require a direct monetary outlay.

On the other hand, implicit costs are the opportunity costs associated with the use of resources. They represent the value of the next best alternative foregone when a particular choice is made. Implicit costs do not involve a direct monetary payment but rather reflect the value of resources used in alternative ways. For example, if an entrepreneur decides to start a business, the implicit costs would include the forgone salary from their previous job or the foregone returns from investing their capital elsewhere. In general terms, when a company chooses to invest in one project, it foregoes the potential returns from investing in alternative projects. This cost is not always explicitly measured in monetary terms but rather in terms of the potential gains or benefits that could have been realised from the next best alternative.

Opportunity cost is a subjective concept, as it depends on individual preferences and the available alternatives. It can vary across individuals and circumstances. Moreover, opportunity costs are not always easily quantifiable, especially when considering non-monetary factors such as personal satisfaction, environmental impact, or social value.

Understanding both explicit and implicit costs is essential for assessing the true cost of production or decision-making. While explicit costs are relatively straightforward to measure, implicit costs require careful consideration and estimation to account for the value of forgone opportunities. By considering both types of costs, individuals and firms can make more accurate economic evaluations and allocate resources effectively to maximise their overall well-being and profitability.

1.8.3 Application to the Cobb-Douglas Production Function

As mentioned before, the Cobb-Douglas production function is often taken as an example in Economy textbooks. We will also use this production function as an example to study the behaviour of the average cost and marginal cost. We recall that the Cobb-Douglas production function is given by $q=f\left(L,K\right)=AK^{\alpha}L^{\beta}$ and that the total cost is given by $TC = \omega_K K + \omega_L L$, where, for simplification, fixed costs are not considered.

To compute the average total cost (Eq. 1.50) and the marginal cost (Eq. 1.51), we need to express the total cost as a function of the output quantity. Recalling Eqs. 1.39 and 1.40, one can write for the total cost:

$$TC = k_1 \left(k_2 + k_3\right) q^{\frac{1}{\alpha+\beta}} \tag{1.54}$$

where:

$$k_1 = \left(\frac{1}{A}\right)^{\frac{1}{\alpha+\beta}} \tag{1.55}$$

$$k_2 = \omega_L \left(\frac{\omega_K \beta}{\omega_L \alpha}\right)^{\frac{\alpha}{\alpha+\beta}} \tag{1.56}$$

$$k_3 = \omega_K \left(\frac{\omega_L \alpha}{\omega_K \beta}\right)^{\frac{\beta}{\alpha+\beta}} \tag{1.57}$$

It is noted that k_1, k_2, k_3 are constant parameters that only depend on the unit prices of the production factors, (L, K), supposed constant in our analysis, and on the Cobb-Douglas function parameters, again supposed constant.

The total cost is an increasing polynomial function of the output quantity. $\alpha + \beta$ determine the type of polynomial (in fact, it is a monomial). Before going further, let us recall that $\alpha + \beta = 1 \overset{\text{yields}}{\rightarrow} CRS$, $\alpha + \beta > 1 \overset{\text{yields}}{\rightarrow} IRS$, $\alpha + \beta < 1 \overset{\text{yields}}{\rightarrow}$ DRS.

- If $\alpha + \beta = 1$, we have a linear function of the output.
- If $\alpha + \beta > 1$, we have a convex function of the output.
- If $\alpha + \beta < 1$, we have a concave function of the output.

The average total cost and the marginal cost are, respectively:

$$ATC = \frac{TC}{q} = k_1 \left(k_2 + k_3\right) q^{\frac{1}{\alpha+\beta}-1} \tag{1.58}$$

$$MC = \frac{dTC}{dq} = k_1 \left(k_2 + k_3\right) \left(\frac{1}{\alpha + \beta}\right) q^{\frac{1}{\alpha+\beta}-1} \tag{1.59}$$

Again, we have polynomial functions. The type of polynomial is determined by $\frac{1}{\alpha+\beta} - 1$. Also, it is interesting to note that:

$$\frac{ATC}{MC} = \alpha + \beta \tag{1.60}$$

- If $\alpha + \beta = 1$, both ATC and MC are constant and equal one to the other.
- If $\alpha + \beta > 1$, both ATC and MC are decreasing functions of the output and $ATC > MC$.

- If $\alpha + \beta < 1$, both ATC and MC are increasing functions of the output and $ATC < MC$.

So far, we have considered constant unit costs of the production factors, ω_L, ω_K. It would be interesting to study how changes in the unit costs of the production factors affect the total, average, and marginal costs, for a constant output level. Let us exemplify with the effect of a change in the capital unit cost, ω_K, bearing in mind that the process for the labour unit cost is the same. For this purpose, let us recall Eqs. 1.39 and 1.40

$$K^* = \left(\frac{w_L \alpha}{w_K \beta}\right)^{\frac{\beta}{\alpha+\beta}} \left(\frac{q}{A}\right)^{\frac{1}{\alpha+\beta}} \tag{1.61}$$

$$L^* = \frac{\omega_K \beta}{\omega_L \alpha} K^* = \left(\frac{w_K \beta}{w_L \alpha}\right)^{\frac{\alpha}{\alpha+\beta}} \left(\frac{q}{A}\right)^{\frac{1}{\alpha+\beta}} \tag{1.62}$$

We now plug them into the total cost equation:

$$TC = \omega_L L^* + \omega_K K^* = (k_4 + k_5)\omega_K^{\frac{\alpha}{\alpha+\beta}} \tag{1.63}$$

where:

$$k_4 = \omega_L \left(\frac{q}{A}\right)^{\frac{1}{\alpha+\beta}} \left(\frac{\beta}{\omega_L \alpha}\right)^{\frac{\alpha}{\alpha+\beta}} \tag{1.64}$$

$$k_5 = \left(\frac{q}{A}\right)^{\frac{1}{\alpha+\beta}} \left(\frac{\omega_L \alpha}{\beta}\right)^{\frac{\beta}{\alpha+\beta}} \tag{1.65}$$

The average total cost is:

$$ATC = \frac{TC}{q} = \frac{(k_4 + k_5)\omega_K^{\frac{\alpha}{\alpha+\beta}}}{q} \tag{1.66}$$

The marginal cost is:

$$MC = \frac{\partial TC}{\partial q} = \frac{1}{\alpha+\beta} \left(\frac{1}{A}\right)^{\frac{1}{\alpha+\beta}} q^{\frac{1}{\alpha+\beta}-1} (k_4 + k_5)\omega_K^{\frac{\alpha}{\alpha+\beta}} \tag{1.67}$$

The optimal quantities of labour and capital can also be computed as a function of the capital unit cost, ω_K as:

$$L^* = \frac{k_4}{\omega_L} \omega_K^{\frac{\alpha}{\alpha+\beta}} \tag{1.68}$$

$$K^* = k_5 \omega_K^{-\frac{\beta}{\alpha+\beta}} \tag{1.69}$$

Example 1.6

Consider a firm with a Cobb-Douglas production function given by $q = 10K^{\alpha}L^{\beta}$. The unit prices of labour and capital are 2 and 1, respectively. We want to study three cases for the Cobb-Douglas parameters, cases 1, 2, and 3, respectively: $1.\ \alpha = \beta = 0.5,\ 2.\ \alpha = 0.2\ and\ \beta = 0.1,\ 3.\ \alpha = 1\ and\ \beta = 2.$

1. For an output level of 10, determine the optimal labour and capital quantities, the marginal cost (MC), and the minimum total cost TC).
2. Plot the graphs of the Total Cost (TC), Average Cost (AC), and Marginal Cost (MC) as a function of the output. Note that the average total cost is often referred to as the average cost.
3. When the capital unit cost doubles, how are the costs affected?
4. For an output level of 10, plot the graphs of the Total Cost (TC), Average Cost (AC), and Marginal Cost (MC) as a function of the capital unit cost.
5. Plot the graphics of the optimal production factors as a function of the capital unit cost, considering an output of 10.

◀

Solution

1.

The required results are computed using the Lagrange method which leads to:

$$K^{*} = \left(\frac{w_L \alpha}{w_K \beta}\right)^{\frac{\beta}{\alpha+\beta}} \left(\frac{q}{A}\right)^{\frac{1}{\alpha+\beta}}$$

$$L^{*} = \frac{\omega_K \beta}{\omega_L \alpha} K^{*}$$

$$\lambda = MC = \frac{\omega_K}{A\alpha K^{*\alpha-1} L^{*\beta}}$$

$$TC = \omega_L L^{*} + \omega_K K^{*}$$

The obtained results are in Table 1.3.

Table 1.3 Results of Example 1.6 1

	RS	K^*	L^*	MC	TC
Case 1	CRS	1.4142	0.7071	0.2828	2.8284
Case 2	DRS	1.5874	0.3969	0.7937	2.3811
Case 3	IRS	1	1	0.1	3

2.

The variation of the Total Cost (TC), Average Cost (AC) and Marginal Cost (MC) with the output level is obtained using:

$$TC = k_1\,(k_2 + k_3)\,q^{\frac{1}{\alpha+\beta}}$$

$$AC = k_1\,(k_2 + k_3)\,q^{\frac{1}{\alpha+\beta}-1}$$

$$MC = k_1\,(k_2 + k_3)\left(\frac{1}{\alpha+\beta}\right)q^{\frac{1}{\alpha+\beta}-1}$$

The results for the three cases are displayed in Table 1.4.

These equations are plotted in the next graphs (Figs. 1.17, 1.18 and 1.19).

Table 1.4 Results of Example 1.6 2

	TC	AC	MC
Case 1	$0.2828q$	0.2828	0.2828
Case 2	$0.0011q^{3.3333}$	$0.0011q^{2.3333}$	$0.0037q^{2.3333}$
Case 3	$1.3925q^{0.3333}$	$1.3925q^{-0.6667}$	$0.4642q^{-0.6667}$

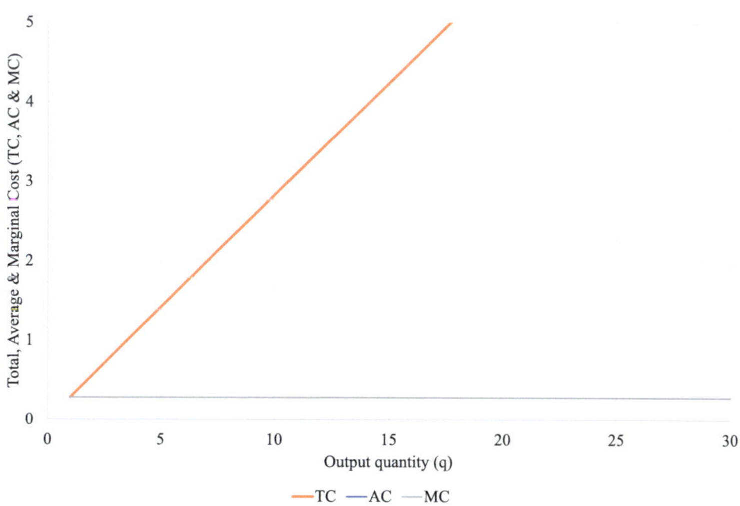

Fig. 1.17 Total Cost (TC), Average Cost (AC) and Marginal Cost (MC) for a CRS Cobb-Douglas production function; Case 1. Example 1.6 2

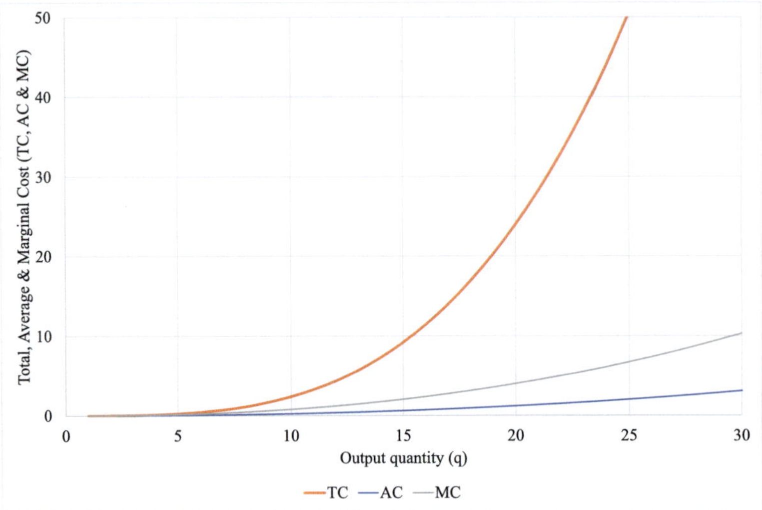

Fig. 1.18 Total Cost (TC), Average Cost (AC) and Marginal Cost (MC) for a DRS Cobb-Douglas production function; Case 2. Example 1.6 2

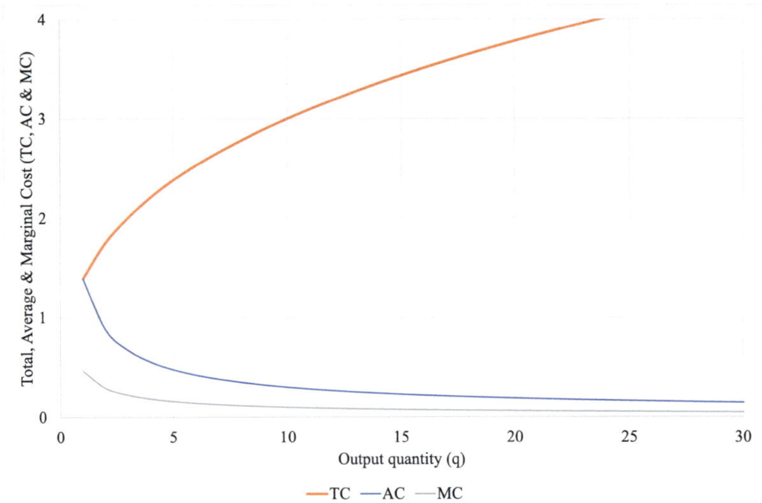

Fig. 1.19 Total Cost (TC), Average Cost (AC) and Marginal Cost (MC) for an IRS Cobb-Douglas production function; Case 3. Example 1.6 2

It is worth noting that:

- For a CRS Cobb-Douglas function, the total cost is a linear function of the output, and the average cost is constant and equal to the marginal cost.

- When we have a DRS Cobb-Douglas function, the total cost is a convex function of the output ($TC = kq^a,\ a > 1$) and both the average cost and the marginal cost are increasing functions, the marginal cost being higher than the average cost (see Fig. 1.16).
- If the Cobb-Douglas function exhibits IRS, the total cost is a concave function of the output ($TC = kq^a,\ a < 1$) and both the average cost and the marginal cost are decreasing functions, the marginal cost being lower than the average cost (see Fig. 1.16).

3.
If the capital unit cost doubles, the curves will shift left, denoting that it will be more expensive to produce the same output quantity. This is shown in Fig. 1.20, taking DRS Case 2 as an example.
4.
The equations to compute the TC, AC and MC when the capital cost changes, keeping the same output, are:

$$TC = (k_4 + k_5)\omega_K^{\frac{\alpha}{\alpha+\beta}}$$

$$AC = \frac{(k_4 + k_5)}{q}\omega_K^{\frac{\alpha}{\alpha+\beta}}$$

$$MC = \frac{1}{\alpha+\beta}\left(\frac{1}{A}\right)^{\frac{1}{\alpha+\beta}} q^{\frac{1}{\alpha+\beta}-1}(k_4 + k_5)\omega_K^{\frac{\alpha}{\alpha+\beta}}$$

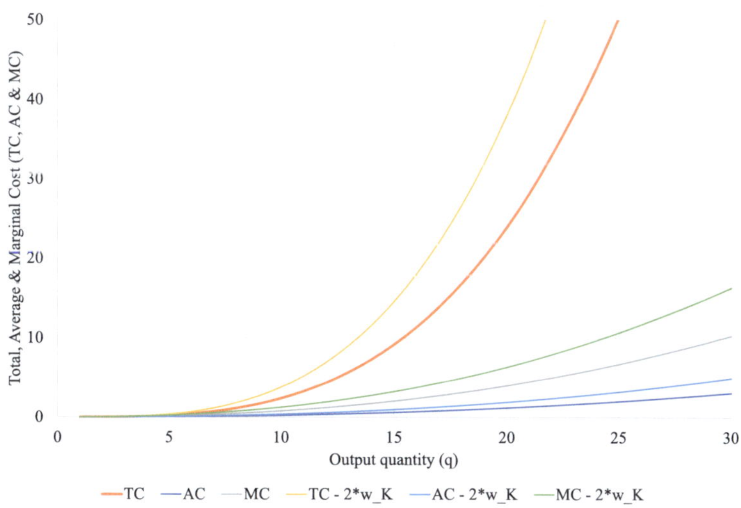

Fig. 1.20 Total Cost (TC), Average Cost (AC) and Marginal Cost (MC) for a DRS Cobb-Douglas production function; Case 2; Comparison with the case of doubling capital unit cost; Example 1.6 3

Table 1.5 Results of Example 1.6 4

	TC	AC	MC
Case 1	$2.8284\omega_K^{0.5}$	$0.2828\omega_K^{0.5}$	$0.2828\omega_K^{0.5}$
Case 2	$2.3811\omega_K^{0.6667}$	$0.2381\omega_K^{0.6667}$	$0.7937\omega_K^{0.6667}$
Case 3	$3\omega_K^{0.3333}$	$0.3\omega_K^{0.3333}$	$0.1\omega_K^{0.3333}$

The results can be seen in (Table 1.5).
The plotting of the results is in Figs. 1.21, 1.22 and 1.23.
All the costs increase when the capital unit cost increases, as would be expected. For Case 1 (CRS), $AC = MC$, for Case 2 (DRS), $AC < MC$ and for Case 3 (IRS), $AC > MC$.
5.
The computation of the change of the production factors with the capital unit costs are:

$$L^* = \frac{k_4}{\omega_L}\omega_K^{\frac{\alpha}{\alpha+\beta}}$$

$$K^* = k_5\omega_K^{-\frac{\beta}{\alpha+\beta}}$$

Table 1.6 shows the results obtained.

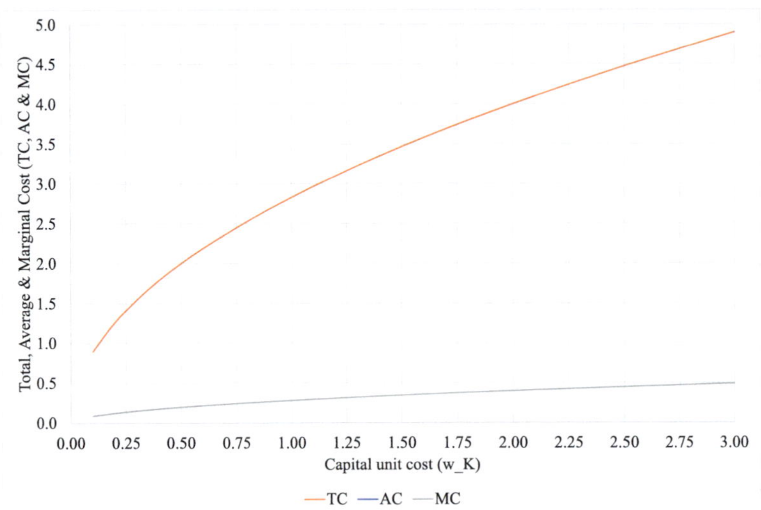

Fig. 1.21 Total Cost (TC), Average Cost (AC) and Marginal Cost (MC) for changing capital unit cost; CRS Cobb-Douglas production function; Case 1. Example 1.6 4

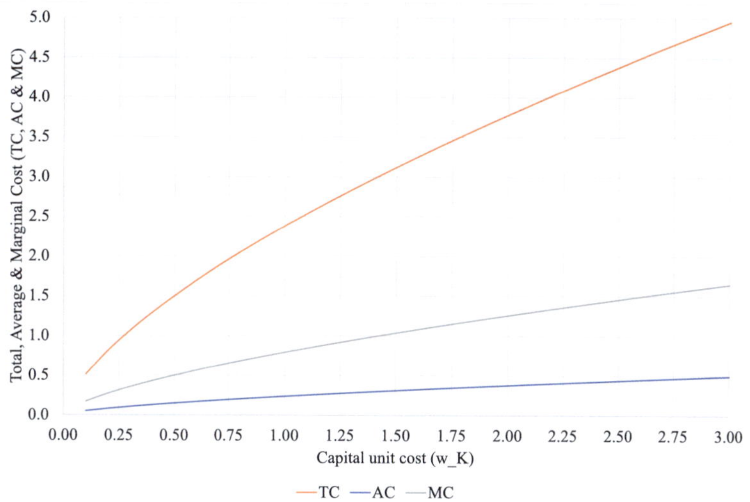

Fig. 1.22 Total Cost (TC), Average Cost (AC) and Marginal Cost (MC) for changing capital unit cost; DRS Cobb-Douglas production function; Case 2. Example 1.6 4

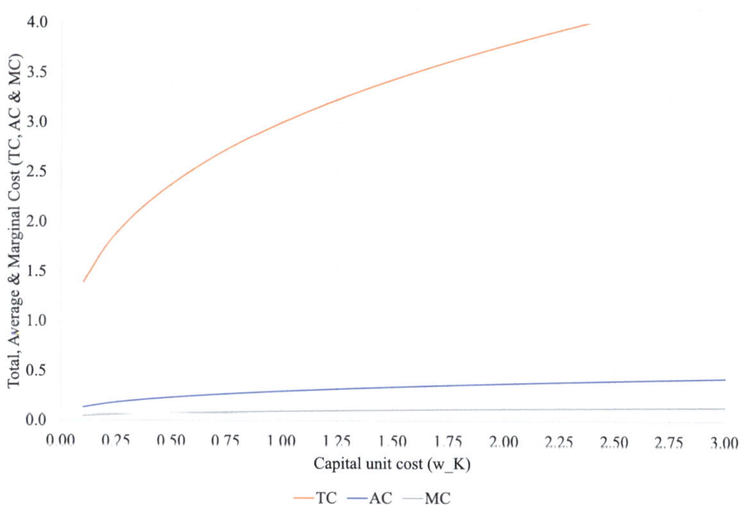

Fig. 1.23 Total Cost (TC), Average Cost (AC) and Marginal Cost (MC) for changing capital unit cost; IRS Cobb-Douglas production function; Case 3. Example 1.6 4

The following plots illustrate the graphical representation Figs. 1.24, 1.25 and 1.26 illustrate the graphical representation. ◀

Table 1.6 Results of Example 1.6 5

	L^*	K^*
Case 1	$0.7071\omega_K^{0.5}$	$1.4142\omega_K^{-0.5}$
Case 2	$0.3969\omega_K^{0.6667}$	$1.5874\omega_K^{-0.3333}$
Case 3	$\omega_K^{0.3333}$	$\omega_K^{-0.6667}$

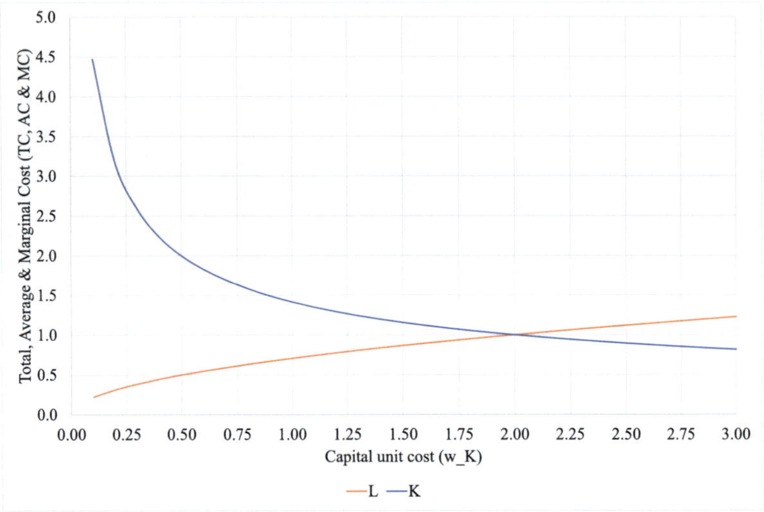

Fig. 1.24 Optimal Labour (L) and Capital (K) for changing capital unit cost; CRS Cobb-Douglas production function; Case 1. Example 1.6 5

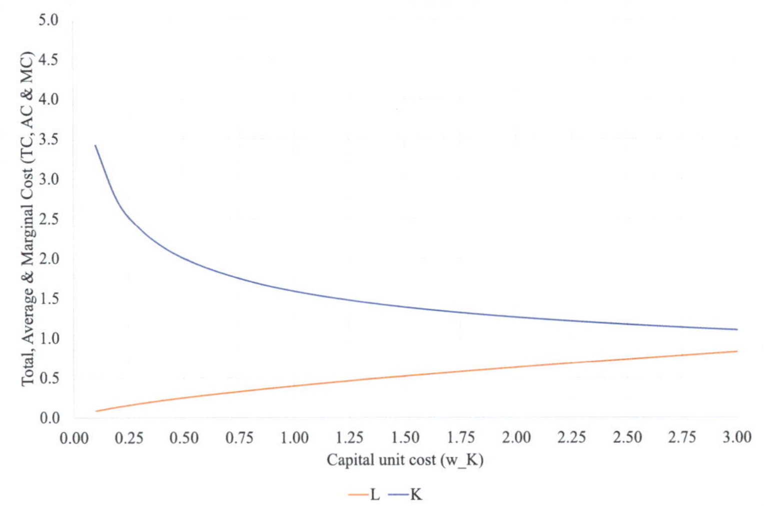

Fig. 1.25 Optimal Labour (L) and Capital (K) for changing capital unit cost; DRS Cobb-Douglas production function; Case 2. Example 1.6 5

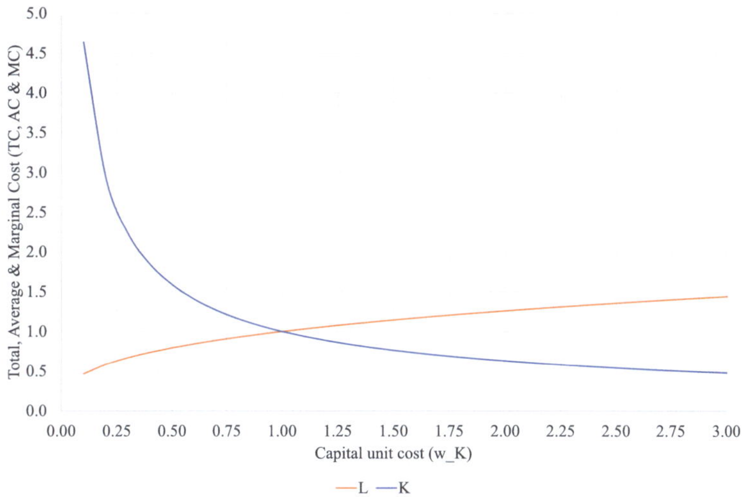

Fig. 1.26 Optimal Labour (L) and Capital (K) for changing capital unit cost; IRS Cobb-Douglas production function; Case 3. Example 1.6 5

1.9 Supply Curve

The supply curve depicts the (unitary) price a supplier is willing to sell a certain quantity of the product. In general, it is a function:

$$p = f(q) \tag{1.70}$$

This means the price is plotted on the y-axis and the quantity on the x-axis. In traditional Economics, supply curves are always upward, meaning that the supplier is willing to sell higher quantities of the product at a higher price each. As the price of a good increases, sellers are willing to supply more of that good. However, this does not hold always true. Supply curves can be downward as we will see later in this book.

Supply curves are based on production costs. The objective of the sellers is to maximise the profit, i.e., the difference between revenues and costs.

Changes in the supply cause the supply curve to shift, leftward or rightward. A leftward shift results in a reduced supply of the good at all prices—for the same price, less quantity is offered. A rightward shift results in an increased supply of the good at all prices—for the same price, more of the good is delivered.

Several factors may affect the quantity supplied. Changes in the production factors' cost, technological advances, prices of substitutes and complements, and perception of future prices are among the factors that may shift the supply curve. For instance, a decrease in the prices of raw materials reduces costs and allows suppliers to supply more of that good at all prices, shifting rightward the supply curve for that good.

1.9.1 Example with the Cobb-Douglas Production Function

To exemplify the process of determining the supply curve, let us consider a producer that sells in a competitive market (we will see later that the price in the competitive market is constant) and shows a Cobb-Douglas production function. Our objective is to determine the supply curve, $p = f(q)$, where p is price and q is quantity. The producer wants to maximise its profit, which is given by:

$$\Pi(p, q, \omega_L, \omega_K, L, K) = pq - \omega_L L - \omega_K K = pf(L, K) - \omega_L L - \omega_K K \quad (1.71)$$

The maximum profit is obtained by setting:

$$\frac{\partial \Pi}{\partial q} = 0 \quad (1.72)$$

We need to express L and K as a function of the output quantity, q. At this point, we recall Eqs. 1.39 and 1.40, which we repeat here for convenience:

$$K^* = \left(\frac{w_L \alpha}{w_K \beta}\right)^{\frac{\beta}{\alpha+\beta}} \left(\frac{q}{A}\right)^{\frac{1}{\alpha+\beta}} \quad (1.73)$$

$$L^* = \frac{\omega_K \beta}{\omega_L \alpha} K^* = \left(\frac{w_K \beta}{w_L \alpha}\right)^{\frac{\alpha}{\alpha+\beta}} \left(\frac{q}{A}\right)^{\frac{1}{\alpha+\beta}} \quad (1.74)$$

Replacing Eqs. 1.73 and 1.74 in Eq. 1.71, we obtain:

$$\Pi = pq - \left(\frac{1}{A}\right)^{\frac{1}{\alpha+\beta}} \left[\omega_L \left(\frac{\omega_K \beta}{\omega_L \alpha}\right)^{\frac{\alpha}{\alpha+\beta}} + \omega_K \left(\frac{\omega_L \alpha}{\omega_K \beta}\right)^{\frac{\beta}{\alpha+\beta}}\right] q^{\frac{1}{\alpha+\beta}} = pq - TC \quad (1.75)$$

which is a result that includes the total cost obtained in Eq. 1.54.

Now, we have to apply Eq. 1.72 and solve for p. The result is $p = MC$, which is a result that we have previously obtained in Eq. 1.59, and we copy it here again.

$$p(q) = \left(\frac{1}{A}\right)^{\frac{1}{\alpha+\beta}} \left[\omega_L \left(\frac{\omega_K \beta}{\omega_L \alpha}\right)^{\frac{\alpha}{\alpha+\beta}} + \omega_K \left(\frac{\omega_L \alpha}{\omega_K \beta}\right)^{\frac{\beta}{\alpha+\beta}}\right] \frac{1}{\alpha + \beta} q^{\frac{1}{\alpha+\beta} - 1} \quad (1.76)$$

For constant costs of the production factors, ω_L and ω_K, we obtain:

$$p(q) = k_1(k_2 + k_3) \frac{1}{\alpha + \beta} q^{\frac{1}{\alpha+\beta} - 1} \quad (1.77)$$

This is a known result from the theory of competitive markets that will become apparent to you in the dedicated chapter, in this book.

To see examples of supply functions, the reader is invited to look again at the MC plotting of Figs. 1.17, 1.18 and 1.19.

1.9.2 Returns to Scale Again

As it can be easily seen from Eq. 1.77, the sign of:

$$\gamma = \frac{1}{\alpha + \beta} - 1 \tag{1.78}$$

determines the type of price dependence on the quantity.

- If $\gamma = 0$, it follows that $\alpha + \beta = 1$. The supply curve is constant in quantity and the producer shows CRS.
- If $\gamma > 0$, it follows that $\alpha + \beta < 1$. The supply curve is upward on quantity (price increases with quantity) and the producer shows DRS. This is the usual situation addressed in traditional Economics textbooks.
- If $\gamma < 0$, it follows that $\alpha + \beta > 1$. The supply curve is downward on quantity (price decreases with quantity) and the producer shows IRS.

1.10 Conclusions

In this chapter, we delved into the theory of producers, also known as the theory of the firm. The producers are one of the most important economic agents, the other being the consumers. Producers play a critical role as economic agents within an economic system, namely within the power system. They are the entities responsible for transforming inputs into goods and services, thereby creating value and driving economic growth. As key participants in the energy production process, producers contribute to the overall functioning of energy markets and shape the dynamics of supply and demand of energy.

The theory of producers provides valuable insights into the behaviour and decision-making of firms in the power generation sector. Throughout this chapter, we have explored key concepts such as returns to scale, production function, the Cobb-Douglas function, optimisation problem, production costs, and the supply function, all of which contributed to our understanding of how producers operate within the power system.

Returns to scale refer to the changes in output resulting from proportional changes in all inputs used in production. It helped us understand how generation firms can achieve economies of scale or face diminishing returns as they increase their production levels. By analysing returns to scale, producers can determine the most efficient scale of production and optimise the size of power plants.

The production function is a mathematical representation of the relationship between inputs and outputs in the power production process. It enables firms to assess the productivity of different combinations of inputs, such as labour, capital, and technology, in generating output energy. By understanding the production function, power system engineers can make informed decisions about resource allocation and production techniques to maximise output energy. We have introduced key concepts related to production function such as marginal product and marginal rate of technical substitution.

The Cobb-Douglas function, a widely used typical production function, incorporates the factors of production, their respective shares, and the elasticity of substitution. It provides a flexible framework for analysing production relationships and estimating the impact of input changes on output. The Cobb-Douglas function helps producers identify the optimal combination of inputs to achieve their production goals.

The optimisation problem lies at the core of producers' decision-making. Firms aim to maximise their profits or minimise their costs, subject to various constraints, such as resource availability and technological limitations. By formulating and solving optimisation problems, producers can determine the optimal levels of inputs, output, and pricing strategies that will yield the desired outcomes. We have introduced the Lagrangian method as a mathematical tool to solve optimisation problems and highlighted the physical meaning of the Lagrange multiplier. Also, the methods to deal with special production functions, such as perfect substitutes and perfect complements, were addressed.

Production costs are a crucial consideration for producers, as they directly impact profitability. Various cost factors, such as labour, raw materials, and capital, must be carefully managed to minimise expenses and increase efficiency. Understanding the cost structure and employing cost-saving measures allow producers to remain competitive in the market. The important concepts of average cost and marginal cost were explained in this chapter.

The supply function represents the relationship between the price of a good or service and the quantity that producers are willing and able to supply. It incorporates factors such as production costs, market conditions, and expectations. The supply function helps firms determine the quantity of goods or services to offer at different price levels, influencing market equilibrium and ensuring efficient allocation of resources. Examples using the Cobb-Douglas function were provided for illustration purposes.

In summary, the theory of producers encompasses a range of concepts and models that shed light on their decision-making process and behaviour. By considering returns to scale, production functions, the Cobb-Douglas function, the optimisation problem, production costs, and the supply function, power system engineers can optimise resource allocation, improve power generation efficiency, manage costs effectively, and ensure the reliable supply of electricity to meet the evolving needs of the energy market while considering economic, environmental, and regulatory factors.

1.11 Proposed Exercises

Problem 1.1

Assume that a manufacturing process is characterised by the production function given by the Cobb-Douglas expression:

$$q = f(K, L) = 0.1K^2L$$

Where K and L (capital and labour) are input variables (production factors) and q represents the output (production).

The production factors have unitary prices given by:

$$\omega_K = 3; \omega_L = 5$$

1. What type of returns to scale has this manufacturing process? Justify.
2. On the K and L axes draw the isoquant for $q = 10^6$. What is the meaning of this curve?
3. Again, on the K and L axes draw a cost line for $TC = 1000$. What is the meaning of this line?
4. Compute the minimum production cost for an output quantity $q = 10^6$. What is the average cost of production?
5. If the unitary costs of the production factors are kept constant and the output production is increased, will the average cost of production change? Justify.

Key Problem 1.1

1.

The production function of the problem is:

$$f(K, L) = 0.1K^2L$$

It is noted that the x and y-axis for K and L are arbitrary. In the theory exposed before, we used the x-axis for the labour and the y-axis for the capital. In this problem, we are going to use the opposite, just to show that the selection of the axis is indifferent.

The general form of a Cobb-Douglas production function is:

$$f(K, L) = AK^{\alpha}L^{\beta}$$

When $\alpha + \beta > 1$, which is the case, the production function shows increasing returns to scale (IRS).

2. and **3.**

A production function represents the maximum output quantity that can be obtained by a given combination of inputs (production factors) quantities.

$$q = 10^6 = 0.1K^2L \overset{\text{yields}}{\rightarrow} L = \frac{10^6}{0.1K^2}$$

Figure 1.27 shows the result obtained. This is called an isoquant, a curve that contains all combinations of inputs which generate the same level of output.

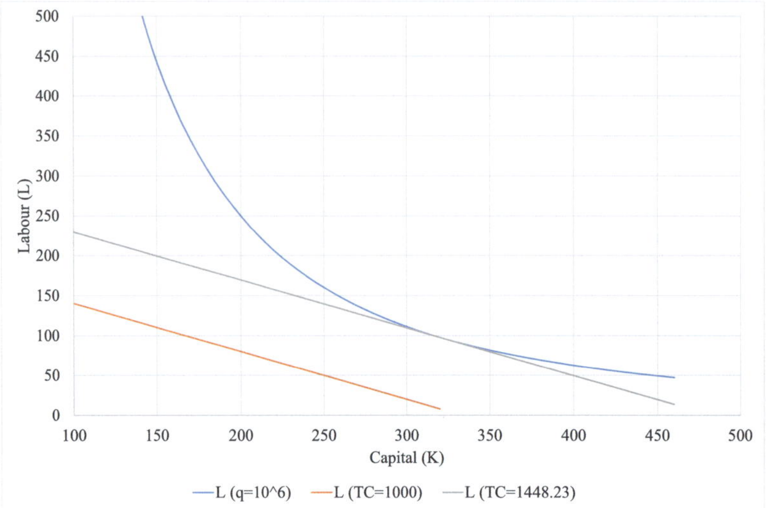

Fig. 1.27 Isoquant (blue) and isocost curves (grey and orange). Problem 1.1

In the same figure, we also represented two cost lines. A cost line represents the total cost of a product given the unitary costs of each input and the input quantities:

$$TC = \omega_K K + \omega_L L \overset{\text{yields}}{\rightarrow} L = \frac{\text{TC}}{\omega_L} - \frac{\omega_K}{\omega_L} K$$

In the figure, the two cost lines represented are for $TC = 1000$ and $TC = 1448.23$. The slope of the cost line is the ratio of the two production factors' unitary costs.

The cost line for $TC = 1000$ does not intercept the isoquant. This means that it is impossible to produce the desired quantity of output, 10^6, at a cost of 1000.

The cost line for $TC = 1448.23$ intercepts the isoquant at a single point (tangency point). The coordinates of this point (K^*; L^*) define the input quantities that correspond to the minimum cost. We will see this in an analytical way, in the next question.

4.

Problem formulation:

$$\min\,[3K + 5L]$$

$$s.t.\ 0.1K^2L - 10^6 = 0$$

Solution method (Lagrangian method):

$$Lag = 3K + 5L - \lambda(0.1K^2L - 10^6)$$

Optimal conditions:

$$\frac{\partial Lag}{\partial K} = 3 - 0.2KL\lambda = 0$$

$$\frac{\partial Lag}{\partial L} = 5 - 0.1K^2\lambda = 0$$

$$\frac{\partial Lag}{\partial \lambda} = 10^6 - 0.1K^2L = 0$$

Most common solution procedure: Divide the first equation by the second and plug in in the third.

$$\frac{3}{5} = 2\frac{L}{K} \overset{\text{yields}}{\rightarrow} L = \frac{3}{10}K$$

$$\frac{0.3}{10}K^3 = 10^6 \overset{\text{yields}}{\rightarrow} K^* = \left(\frac{10^6 10}{0.3}\right)^{\frac{1}{3}} = 321.83$$

$$L^* = 96.55$$

$$\lambda = \frac{3}{0.2K^*L^*} = 4.83 \times 10^{-4}$$

$K^*,\ L^*$ are the production factors quantities that minimize the total cost. The minimum total cost is:

$$TC^* = 3K^* + 5L^* = 1448.23$$

$\lambda = \frac{\partial TC}{\partial q}$ is the marginal cost, i.e., the cost of producing the last unit. Note that, in general, the λ parameter is the derivative of the objective function with respect to the restriction function.

The average cost is:

$$AC = \frac{TC}{q} = 0.0014$$

5.

$$TC = \omega_K K + \omega_L L = 3K + 5\left(\frac{3K}{10}\right) = \frac{9}{2}K = \frac{9}{2}\left(\frac{10q}{0.3}\right)^{\frac{1}{3}}$$

$$AC = \frac{\text{TC}}{q}$$

$$MC = \lambda = \frac{\partial TC}{\partial q} = \frac{50}{\left(\frac{10q}{0.3}\right)^{\frac{2}{3}}}$$

In Figs. 1.28 and 1.29, we show the total cost as a function of the quantity, and both the average cost and the marginal cost as a function of the output quantity, respectively. Both figures are plotted keeping the production factors' unitary costs ($\omega_K;\ \omega_L$) constant and the same production function. We can see that both the marginal and average costs depend on the output quantity. Also, the results for the marginal cost and average cost are compatible with an IRS production function. Both

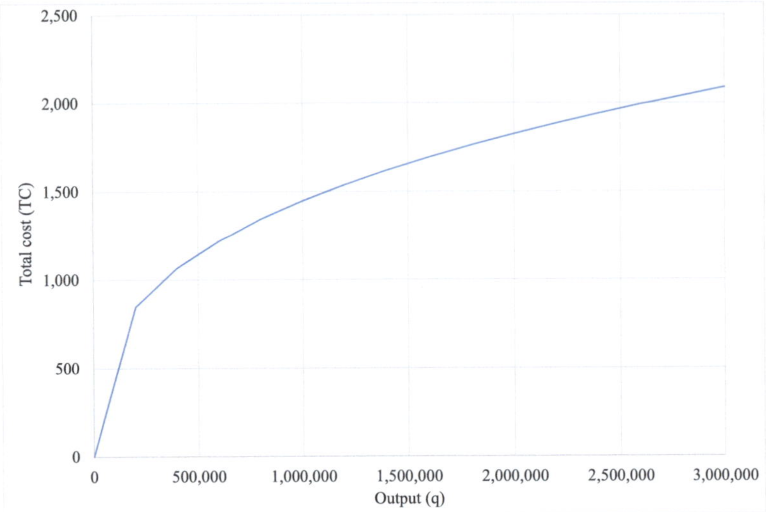

Fig. 1.28 Total cost as a function of the output quantity. Problem 1.1

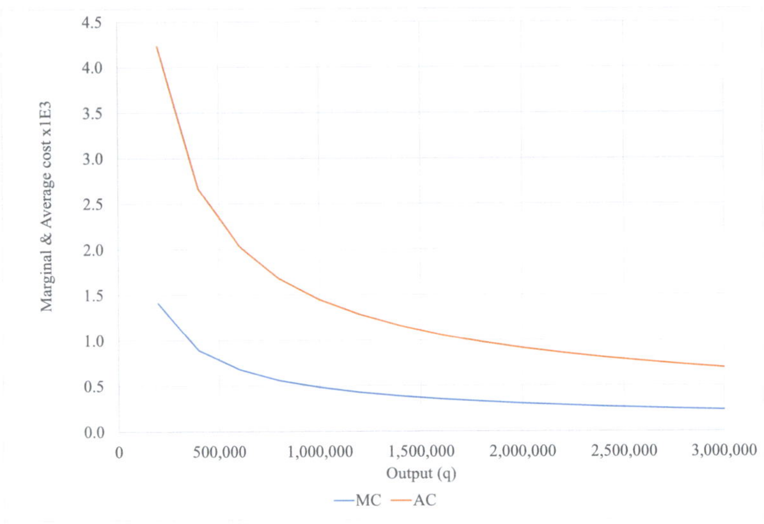

Fig. 1.29 Average cost (orange—AC) and marginal cost (blue—MC) as a function of output quantity. Problem 1.1

are decreasing functions of the output quantity, the marginal cost being lower than the average cost.

We could formulate the dual problem as:

Compute the maximum quantity that can be produced at a cost of 1448.23.

The problem is formulated as follows:

$$\max\left[0.1K^2L\right]$$

$$s.t.\ 3K + 5L - 1448.23 = 0$$

The reader is invited to solve this problem using the Lagrangian method.

Of course, the optimal input quantities are identical to those before and the maximum quantity is $q = 10^6$. The λ parameter is $\lambda = \frac{\partial q}{\partial TC} = 2071.48$ and represents how many units of output are produced when we increase the total cost in one unit.

Theory of Consumers

2

Abstract

In this chapter, we address the theory of consumers, which is a fundamental concept in the field of engineering of power systems economics. The utility function is key in the development of the theory of consumers. It represents the preferences and satisfaction of consumers for different combinations of goods and services. By quantifying utility, energy engineers can understand consumers' preferences and make informed decisions regarding resource allocation and pricing strategies in the power systems context. The budget line reflects the budgetary constraints faced by consumers. It represents the combinations of goods and services that consumers can afford given their income and the prices of goods. Energy analysts assess the budget line to understand consumers' purchasing power and the trade-offs they face when allocating their limited resources. The optimisation problem, from the consumers' viewpoint, involves maximising utility subject to the constraints represented by the budget line. Engineers can employ this method to determine the optimal consumption bundle that maximises consumer satisfaction within their budgetary limitations. The demand function is derived from the optimisation problem and represents the relationship between the price of a good or service and the quantity demanded by consumers. Understanding the demand function enables energy engineers to analyse the responsiveness of consumers to price changes and make predictions about consumer behaviour in power systems markets. This knowledge contributes to the efficient operation and management of power systems, ultimately benefiting both consumers and providers of electricity.

2.1 Introduction

The consumers are, alongside the producers, a main economic agent, namely in the power systems industry. The theory that explains their behaviour is similar to the theory of the producers, with the necessary adaptations.

R. Castro, *Engineering of Power Systems Economics*,
https://doi.org/10.1007/978-3-031-55251-9_2

The principal assumption upon which the theory of consumer behaviour and demand is built is: “A consumer attempts to allocate his/her limited money income among available goods and services so as to maximise his/her utility (satisfaction)”. In what concerns the producers, a similar basic principle could be formulated: “A producer attempts to allocate the minimum possible cash to produce a given quantity of output”.

A consumer could be defined as an economic agent that chooses and executes a consumption plan. And what is the definition of a consumption plan? It is a consumer action that specifies the quantities of the products that allow said consumer to obtain maximum satisfaction. Surely, the consumption plan should consider several limitations, the main ones being, perhaps, of a budgetary and physiological nature. We can agree that maximum satisfaction is the objective of the consumer, but to attain such an objective, several restrictions should be considered, and a few procedures have to be followed.

The theory of consumers is a fundamental concept in economics that explores how individuals make choices and allocate their limited resources to maximise their well-being or utility. It provides a framework for understanding how consumers make rational decisions based on their preferences, budget constraints, and market prices.

The utility function is a mathematical representation of an individual’s preferences over different goods and services. It assigns a level of satisfaction or utility to different combinations of goods consumed. The utility function allows engineer-s and economists to quantify and analyse consumer preferences and choices in a meaningful way.

The budget line represents the combinations of goods and services that a consumer can afford given their income and the prevailing market prices. It illustrates the trade-off between different goods due to limited resources.

Consumers face an optimisation problem: how to allocate their limited income among different goods to maximise their utility? The objective is to choose the combination of goods that provides the highest level of utility within the budget constraint. Mathematically, this problem can be formulated as maximising the utility function subject to the budget constraint. It can be solved using the Lagrangian method, in a similar way we have seen in the theory of producers.

The demand function represents the relationship between the price of a good and the quantity demanded by a consumer. It expresses how consumers respond to changes in prices while aiming to maximise their utility. The demand function shows the quantity of a good that a consumer is willing and able to purchase at different price levels, holding other factors constant, such as income and the prices of other goods.

By leveraging these concepts, engineers can gain insights into consumer preferences, optimise resource allocation, design effective pricing strategies, and understand the dynamics of demand in power systems markets.

This chapter is devoted to the theory of the consumers, which, alongside the producers, compose the two main agents that operate the market. It is going to be shown that most of the related concepts have equivalent concepts in the theory of the producers. To begin with, the concept of *utility function*, the equivalent of the production function, is introduced, involving the *total utility* and *marginal utility*, *indifference curves* and *marginal rate of substitution*. Also, the *budget line* is addressed. A similar *optimisation problem* can be formulated as *what is the maximum utility a consumer may derive from the consumption of a bundle of goods?*. Again, the *Lagrangian method* is used to solve the optimisation problem, with the aid of the *Lagrange multiplier*, here known as *marginal utility of wealth*. In the optimisation problem context, the *substitution effect* and the *income effect* are discussed. Next, typical utility functions, the most known being the *Cobb-Douglas utility function*, are dealt with, emphasising their main features. The chapter ends with an explanation about *demand functions* and the important concept of *elasticity of demand* is introduced. As usual, at the very end, a set of application exercises is proposed.

2.2 Consumer Related Concepts

For the consumer, the products and services are the available objects of choice. We define a product vector as a list of the quantities of the various products, for instance:

$$a = [a_1, a_2, a_3, ..., a_i, ..., a_L] \tag{2.1}$$

where a_i is the quantity of product i.

A consumption set is a subset of the products space, whose elements are the consumption bundles that the individual can consume.

The unitary prices are represented by a price vector:

$$p = [p_1, p_2, p_3, ..., p_i, ..., p_L] \tag{2.2}$$

where p_i is the unitary price of product i.

It is assumed that consumers are price-takers, meaning that they cannot influence the price at which they buy the product. The price is defined elsewhere, using mechanisms that will be detailed later in this book.

The wealth of a particular consumer is designated by W. It is said that a consumption bundle is affordable for a particular consumer if:

$$p_1 \times a_1 + p_2 \times a_2 + ... + p_i \times a_i + ... + p_L \times a_L \leq W \tag{2.3}$$

By this time, we should already have acknowledged that a bundle of goods is a combination of goods quantities. For instance, if X is the number of slices of pizza (quantity) and Y is the number of glasses of juice (quantity), examples of bundles are the sets P, Q, R, S defined as:

- $P : X = 1, Y = 1$
- $Q : X = 3, Y = 0$
- $R : X = 0, Y = 3$
- S: $X = 2, Y = 1$

In many cases, it is convenient to have a way of expressing the consumer's preferences over bundles of goods. Preferences are a nuclear concept in the theory of consumers. The preferences of a particular consumer over the bundles of goods we have introduced before could be stated, for instance, as: P is preferred to both Q and R; S is preferred to P.

This works fine because there are only 4 bundles. However, this way of representing preferences, by enumerating all the possibilities, would be very messy if we had many bundles. In this section, we will come up with a simple way of representing preferences over bundles of goods. This proves to be very useful because, in the real world, there are many bundles of goods.

Before we proceed, we need to define some requirements the preferences should obey to ensure the preferences are expressed in an organised way, therefore allowing the method to work. These requirements are the three Axioms of Rational Choice.

- Completeness—if A and B are any two bundles, an individual can always specify exactly one of these possibilities: - A is preferred to B; - B is preferred to A; - A and B are equally attractive.
- Transitivity—if A is preferred to B, and B is preferred to C, then A is preferred to C. This is to ensure the individual's choices are internally consistent.
- Continuity—if A is preferred to B, then bundles suitably "close to" A must also be preferred to B.

2.3 The Utility Function

Given the rules defined by the axioms of rational choice, it is possible to show that consumers can rank all possible bundles from the least desirable to the most.

In economic theory, this is called ranking utility. This may be represented by a function, whose meaning is "if bundle A is preferred to bundle B, then the utility assigned to A exceeds the utility assigned to B". Mathematically, we can write:

$$U(A) > U(B) \tag{2.4}$$

where U is the utility function.

In the same way, if the consumer is indifferent to the choice between bundle A and bundle B, then:

$$U(A) = U(B) \tag{2.5}$$

The point behind the establishment of a utility function is that the consumer is assumed to have preferences on the consumption bundles so that he/she can compare and rank various goods available in the economy. We will see now whether this can be accomplished either by using numbers or not.

We note that the concept of the utility function in the theory of consumers is equivalent to the production function concept in the theory of producers.

2.3.1 Cardinal Utility and Ordinal Utility

Numbers can be assigned to measure utility. These numbers are called "utils". Respected economists, such as William Jevons (1835–1882), an expounder of the marginal utility theory of value, Léon Walras (1834–1910), who formulated the marginal theory of value, Alfred Marshall (1842–1924), one of the founders of neoclassical economics and who brought the ideas of supply and demand together, followed this approach. Still, in many introductory microeconomics textbooks, this approach to measuring utility is still considered effective for teaching purposes. For instance, one can state that "Jack derives 10 utils from having one slice of pizza but only 5 utils from having a burger". This is called cardinal utility.

There is another trend that advocates that utility should be measured by comparison (ordinal utility). Another set of renowned economists, such as Vilfredo Pareto (1848–1923), who made important contributions to income distribution, and introduced the concept of Pareto efficiency, John Hicks (1904–1989), responsible for stating the consumer demand theory, and Eugen Slutsky (1880–1948), developer of the Slutsky equation widely used in consumer theory, were in favour of this approach. For instance, it could be said in a simple way that "Jill prefers a burger to a slice of pizza and a slice of pizza to a hotdog", but more precision could be added if one says, "Jill is willing to trade a burger for four hotdogs, but she will give up only two hotdogs for a slice of pizza". Translating in ordinal utility, it would become "To Jill, a burger has twice as much utility as a slice of pizza, and a slice of pizza has twice as much utility as a hotdog".

For the supporters of this approach, the difference in the utility numbers is meaningless. The important thing about the numbers is that they can be used to represent rankings (orderings). Because utility measures are not unique, it makes no sense to consider how much more utility is gained from A than from B. This gain in utility will depend on the scale, which is arbitrary—any transformation that preserves the ordering (multiply by a positive number, take it at a power of a positive number, take log) will give the same ordering.

As far as adherence to reality is concerncd, it is fair to say that ordinal utility is closer to the real world than cardinal utility. However, as said before, for teaching purposes, the cardinal utility approach is more pedagogic, which is why we are going to use it further in this book.

2.3.2 Total Utility and Marginal Utility

Total utility is the utility a consumer derives from the consumption of all the units of a good or a bundle of goods.

Table 2.1 Example of total utility and marginal utility for ice cream

Quantity	Total utility	Marginal utility
0	0	
1	40	40
2	85	45
3	120	35
4	140	20
5	150	10
6	157	7
7	160	3
8	160	0
9	155	−5
10	145	−10

Marginal utility is the utility a consumer derives from the last unit of a good or a bundle of goods. The marginal utility can be mathematically expressed as:

$$MU = \frac{\partial U}{\partial a_1} \approx \frac{\Delta U}{\Delta a_1} \tag{2.6}$$

where U is the utility function and a_1 is the quantity of good 1. The equivalent concept in the theory of producers is the marginal product.

Of course, both definitions hold for a given consumption period.

In a similar way to the theory of the producers, where we have introduced the law of diminishing marginal productivity, we can now present the *Law of Diminishing Marginal Utility*. It can be stated as follows: the more of a good (or bundle of goods) a consumer has consumed, the less marginal utility an additional unit contributes to his or her overall satisfaction (total utility). Alternatively, we could say that as more and more of a good is consumed by a consumer, beyond a certain point, the marginal utility of additional units begins to fall. In the example (Table 2.1), this is apparent.

2.3.3 Indifference Curves

We recall that in the theory of producers, we had the concept of isoquants, defined as the combinations of inputs' quantities that give rise to the same output quantity (production function). Here, we now introduce the *Indifference Curves*, as the combination of goods quantities that lead to the same utility. Therefore, the consumer is indifferent between the choice of any two pairs of goods quantities located in the same indifference curve (see Fig. 2.1 where a_1 is the quantity of good 1 and a_2 is the quantity of good 2).

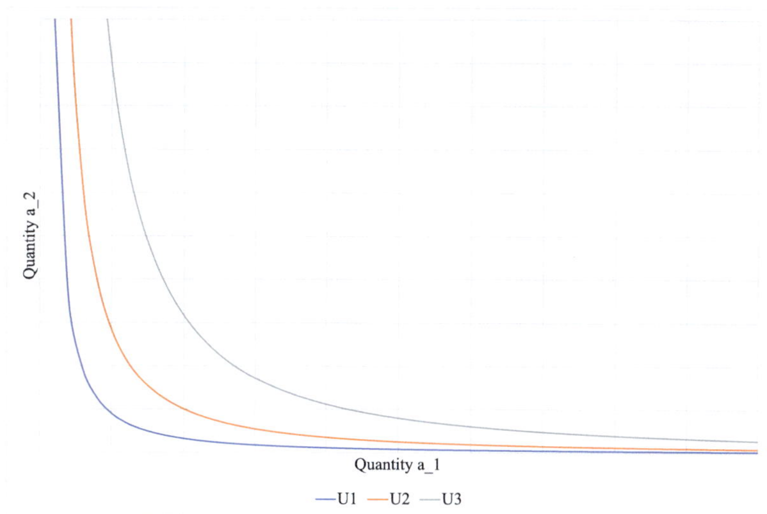

Fig. 2.1 Indifference curves. $U(a_1, a_2) = constant$

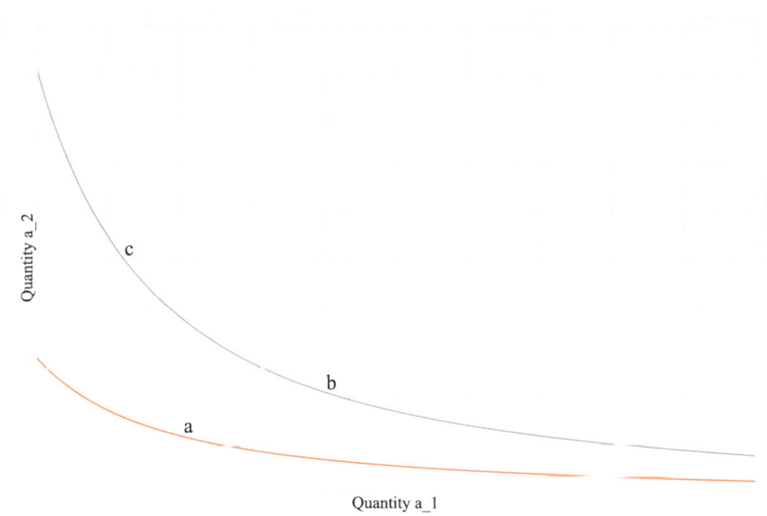

Fig. 2.2 Example with two indifference curves

Consider two indifference curves as in Fig. 2.2. Every point in the higher indifference curve corresponds to a higher preference than any point in the lower indifference curve. For instance, point b is preferred to point a; points b and c are indifferent, and, therefore, c is preferred to a.

2.3.4 Marginal Rate of Substitution

Indifference curves generally show negative slopes, illustrating the common-sense fact that, in a bundle of two goods, to achieve the same level of utility, increasing the consumption of one good implies decreasing the consumption of the other good. We now define the *Marginal Rate of Substitution* (MRS) as the rate at which a consumer is willing to substitute one good for the other while receiving the same level of utility. We note the similarity with the Marginal Rate of Technical Substitution ($MRTS$) we have defined in the theory of producers. Mathematically, we write:

$$MRS = \frac{\partial a_2}{\partial a_1} = \frac{\frac{\partial U}{\partial a_1}}{\frac{\partial U}{\partial a_2}} \approx \frac{\Delta a_2}{\Delta a_1} \tag{2.7}$$

As stated before, a negative MRS means that to increase the consumption of one product, the consumer is prepared to decrease the consumption of another product. Also, we can see that the MRS is the ratio of the marginal utility of one good to the marginal utility of the other, in the same way, the $MRTS$ is the ratio of one marginal product to the other. A graphical interpretation of the MRS points to the MRS being the slope of the indifference curve at any single point along the curve. We know that, for common utility functions, the MRS is negative, but oftentimes we drop the negative sign for simplification.

2.4 The Budget Line

So far, we have assumed that any amount of goods and services is readily available for consumption. In reality, consumers face constraints: the income, or in a more technical way, the budget, is limited and goods are obtained at a price.

This leads to the definition of *Budget Line* as the representation, in the plan of the quantities of both goods (two-dimensional space), of all combinations of goods quantities that are available to the consumer given his budget (money income) and the prices of the goods he/she purchases. In mathematical terms, for two goods, we write:

$$p_1 a_1 + p_2 a_2 = W \tag{2.8}$$

where p_i is the unitary price of good i, a_i is the quantity of good i and W is the wealth, or income, of the consumer. The plotting of Eq. 2.8 is a straight line, given by:

$$a_2 = \frac{W}{p_2} - \frac{p_1}{p_2} a_1 \tag{2.9}$$

We note that the slope is $-\frac{p_1}{p_2}$, the ratio of one price to the other. The intercept in the y-axis is $\frac{W}{p_2}$ and the intercept in the x-axis is $\frac{W}{p_1}$.

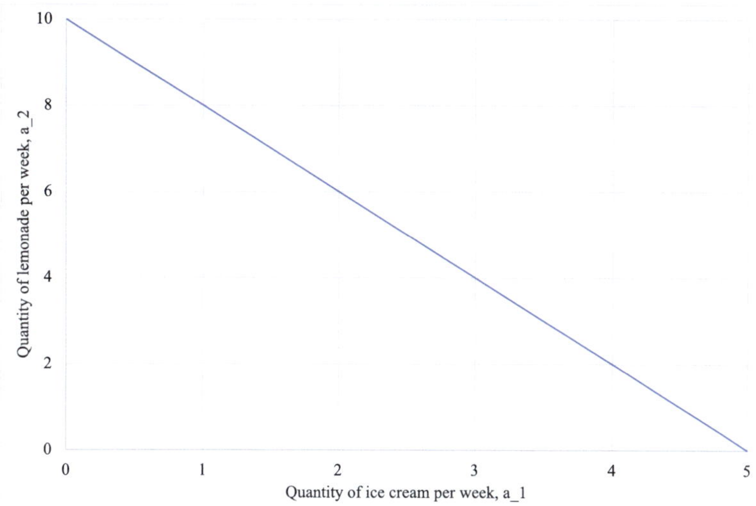

Fig. 2.3 Example of a budget line

Figure 2.3 displays an example of the graphical representation of the budget line. The quantities of the two goods are a_1 (x-axis)—quantity of ice cream per week; a_2 (y-axis)—quantity of lemonade per week. The wealth is $W = 10$ and the (unitary) prices are $p_1 = 2$ and $p_2 = 1$. The budget line equation is $a_2 = 10 - 2a_1$.

To exhaust the income, the consumer can, for instance, buy 3 ice-creams and 4 lemonades, or 5 ice-creams and no lemonade.

2.5 The Optimisation Problem

A similar optimisation problem to the one we have devised in the theory of producers can be formulated in the theory of consumers. By now, the reader should be able to easily understand that the optimisation problem is *What is the maximum utility—objective function—a consumer may derive from the consumption of a bundle of goods, taking into account the limited budget—restriction function—he has available to spend?*. Doing the maths, this becomes, for a bundle of two goods:

$$\max[U(a_1, a_2)] \tag{2.10}$$

s.t.

$$p_1 a_1 + p_2 a_2 = W \tag{2.11}$$

We note that the objective is to compute the optimal quantities, a_1, a_2, and the λ parameter, called the *Lagrange Multiplier*. The utility function, U, the unitary prices of the goods, p_1 and p_2, and W, the consumer wealth, are given.

2.5.1 The Lagrangian Method

The solution of this optimisation problem is carried out exactly in the same way as we did for the producers' optimisation problem, i.e., the Lagrange method is to be used. We begin by writing the Lagrangian function:

$$Lag = U(a_1, a_2) - \lambda(p_1 a_1 + p_2 a_2 - W) \tag{2.12}$$

Then, we know that, at the optimal point, the following conditions hold:

$$\begin{aligned} \frac{\partial Lag}{\partial a_i} &= 0 \\ \frac{\partial Lag}{\partial \lambda} &= 0 \end{aligned} \tag{2.13}$$

For a bundle of 2 goods, Eq. 2.13 results in a system of 3 equations and 3 unknowns—a_1, a_2 and λ. The usual solution method involves dividing the first equation by the second, to eliminate the λ parameter, obtain a relationship between the two quantities and replace this relationship in the third equation, which does not depend on the Lagrange multiplier, to obtain the value of one unknown quantity. The other unknown quantity is obtained from the relationship previously mentioned. The λ parameter is computed from the first or second equation, whatever is simpler. The reader is invited to recall the optimisation section of the theory of producers chapter.

Based on the development of the Lagrangian method, it is possible to come up with a more straightforward process of computing the 3 unknowns. The first equation of Eq. 2.13 is equivalent to:

$$\frac{\partial U}{\partial a_i} = \lambda p_i \tag{2.14}$$

For 2 products, Eq. 2.14 results in:

$$\frac{\frac{\partial U}{\partial a_1}}{\frac{\partial U}{\partial a_2}} = MRS = \frac{MU_1}{MU_2} = \frac{p_1}{p_2} \tag{2.15}$$

From Eq. 2.15, it is possible to derive a relationship between the two quantities, a_1 and a_2. This relationship is plugged in the restriction Eq. 2.11 and the value of the quantity of one good is obtained. The quantity of the other good comes from the relationship just derived.

2.5.2 The Lagrange Multiplier

To obtain the Lagrange multiplier, one notes that from Eq. 2.14 it is possible to write:

$$\frac{\frac{\partial U}{\partial a_1}}{p_1} = \frac{\frac{\partial U}{\partial a_2}}{p_2} = \lambda \tag{2.16}$$

This allows obtaining the unknown λ parameter. In short, Eqs. 2.15, 2.11 and 2.16 allow for the computation of the 3 unknowns—a_1, a_2 and λ.

Now, it is time to take a closer look at the λ parameter, the Lagrange multiplier. Developing Eq. 2.14 for 2 products, results in:

$$\begin{aligned}\frac{\partial U}{\partial a_1} &= \lambda p_1 \\ \frac{\partial U}{\partial a_2} &= \lambda p_2\end{aligned} \tag{2.17}$$

Taking the differentials, we obtain:

$$\frac{\partial U}{\partial a_1} da_1 + \frac{\partial U}{\partial a_2} da_2 = \lambda(p_1 da_1 + p_2 da_2) \tag{2.18}$$

which results in:

$$dU = \lambda dW \overset{\text{yields}}{\rightarrow} \lambda = \frac{dU}{dW} \tag{2.19}$$

The λ parameter is called the *Marginal Utility of Wealth* and measures the change in utility due to a change of one monetary unit in the consumer's income. We recall that, in general, the Lagrange multiplier describes the change in the objective function when the restriction function is varied in one unit.

We have just explained the analytical solution to the utility maximisation problem. The graphical interpretation follows in Fig. 2.4. The consumer's utility is maximised at point A where the indifference curve is tangent to the budget line.

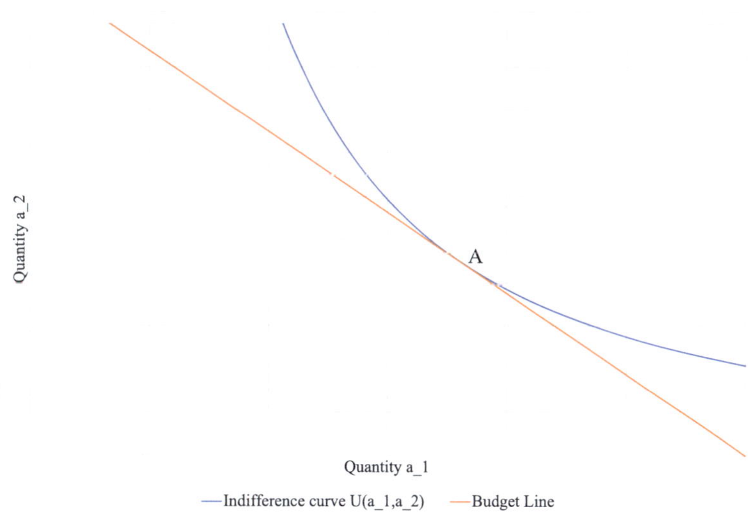

Fig. 2.4 Graphical interpretation of the consumer's optimisation problem

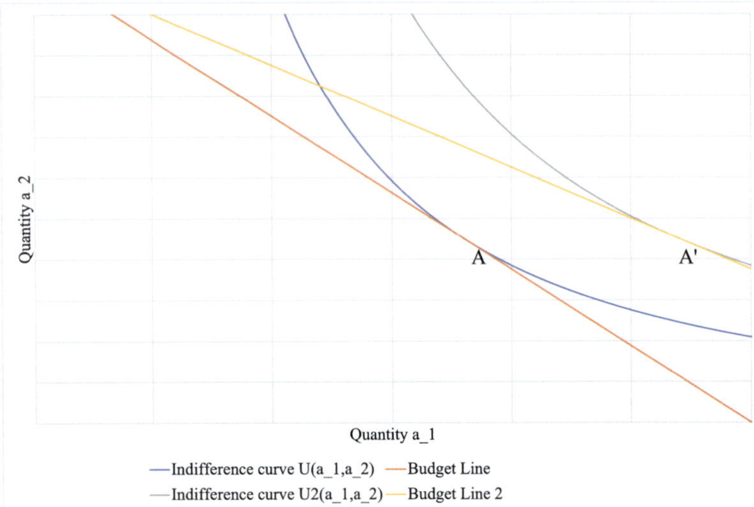

Fig. 2.5 Changing the budget line. Example with decreasing p_1, while keeping W and p_2 constant

2.5.3 Changes in the Budget Line

What happens to the optimal point (a_1^*, a_2^*) if the budget line changes? Let us suppose the price of good 1 decreases from p_1 to p_1', as represented in Fig. 2.5, while the wealth and price of good 2 are kept constant. The absolute value of the budget line slope, $\frac{p_1}{p_2}$, decreases, so the new budget line is less sloped than the original. The y-intercept, $\frac{W}{p_2}$, remains unchanged. This implies the optimal point moving to a higher indifference curve (from point A to point A'). The consumer can now buy more units of good 1 while receiving a higher utility.

2.5.4 Substitution Effect and Income Effect

In fact, the situation that occurs when the price changes can be further explained. When the price of a product changes, there are two effects to be considered. On one side, the price ratio changes, and, on the other side, the real income of the consumer changes. The first effect is called *Substitution effect* and the second is called *Income effect*. Let us look at an example to better explain the concept. In Fig. 2.6, initially, the consumer is consuming 18 units of good 1 (point A) and is located in the indifference curve IC_1 and budget line BL_1.

Suppose, the price p_1 decreases (we recall that the x-intercept is $\frac{W}{p_1}$), while keeping the same p_2 (the y-intercept is $\frac{W}{p_2}$) and the same wealth W. The final operating point corresponds to 27 units consumed (point B) and the consumer is now located in indifference curve IC_2 and budget line BL_2. This movement may be viewed as reflecting two separate effects. The first effect—substitution effect (also known as

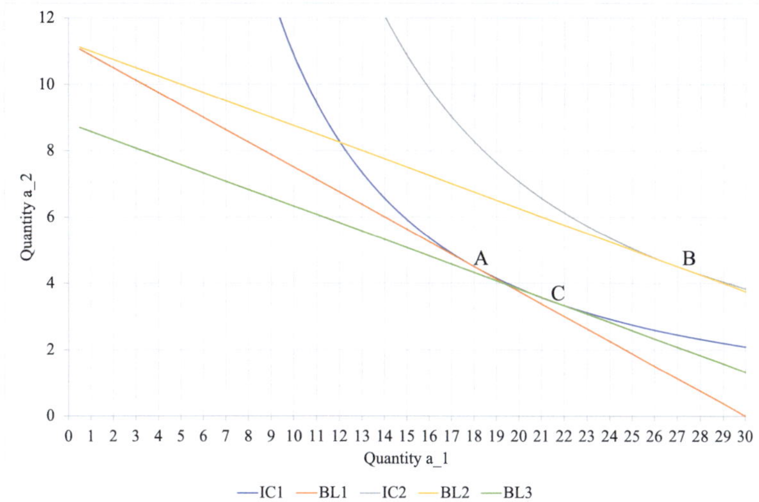

Fig. 2.6 Substitution effect and income effect

the Hicksian substitution effect after the economist John Hicks)—is computed by considering solely the price ratio change. It is determined as the point where the initial indifference curve (IC_1) meets an imaginary budget line ($BL3$) with the new price ratio and the income that would be necessary to keep the consumer in the original indifference curve. This "intermediate" budget line is attempting to hold real income fixed so we can isolate the substitution effect. In this hypothetical situation, the consumer would buy 21 units of good 1 (point C).

Then, the other effect—income effect—comes into place. The real income has been increased because when the price goes down the consumer is able to buy more units of goods 1 than he/she could have purchased before. In real terms, the consumer has more purchasing power. Hence, the consumer moves to a higher indifference curve, due to the increase in the individual's purchasing power brought about by one of the goods becoming cheaper. The final budget line is plotted, allowing us to determine the final consumption bundle as the tangency point with the higher indifference curve. The income effect is measured as the difference between the imaginary quantity and the final quantity (27 units). In this case, the variation in the consumption of a_1 is $27 - 18 = 9 = 27 - (21 - 3)$.

Example 2.2

Let us retake the example depicted in Fig. 2.6. The utility function is $U(a_1, a_2) = a_1^3 a_2^2$. The initial prices are $p_1 = 1.5$ and $p_2 = 4$. The final prices are $p_1' = 1$ and $p_2 = 4$. The wealth is 45. Determine points A, B and C, using the analytical method. ◄

Solution

Point A is the initial optimal point, i.e., the optimal quantities of goods 1 and 2 that maximise the utility. Recalling what we have learned, and as a consequence of the Lagrangian method, we have:

$$a_1^* = \frac{\alpha}{\alpha + \beta} \frac{W}{p_1} = 18$$

$$a_2^* = \frac{\beta}{\alpha + \beta} \frac{W}{p_2} = 4.5$$

$$U = 18^3 \times 4.5^3 = 118{,}098$$

Point B is the final optimal point, after the price change. As the price of good 1 decreases, more units of this good can be bought and a higher utility is achieved.

$$a'^*_1 = \frac{\alpha}{\alpha + \beta} \frac{W}{p'_1} = 27$$

$$U' = 27^3 \times 4.5^3 = 398{,}580.75$$

Quantity a_2 remains unchanged.
Point C represents the Hicksian substitution effect. It is determined in the initial indifference curve by considering an imaginary budget line with a slope equal to the final price ratio. We need to compute the income, W', that would be necessary to keep the consumer in the initial indifference curve with the new price ratio.

$$U(W') = \left(\frac{\alpha}{\alpha + \beta} \frac{W'}{p'_1}\right)^3 \left(\frac{\beta}{\alpha + \beta} \frac{W'}{p_2}\right)^2$$

$$118{,}098 = \left(\frac{3}{5} \frac{W'}{1}\right)^3 \left(\frac{2}{5} \frac{W'}{4}\right)^2 \overset{\text{yields}}{\rightarrow} W' = 35.2824$$

$$a''^*_1 = \frac{3}{5} \frac{35.2824}{1} = 21.1694$$

$$a''^*_2 = \frac{2}{5} \frac{35.2824}{4} = 3.5282$$

◄

2.6 Typical Utility Functions

2.6.1 Cobb-Douglas

The Cobb-Douglas utility function is perhaps the most known utility function. The general formula of the Cobb-Douglas utility function is (we will stick to examples with 2 products:

$$U(a_1, a_2) = a_1^{\alpha} a_2^{\beta} \tag{2.20}$$

where a_1 and a_2 are the quantities of the 2 products and α and β are parameters and $\alpha, \beta \geq 0$.

A typical representation of a Cobb-Douglas utility function is shown in Fig. 2.7. It is observed that the indifference curves never intercept the axis: utility is zero for either a_1 or a_2 equal to zero. Also, the indifference curves are convex, because as the consumer consumes more of one good, he/she will consume less of the other.

The marginal utilities are:

$$\begin{aligned} MU_1 &= \alpha a_1^{\alpha-1} a_2^{\beta} \\ MU_2 &= \beta a_2^{\beta-1} a_1^{\alpha} \end{aligned} \tag{2.21}$$

We point out that the marginal utilities are positive and, taking, for instance, MU_1 and $\alpha < 1$, when a_1 goes to infinity, MU_1 goes to zero. This behaviour is explained by the law of diminishing marginal utility, which states that, beyond a certain quantity, the utility of consuming an additional unit of the good declines.

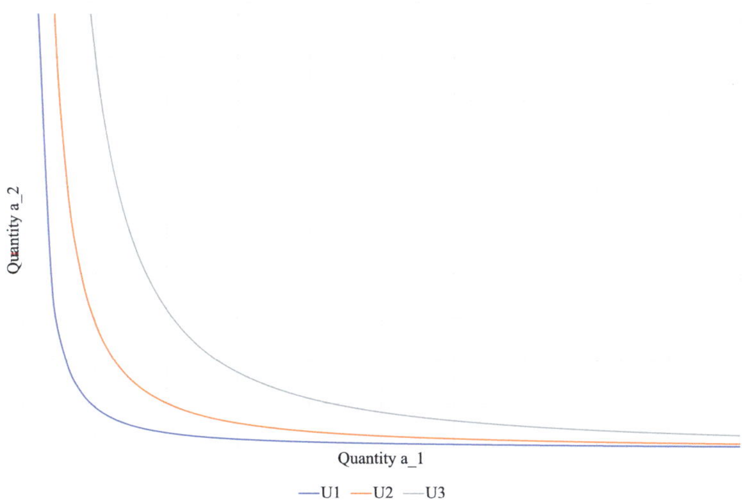

Fig. 2.7 Typical Cobb-Douglas utility function

The marginal rate of substitution of good 1 by good 2 is:

$$MRS_{1,2} = \frac{MU_1}{MU_2} = \frac{\alpha a_2}{\beta a_1} \tag{2.22}$$

Again, when a_1 goes to infinity, $MRS_{1,2}$ goes to zero ($MRS_{1,2}$ is decreasing) because the utility function is convex, in this case.

Based on the relationships we have derived when solving the optimisation problem, we have seen that to find the optimal quantities, we need to solve the following system of two equations:

$$\begin{aligned} \frac{MU_1}{MU_2} &= \frac{p_1}{p_2} \\ p_1 a_1 + p_2 a_2 &= W \end{aligned} \tag{2.23}$$

Applying to the case of the Cobb-Douglas utility function, where $\frac{MU_1}{MU_2} = \frac{\alpha a_2}{\beta a_1}$, from the first equation of Eq. 2.23 we obtain:

$$p_1 a_1 = \frac{\alpha p_2 a_2}{\beta} \tag{2.24}$$

and replacing in the second equation of Eq. 2.23 (budget line) leads to:

$$\begin{aligned} a_1^* &= \frac{\alpha}{\alpha+\beta}\frac{W}{p_1} \\ a_2^* &= \frac{\beta}{\alpha+\beta}\frac{W}{p_2} \end{aligned} \tag{2.25}$$

This means that, with a Cobb-Douglas utility function, the consumer spends $\frac{\alpha}{\alpha+\beta}$ of the income on good 1 and $\frac{\beta}{\alpha+\beta}$ of the income on good 2.

To find the marginal utility of wealth, we must resort to the Lagrangian method. At this stage, it should be apparent to the reader that, at the optimal point, the λ parameter can be computed from:

$$\lambda = \frac{\alpha}{p_1} a_1^{*\alpha-1} a_2^{*\beta} \tag{2.26}$$

Example 2.3

Consider two products, X and Y, available for a consumer to buy. The utility function for these products is $U(X, Y) = X^{0.5}Y^{0.5}$. The consumer has a wealth of 800 and the unitary prices of the products are $p_X = 4$ and $p_Y = 1$.

1. In the plane (X, Y) draw the indifference curves for $U = 50$, $U = 100$ and $U = 200$ and the budget line.
2. Compute the maximum utility the consumer can derive from the consumption of these products, given the income he/she has available.
3. How will the utility change if the income is increased in one unit?

4. Find the marginal utility of good X at the optimal point.
5. Compute the marginal rate of substitution of good X by good Y at the optimal point.

◀

Solution

1.

The indifference curves obey the equation:

$$Y = \left(\frac{U}{X^{0.5}}\right)^2, U = 50, 100, 200$$

and the budget line is:

$$Y = 800 - 4X$$

The results are displayed in Fig. 2.8.

It can be seen graphically that, for the available income, the consumer should consume 100 units of good X and 400 units of good Y, receiving a utility of 200.

2.

Now, let us prove it in an analytical way. We will solve this example by using the corollary of the Lagrange method.

$$\frac{MU_X}{MU_Y} = \frac{p_X}{p_Y} \overset{\text{yields}}{\rightarrow} \frac{0.5X^{-0.5}Y^{0.5}}{0.5X^{0.5}Y^{-0.5}} = \frac{Y}{X} = 4 \overset{\text{yields}}{\rightarrow} Y = 4X$$

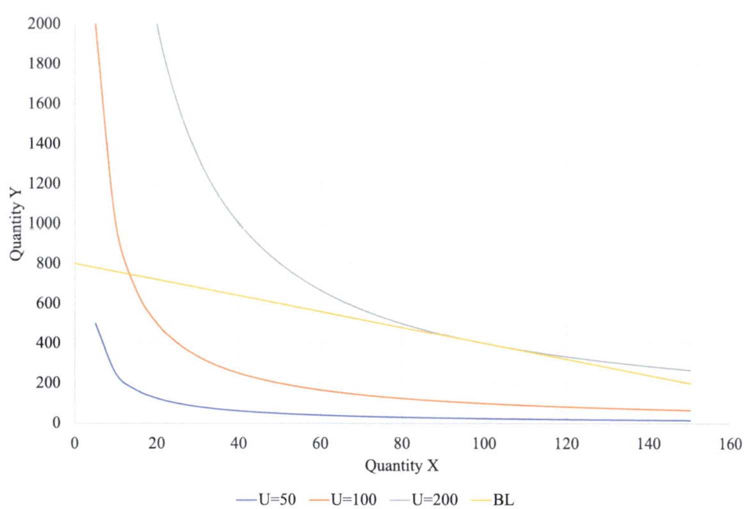

Fig. 2.8 Indifference curves and budget line for Example 2.3

Plugging in the budget line equation, $4X + Y = 800$, one obtains $X^* = 100$ and $Y^* = 4X^* = 400$. These are the optimal quantities the consumer should buy in order to achieve maximum utility. The maximum utility is $U^* = X^{*0.5}Y^{*0.5} = 200$

3.

The Lagrange multiplier is called the marginal utility of wealth in the theory of consumers. It represents how much the utility will change when the wealth is increased in one unit. So, in this question, the objective is to find the λ parameter. We recall that one way of computing it is:

$$\lambda = \frac{MU_Y}{p_Y} = 0.5X^{*0.5}Y^{*-0.5} = 0.25$$

It is concluded that, when the wealth is increased by one unit, the utility increases 0.25 units. The reader is invited to compute the utility for an income of 801.

4.

At the optimal point, the marginal utility of X is:

$$MU_X = 0.5X^{*-0.5}Y^{*0.5} = 1$$

This means that, when the quantity of good X is increased in one unit, the utility is also increased in one unit. This applies at the optimal point. Therefore, the reader is invited to show that $U(101, 400) = 201$.

5.

At the optimal point, the MRS is:

$$MRS_{X,Y} = \frac{p_X}{p_Y} = 4$$

The physical interpretation is that, at the optimal point, the consumer can change one unit of X by four units of Y, while receiving the same utility. Therefore, it is true that $U(99, 404) = 200$. ◀

2.6.2 Perfect Substitutes

The general equation for a perfect substitutes utility function is:

$$U(a_1, a_2) = \alpha a_1 + \beta a_2 \tag{2.27}$$

The typical aspect of perfect substitutes utility functions is depicted in Fig. 2.9.

Typical examples of perfect substitutes are gold from different mines, two banknotes of ten Euros, in general, any commodity from different providers. For instance, an individual is ready to change a banknote of 10 Euros for another one. This is why the indifference curves are straight lines with a negative slope.

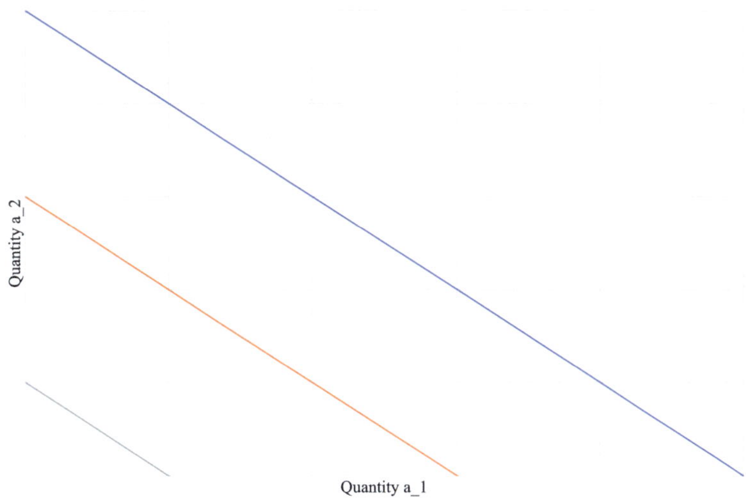

Fig. 2.9 Typical perfect substitutes utility function

The two marginal utilities are now constant:

$$\begin{aligned} MU_1 &= \alpha \\ MU_2 &= \beta \end{aligned} \tag{2.28}$$

The same goes for the marginal rate of substitution:

$$MRS_{1,2} = \frac{\alpha}{\beta} \tag{2.29}$$

As the two goods are perfect substitutes, the consumer shows a constant rate at which he/she is willing to replace one good with the other. This also explains that the indifference curves are straight lines.

It is observed that except for the particular case $\frac{\alpha}{\beta} = \frac{p_1}{p_2}$, the MRS is not equal to the price ratio. In fact, for this case, we need to compare the MRS with the price ratio. We recall that the price ratio is the slope of the budget line. Before we go on, we introduce the following concepts with respect to good 1:

$$\begin{aligned} MB_1 &= \frac{MU_1}{p_1} \\ MC_1 &= \frac{MU_2}{p_2} \end{aligned} \tag{2.30}$$

MB_1 is the *Marginal Benefit* of good 1, i.e., the marginal utility per monetary unit spent on buying good 1. MC_1 is a measure of the *Marginal Cost* of good 1, i.e., the marginal utility per monetary unit that the consumer is not receiving by buying product 2, instead of product 1.

Returning to the question at hand, if $MRS > \frac{p_1}{p_2}$, it follows that the marginal benefit of good 1 always exceeds its marginal cost, therefore, the consumer will spend all the wealth on buying product 1 and zero units of good 2 will be bought. The inverse happens if $MRS < \frac{p_1}{p_2}$, only good 2 will be chosen. If, by any chance, $MRS = \frac{p_1}{p_2}$, the marginal benefit is equal to the marginal cost and, hence, any consumption bundle located in the indifference curve will maximise the utility. As long as all money is spent, any split of a_1 and a_2 is optimal.

Mathematically, we write:

$$\begin{aligned} &\text{If } MRS > \frac{p_1}{p_2} \rightarrow a_1^* = \frac{W}{p_1}; a_2^* = 0 \\ &\text{If } MRS < \frac{p_1}{p_2} \rightarrow a_1^* = 0; a_2^* = \frac{W}{p_2} \end{aligned} \tag{2.31}$$

The perfect substitutes utility function portrays a case of corner solutions. Depending on the relative position of the indifference curves and the budget line, the optimal solution is located either on the y-axis or the x-axis.

Example 2.4

A consumer consumes two goods, whose quantities are x and y, according to the utility function $U(x, y) = 3x + 4y$. The wealth of the consumer is 150.

1. Draw the indifference curves corresponding to the utility levels of 25, 75 and 125.
2. Consider the following budget lines: 1. BL1: $p_{1x} = 6$; $p_{1y} = 8$; 2. BL2: $p_{2x} = p_{1x}$; $p_{2y} = 2p_{1y}$; 3. BL3: $p_{3x} = 2p_{1x}$; $p_{3y} = p_{1y}$; 4. BL4: $p_{4x} = 2p_{1x}$; $p_{4y} = 0.5p_{1y}$. Draw the indifference curve for $U = 75$ and the four budget lines on the same graphic. Use the graphical method to conclude about the optimal bundle of goods for each budget line.

◄

Solution

1.

The indifference curves follow the general equation:

$$y = \frac{U}{4} - \frac{3}{4}x, \ U = 25, 75, 125$$

Figure 2.10 shows the result of the plotting of the three indifference curves. It is noted that they are parallel straight lines, with a decreasing slope.

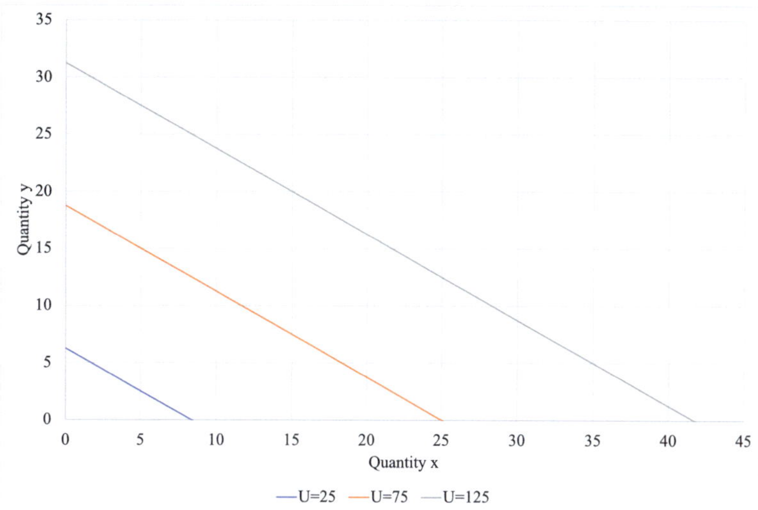

Fig. 2.10 Indifference curves for Example 2.4

2.

The four equations for the budget lines are:

$$BL1: y_1 = 18.75 - 0.75x_1$$

$$BL2: y_2 = 9.375 - 0.375x_2$$

$$BL3: y_3 = 18.75 - 1.5x_3$$

$$BL4: y_4 = 37.5 - 3x_4$$

The plotting of the indifference curve and the four budget lines is depicted in Fig. 2.11.

The indifference curve for $U(x, y) = 75$ is displayed in blue. We can see that $BL1$ (dotted orange line) overlaps the indifference curve. This means that, on one side, the maximum utility is 75, and, on the other side, all the points of the indifference curve are optimal consumption bundles. Therefore, the solution is:

$$x^* \in [0; 25]$$

$$y^* \in [0; 18.75]$$

In what regards $BL2$ (grey line), we also note that the maximum utility is 75 and the consumption basket that maximises the utility is $x^* = \frac{W}{p_x} = 25$, $y^* = 0$ (corner solution). This is because $MRS = 0.75 > \frac{p_x}{p_y} = 0.375$. We recall that, in this case, the price of good y increased twice and the price of good x was kept constant.

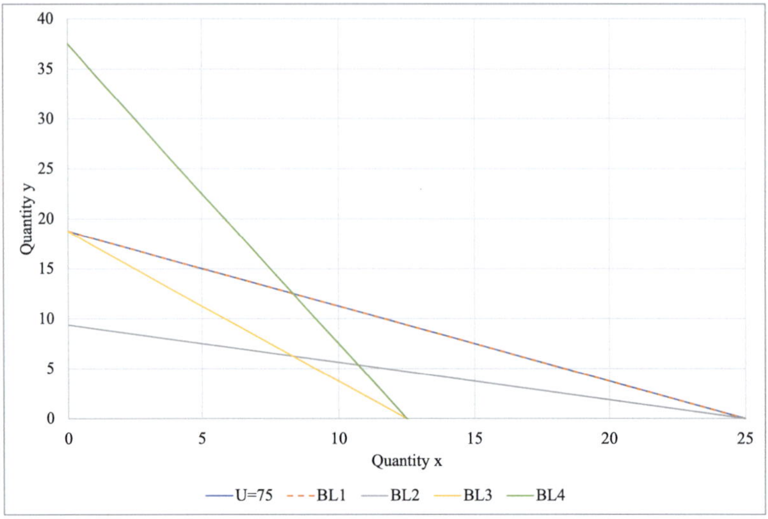

Fig. 2.11 Indifference curve and budget lines for Example 2.4

For the budget line $BL3$ (yellow line), still the maximum utility is 75, this maximum being achieved at $x^* = 0$, $y^* = \frac{W}{p_y} = 18.75$ (corner solution). This is explained by $MRS = 0.75 < \frac{p_x}{p_y} = 1.5$. This conclusion is justified by the fact that the price of good x increased twice, while good x kept the same price.
Finally, in what concerns $BL4$ (green line), it is apparent that the maximum utility is not 75, it is higher than that. The indifference curve should intercept the y-axis at (0; 37.5). Therefore the maximum utility is $U(0; 37.5) = 3 \times 0 + 4 \times 37.5 = 150$. We observe that the other possible corner solution, the x-intercept, corresponds to a lower utility. In this case, the price of x increased twice and the price of good y decreased to one-half. Hence, it is expected that the utility would increase and that the consumption basket only includes good y. ◀

2.6.3 Perfect Complements

The expression for a perfect complements utility function is:

$$U(a_1, a_2) = \min\{\alpha a_1, \beta a_2\} \tag{2.32}$$

Figure 2.12 offers a plotting of typical perfect complements utility function.

The indifference curves are "L-shaped". In this case, we do not have corner solutions, because the indifference curves do not intercept the axis (utility is zero if one good is consumed at zero). Also, no tangency points can be found, because the slope of the indifference curves is zero or infinity. Instead, we say we have a kink solution.

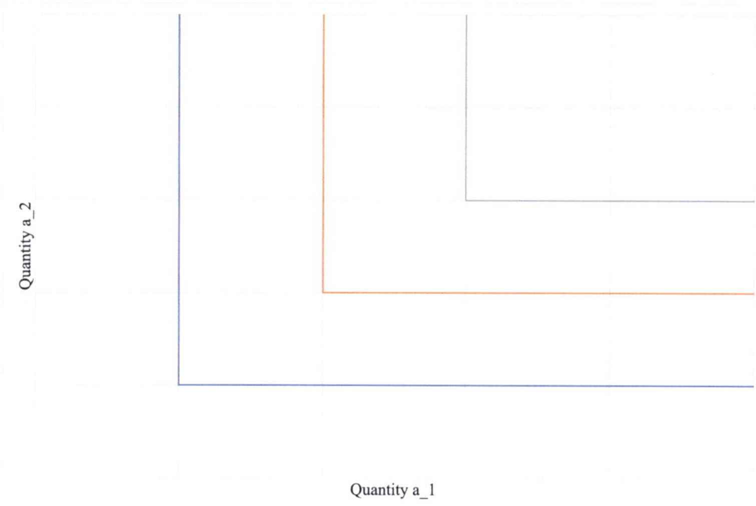

Fig. 2.12 Typical perfect complements utility function

A typical example of perfect complements is left shoes and right shoes. The same utility is obtained for the combination of one left shoe with any number of right shoes greater or equal to one (the plotting is a vertical line if the number of right shoes is plotted on the y-axis). The same applies to one right shoe and any number of left shoes greater or equal to one (the plotting is a horizontal line if the number of left shoes is plotted on the x-axis). This graphic is "L-shaped". The kink is located at point (1; 1). The utility will increase if you have two left shoes, hence we move to a higher indifference curve, but the reasoning remains the same. The kink is now located at point (2, 2).

The utility function is not differentiable, hence we cannot use analytical methods to compute marginal utility and marginal rate of substitution. Moreover, we are unable to represent the utility as a function $a_2 = f(a_1)$. However, we know that the optimal bundle is located at the kink. The kink is determined by setting $\alpha a_1 = \beta a_2$. This means it lies on a straight line whose equation is:

$$a_2 = \frac{\alpha}{\beta} a_1 \tag{2.33}$$

The consumer will always choose a bundle with fixed proportions, $\alpha a_1 = \beta a_2$, as these are the optimal proportions of each product. To find the optimal bundle for each budget line, we need to find the point where the budget line, $W = p_1 a_1 + p_2 a_2$, crosses the straight line given by $a_2 = \frac{\alpha}{\beta} a_1$. This is accomplished by replacing this last equation in the first. The result is:

$$a_1^* = \frac{W}{p_1 + \frac{\alpha}{\beta} p_2}$$
$$a_2^* = \frac{W}{p_2 + \frac{\beta}{\alpha} p_1} \tag{2.34}$$

Example 2.5

The utility function a consumer faces when consuming two goods w and z is $U(w, z) = \min\{2w, 5z\}$. The unitary price of w is 2 and the unitary price of z is 10. The available income to be exhausted in the consumption of the goods is 72.

1. Draw the indifference curves corresponding to the utility levels of 10, 24 and 30, the budget line. Graphically show that the kinks are located across a straight line with equation $z = \frac{2}{5}w$. Based on the graphics, conclude about the optimal consumption basket and the corresponding utility.
2. Compute, using the analytical method, the optimal consumption basket, the maximum utility and the marginal rate of substitution.

◀

Solution

1.

This is a perfect complements utility function. We know that the indifference curves are "L-shaped" around a kink, and we also know that the kinks are located in the straight line obtained after $2w = 5z$. The straight line is $z = \frac{2}{5}w$. For instance, for $U = 10$, at the kink, we have $2w = 5z$, therefore, $w = 5$ and $z = 2$ is the kink, i.e., belongs to the straight line, and corresponds to a utility $U(5, 2) = \min\{2 \times 5, 5 \times 2\} = 10$. The indifference curve for $U = 10$ has two components: the horizontal line $z = 2$ and the vertical line $w = 5$.
The graphics are displayed in Fig. 2.13.
For the given budget line, the interception point of the kinks line and the budget line is (12; 4.8), which defines the optimal bundle. The maximum utility is $U(12, 4.8) = \min\{2 \times 12, 5 \times 4.8\} = 24$.

2.

To compute the optimal bundle we need to solve the system of equations:

$$2w = 5z$$

$$2w + 10z = 72$$

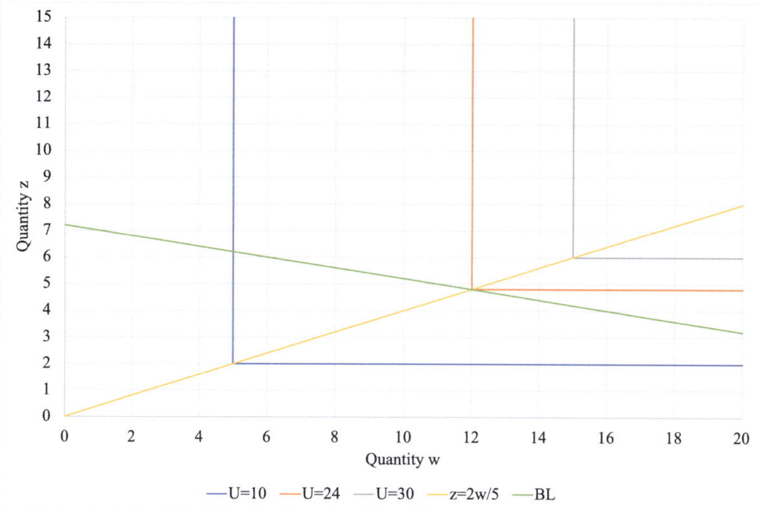

Fig. 2.13 Indifference curves, kinks line and budget line for Example 2.5

The solution is of course $w^* = 12$; $z^* = 4.8$ and the maximum utility is $U = 24$. The same result would be obtained if Eq. 2.34 is used:

$$w^* = \frac{72}{2 + \frac{2}{5} \times 10} = 12$$

$$z^* = \frac{72}{10 + \frac{5}{2} \times 2} = 4.8$$

As far as the marginal rate of substitution is concerned, the concept has no application in this case, because the goods are perfect complements, they are consumed in fixed proportions, and one good is not substitutable by the other. ◀

2.6.4 Quasi-Linear

A quasi-linear utility function can be of two types:

$$\begin{aligned} U(a_1, a_2) &= \alpha a_1 + f(a_2) \\ U(a_1, a_2) &= g(a_1) + \beta a_2 \end{aligned} \tag{2.35}$$

There are two terms: one is linear and the other is non-linear. A typical quasi-linear utility function is in Fig. 2.14. It is observed that the indifference curves are parallel, i.e., vertical shifts one from the other, and can intercept the axis.

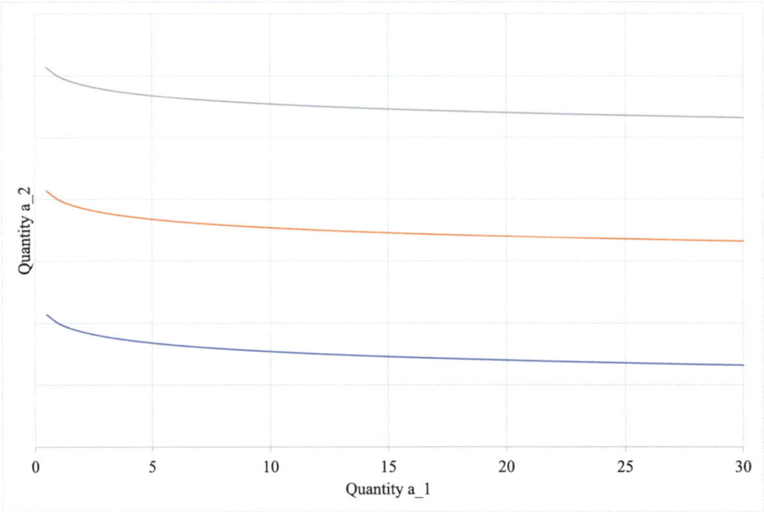

Fig. 2.14 Typical quasi-linear utility function

In this type of preferences, the consumer has a limit on one of the goods he/she can consume, meaning that an increase in the income will be exhausted in the consumption of the other good. In comparison with the Cobb-Douglas function, this is a different feature, because the increase in wealth will not always result in an increase in the consumption of the two goods.

The marginal utilities are (from now on, we focus only on the first type of quasi-linear utility function—first equation of Eq. 2.35 as the reasoning is the same for the second type):

$$\begin{aligned} MU_1 &= \alpha \\ MU_2 &= \frac{df}{da_2} \end{aligned} \tag{2.36}$$

and the marginal rate of substitution is:

$$MRS = \frac{\alpha}{\frac{df}{da_2}} = \frac{\alpha}{f'(a_2)} \tag{2.37}$$

It can be seen that the MU of the variable that enters linearly is constant and the MRS only depends on the non-linear variable.

To solve the optimisation problem, we recall that the fundamental equation is:

$$MRS = \frac{p_1}{p_2} \tag{2.38}$$

As the (unitary) prices of the goods are known and the MRS only depends on a_2, the optimal quantity of good 2 is immediately defined. To compute the optimal quantity of the other good, we use the budget line equation. It happens that the problem has been easily solved, but it is not the case. What if the price of good 2 exceeds the available wealth, i.e., $W < p_2 a_2$? The resulting optimal quantity of good 1 would be negative, which makes no sense at all. To overcome this difficulty, the consumer would buy as many units of good 2 as he/she can afford and the leftovers would be spent in good 1, even if this implies the consumption of zero units of good 1.

Now, it is time to look into the Lagrange multiplier (marginal utility of wealth). To this aim, let us apply the old-school Lagrangian method. The three optimal conditions are (the reader is invited to write the Lagrangian):

$$\begin{aligned}\frac{\partial Lag}{\partial a_1} &= \alpha - \lambda p_1 = 0\\ \frac{\partial Lag}{\partial a_2} &= f'(a_2) - \lambda p_2 = 0\\ \frac{\partial Lag}{\partial \lambda} &= W - p_1 a_1 - p_2 a_2 = 0\end{aligned} \tag{2.39}$$

From the first equation, we immediately conclude:

$$\lambda = \frac{\alpha}{p_1} \tag{2.40}$$

This is interesting because, for a Cobb-Douglas utility function, λ can only be computed after solving for a_1^* and a_2^*. Moreover, in this case, the marginal utility of wealth does not depend on wealth. In fact, it is a constant, depending only on the ratio $\frac{\alpha}{p_1}$. A consequence of this is, as we have already pointed out, when the income increases, the consumer buys more of good 1, because the effect on the utility of an increase in the income is constant.

When we plug in the result of the first equation in the second, we obtain:

$$f'(a_2) = \alpha \frac{p_2}{p_1} \tag{2.41}$$

This allows us to determine (a_2^*). It is verified that the optimal quantity of product 2 also does not depend on wealth, hence, when the income increases, the consumed quantity of good 1 increases. The quantity of good 1 is obtained by resorting to the budget line equation (third equation of Eq. 2.39). If this quantity turns out to be negative, it follows that we are trying to buy more of good 2 than we can afford. Hence, the quantity of good 2 is fixed at the maximum possible and the leftover is spent on buying good 1.

Example 2.6

A consumer's preferences can be described by the utility function $U(a, b) = a^{0.5} + b$, where a and b are the two goods quantities.

1. Draw the indifference curves corresponding to $U = 2$, $U = 2.85$ and $U = 4$, and the budget line $a + 5b = 8$. Based on the graphics you drew, conclude

about the optimal basket and the maximum utility received by the consumption of this bundle.

2. Using the analytical method, show that the optimal basket and the utility are the ones you have just determined in the previous question. Compute the marginal rate of substitution at the optimal point and the marginal utility of wealth.
3. Suppose the income is reduced from 8 to 4. Recalculate the optimal consumption basket and the maximum utility the consumer can derive.

◀

Solution

1.

The indifference curves follow the equation:

$$b = U - a^{0.5};\ U = 2, 2.85, 4$$

The plotting of the indifference curves and the budget line, $b = \frac{8-a}{5}$, is depicted in Fig. 2.15.

From the plotting, it is not obvious, but if one looks at the numbers behind the curves, it is possible to conclude that the budget line is tangent to the indifference curve for $U = 2.85$ at point $(6.25, 0.35)$. This defines the optimal basket and maximum utility.

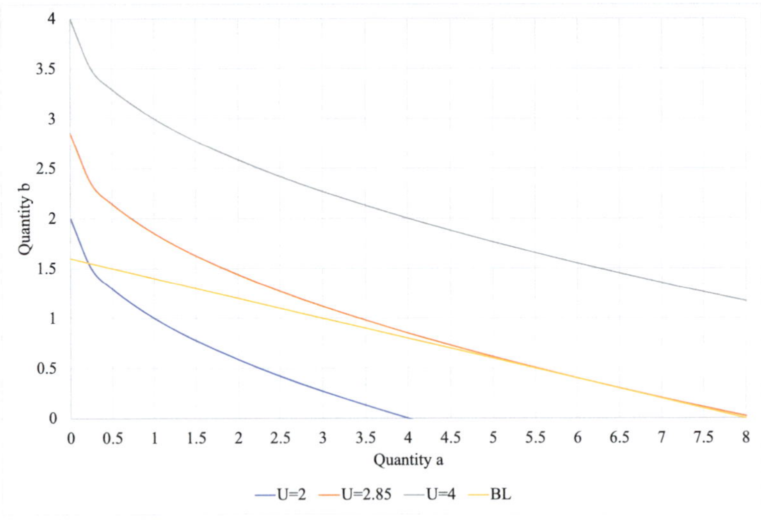

Fig. 2.15 Indifference curves and budget line for Example 2.6

2.
Using the analytical method, we come to:

$$\frac{MU_a}{MU_b} = \frac{p_a}{p_b} \xrightarrow{\text{yields}} 2.5a^{-0.5} = 1 \xrightarrow{\text{yields}} a^* = 6.25$$

The quantity of good a is immediately set because the MRS only depends on a. The remaining income will be used for the consumption of good b. To find this quantity, we use the budget line equation, which leads to $b^* = \frac{8-a^*}{5} = 0.35$. The marginal rate of substitution is:

$$MRS = \frac{MU_a}{MU_b} = 0.5a^{-0.5}$$

At the optimal point, $MRS = 0.2$. In what regards the Lagrange multiplier, we use the equation:

$$\lambda = \frac{MU_b}{p_b} = \frac{1}{5} = 0.2$$

The marginal utility of wealth does not depend on wealth. When the wealth increases, the consumer buys more of good b, because the quantity of good a is fixed beforehand.

3.
The quantity of good a is fixed and does not change with the income reduction ($a^* = 6.25$). If we solve to $b^* = \frac{4-a^*}{5} = -0.45$. This result makes no sense because it leads to a negative quantity consumed. It means that the available income is insufficient to buy all the required units of good a. This is apparent as the income is 4 and the price of the 6.25 units of good a is 6.25. In this case, the consumer would buy all the units of good a he/she can afford ($a_2^* = 4$) and 0 units of good b ($b^* = 0$) will be consumed (corner solution). The utility is $U = 4^{0.5} + 0 = 2$. ◀

2.6.5 Constant Elasticity of Substitution (CES) Functions

The general equation of a *Constant Elasticity of Substitution* (CES) function is:

$$U(a_1, a_2) = \left[\alpha a_1^{\rho} + \beta a_2^{\rho}\right]^{\frac{1}{\rho}} \tag{2.42}$$

In the next section, we will learn more about the concept of elasticity. For the time being, the reader should take note that *Elasticity of Substitution* is defined as the elasticity of the ratio of consumption of two goods to the MRS. We can say that it measures how easily two goods are substitutable along the indifference curve.

$$\varepsilon_S = \frac{\partial \frac{a_2}{a_1}}{\partial MRS} \frac{MRS}{\frac{a_2}{a_1}} \tag{2.43}$$

The elasticity of substitution is constant for CES functions.

It is possible to demonstrate (the reader is invited to do it) that the optimal consumption bundle is given by:

$$a_1^* = \frac{W}{p_1\left(1+\left(\frac{\alpha}{\beta}\right)^{\frac{1}{\rho-1}}\left(\frac{p_2}{p_1}\right)^{\frac{\rho}{\rho-1}}\right)}$$
$$a_2^* = \frac{W}{p_2\left(1+\left(\frac{\beta}{\alpha}\right)^{\frac{1}{\rho-1}}\left(\frac{p_1}{p_2}\right)^{\frac{\rho}{\rho-1}}\right)} \tag{2.44}$$

Moreover, it is also possible to demonstrate that:

- A Cobb-Douglas utility function is obtained when $\rho = 0$.
- Perfect substitutes and perfect complements are obtained when $\rho \to 1^-$ and $\rho \to -\infty$, respectively.

Example 2.7

Consider the utility function given by:

$$U(x, y) = \left[x^{0.5} + y^{0.5}\right]^2$$

Find the optimal consumption bundle, the indirect utility function and the expenditure function. ◀

Solution

At this time, the reader should have learned the methodology to solve this example. We will give only the solutions and invite the reader to come to them.

$$x^* = \frac{W}{p_1 + \frac{p_1^2}{p_2}}$$

$$y^* = \frac{W}{p_2 + \frac{p_2^2}{p_1}}$$

The *indirect utility function* is obtained by expressing the utility as a function of the income and prices. To this aim, we substitute the optimal bundle in the equation of the utility function and obtain:

$$U(W, p_1, p_2) = \left[\left(\frac{W}{p_1 + \frac{p_1^2}{p_2}}\right)^{0.5} + \left(\frac{W}{p_2 + \frac{p_2^2}{p_1}}\right)^{0.5}\right]^2$$

The *expenditure function* is defined as the inverse of the indirect utility function, i.e., we need to solve the indirect utility function for the income.

$$W(U, p_1, p_2) = U\left[\left(\frac{1}{p_1 + \frac{p_1^2}{p_2}}\right)^{0.5} + \left(\frac{1}{p_2 + \frac{p_2^2}{p_1}}\right)^{0.5}\right]^{-2}$$

The expenditure function outputs the needed minimum income to obtain a certain utility level, at the given prices of the buying products. ◀

2.7 The Demand Function

Generally speaking, we can say that a demand function is a representation of how the quantity of good changes with its own price, the price of the other goods, the income and preferences. More complex models include other factors that affect the demand, for instance, the prices of related goods, either complementary, i.e., purchased along with the particular item, or substitutes, i.e., bought instead of the product, consumer expectations about whether prices for the product will rise or fall in the future. In a simpler model, and for a bundle of two goods, the demand function is:

$$\begin{aligned} a_1 &= f_1(p_1, p_2, W, U) \\ a_2 &= f_2(p_2, p_1, W, U) \end{aligned} \tag{2.45}$$

Put this way, this is a highly complex problem because there are many determinants (the said factors that influence the quantity consumed) changing at the same time. So, normally, a simplification is made and it is assumed that all the rest remains constant, except the own price. We represent, in a two-dimensional plane, the variation of the quantity demanded with the price of the good under analysis.

$$\begin{aligned} a_1 &= f_3(p_1) \\ a_2 &= f_4(p_2) \end{aligned} \tag{2.46}$$

Indeed, economists usually represent demand functions the other way around, i.e., they plot the representation of the price of the good as a function of the quantity demanded—price in the vertical y-axis and quantity in the horizontal x-axis, as seen in Fig. 2.16 and Eq. 2.47.

$$\begin{aligned} p_1 &= f_5(a_1) \\ p_2 &= f_6(a_2) \end{aligned} \tag{2.47}$$

2.7.1 Marshallian Demand Function

Let us take our well-known Cobb-Douglas function as an example. For this type of utility function, we recall Eq. 2.25 where we have defined the optimal quantities of

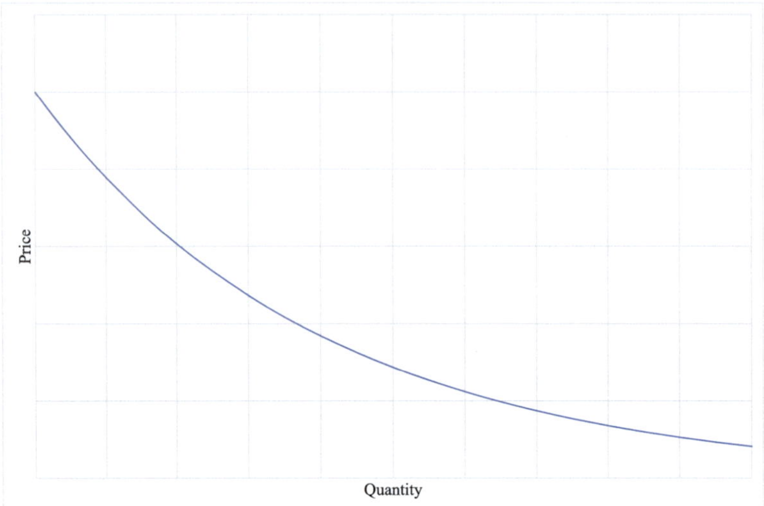

Fig. 2.16 Example of a typical demand function

goods 1 and 2 as a function of the prices, income, and preferences as:

$$\begin{aligned} a_1^* &= \frac{\alpha}{\alpha+\beta}\frac{W}{p_1} \\ a_2^* &= \frac{\beta}{\alpha+\beta}\frac{W}{p_2} \end{aligned} \tag{2.48}$$

Here we have, as targeted, equations that relate the quantity of good 1 (2) with prices, income, and preferences. In this case, it is observed that the quantity a_1 (a_2) does not depend on price p_2 (p_1). This is because, for a Cobb-Douglas function, for instance, the price of good 1 is:

$$p_1 a_1 = \frac{\alpha}{\alpha+\beta} W \tag{2.49}$$

meaning that the consumer spends a fixed proportion of the income on good 1.

As mentioned before, the usual representation of the demand function is not the one displayed in Eq. 2.48. Therefore, solving for p_1 and p_2:

$$\begin{aligned} p_1 &= W\frac{\alpha}{\alpha+\beta}\frac{1}{a_1} = \frac{k_1}{a_1} \\ p_2 &= W\frac{\beta}{\alpha+\beta}\frac{1}{a_2} = \frac{k_2}{a_2} \end{aligned} \tag{2.50}$$

which is called the *Marshallian demand function of good* 1 *(*2*)*, because Alfred Marshall plotted it this way a long time ago.

Taking a_1 as an example, this function is downward-sloping because:

$$\frac{\partial p_1}{\partial a_1} = -W \frac{\alpha}{\alpha + \beta} \frac{1}{a_1^2} < 0 \tag{2.51}$$

This is in good agreement with our common sense that tells us when the price of a good decreases, we tend to consume more of it.

2.7.2 Engel Curve

Another way of looking at Eq. 2.50 is writing the income as a function of the quantity, keeping the two prices and the parameters of the utility function constant. Let us take a_1 as an example.

$$W = \frac{\alpha + \beta}{\alpha} p_1 a_1 \tag{2.52}$$

This is an upward sloping curve, because $\frac{\partial W}{\partial a_1} = \frac{\alpha+\beta}{\alpha} p_1 > 0$. It is called *Engel curve* and it depicts how the quantity changes when the income changes. For a Cobb-Douglas utility function, when the income increases, the consumer consumes more of the good.

2.7.3 Elasticity of Demand

We are now going to introduce the concept of *Elasticity*. Suppose a store manager wants to run a sale. Should he lower the price of a given item? If yes, by how much? The answer will depend on whether the increase in purchases by customers will be larger than the price decrease, (both calculated on a relative basis). This change in purchases by customers and prices is what elasticity is intended to measure.

Price elasticity of demand is defined as the ratio of the percentage change in quantity demanded to a percentage change in price.

$$\varepsilon_p = \frac{\% \Delta a_1}{\% \Delta p_1} = \frac{\frac{\Delta a_1}{a_1}}{\frac{\Delta p_1}{p_1}} = \frac{\Delta a_1}{\Delta p_1} \frac{p_1}{a_1} \tag{2.53}$$

For infinitesimal changes, we write:

$$\varepsilon_p = \frac{\partial a_1}{\partial p_1} \frac{p_1}{a_1} \tag{2.54}$$

As the demand function usually is downward sloping, the price elasticity of demand is usually negative, although the minus sign is often omitted. For example, if the price elasticity of demand is -3, it means that when the price of the product increases by 1%, the consumers are prepared to decrease the consumption level by 3%. However, there are goods that do not obey this general rule. We will see later, that this is called the *Law of Demand*. These goods are called *Giffen goods*, wherein the consumption increases when the price rises, largely because there are few substitutes or alternatives for them. A classical example is food staples consumed by poor people. If consumers have no alternative but to purchase the staple, its consumption will increase, even though it becomes more expensive. In this case, the demand function is upward-sloping.

If the demand is *elastic*, it means that a small price change results in a large quantity change. This would generally take place on the upper portion of the demand curve. If the demand is perfectly elastic (which means that the smallest possible price change results in a virtually infinite quantity change), the demand curve is then horizontal ($\varepsilon_p = \infty$). A classical example of elastic demand is pastry.

If the demand is *inelastic*, it means that even a substantial change in price leads to hardly any change at all in quantity demanded. This generally occurs in the lower portion of the demand curve. If the demand is perfectly inelastic, quantity does not change at all. Perfectly inelastic demand is vertical ($\varepsilon_p = 0$). Classical examples of inelastic demand are medicine and basic foods.

Example 2.8

Consider a demand function given by $p_D(q) = 800 - 2q$. Using this example show that, for a linear demand curve, the demand is elastic in the upper portion of the curve and inelastic in the lower portion. Plot illustrative graphics. ◄

Solution

The variation of the price elasticity of demand with the quantity is given by:

$$\varepsilon_p(q) = \frac{\partial q}{\partial p_D(q)} \frac{p_D(q)}{q} = -0.5 \frac{800 - 2q}{q} = -0.5 \left(\frac{800}{q} - 2 \right)$$

The plot is in Fig. 2.17 where we can see that, for small quantities, the price elasticity of demand is high and, for high quantities, it goes to zero. ◄

In the same way, *Income elasticity of demand* is defined as:

$$\varepsilon_W = \frac{\% \Delta a_1}{\% \Delta W} = \frac{\frac{\Delta a_1}{a_1}}{\frac{\Delta W}{W}} = \frac{\Delta a_1}{\Delta W} \frac{W}{a_1} \tag{2.55}$$

or using calculus:

$$\varepsilon_W = \frac{\partial a_1}{\partial W} \frac{W}{a_1} \tag{2.56}$$

As can be seen, it is the ratio of the percentage change in quantity demanded to a percentage change in income. For instance, if $\varepsilon_W = 2$, it implies that for every 1% increase in income, people will demand more 2% of the quantity of the goods. Thus, if the income is 100 and, at that level of income, the consumer takes 100 units of the product, he/she would demand 102 units, if income rose to 101.

The magnitude of ε_W defines several types of goods.

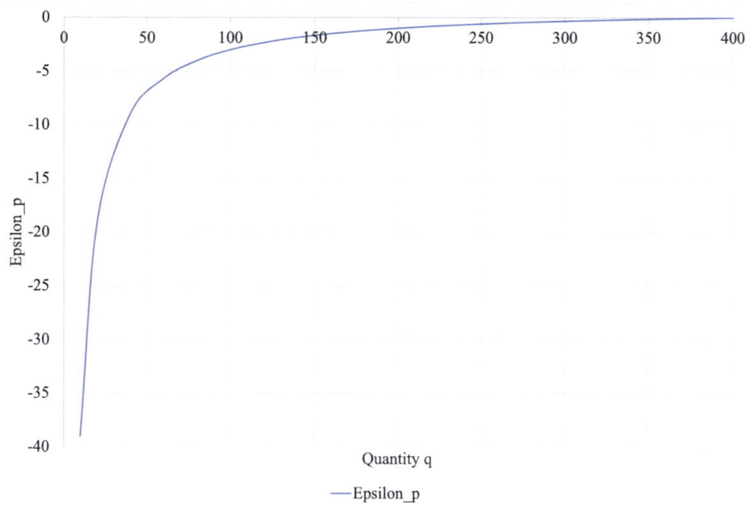

Fig. 2.17 Price elasticity of demand as a function of quantity. Example 2.8

- If $\varepsilon_W > 0$, a good is a *normal good*, meaning its consumption increases as the income increases. The Engel curve is upward-sloping. Examples: clothing, household appliances.
- If $\varepsilon_W < 0$, a good is an *inferior good*, meaning its consumption decreases as the income increases. The Engel curve is downward sloping. Examples: public transportation, generic brand products.
- If $\varepsilon_W > 1$, a good is a *luxury good*, meaning its consumption increases more in proportion of the income increase. Examples: books, restaurant meals, automobiles.
- If $0 \leq \varepsilon_W \leq 1$, a good is a *necessity*. These are the last goods the consumer stops buying when their income declines, and when the income increases, people buy them in a lower proportion with respect to the income increase. Examples: bread, coffee.

These definitions are subjective. The same good can be normal for some people and inferior for other people. Furthermore, goods can be both inferior and normal, depending on the individual's income level.

Another definition applicable to the goods is *Substitutes* and *Complements*. Two goods are said to be substitutes when, if the price of one increases, the quantity demanded of the other good rises. Conversely, two goods are complements when the price of one increases implying the quantity demanded of the other good decreases. It is measured by the *Cross Price Elasticity of Demand*.

$$\varepsilon_{cp} = \frac{\partial a_1}{\partial p_2}\frac{p_2}{a_1} \approx \frac{\Delta a_1}{\Delta p_2}\frac{p_2}{a_1} \tag{2.57}$$

If it is positive, the two goods are substitutes. If it is negative, the two goods are complements.

Example 2.9

The preferences of a consumer are driven by the utility function $U(a, b) = a^3 b$.

1. Derive the demand function for each of the two goods a and b, considering an income of 4. Plot the respective graphics.
2. Find the equation of the Engel curve for good b, assuming its price is 1.
3. Compute the price elasticity of demand of good b at the optimal quantity. In the sequence, consider income and price as defined before whenever appropriate.
4. Figure out the income elasticity of demand of good b at the optimal quantity. Classify good b according to the result obtained.
5. Calculate the cross-price elasticity of demand of good b at the optimal quantity and classify this good accordingly.

◀

Solution

1.

Based on what we have learned, and applying Eq. 2.48, the optimal bundle for this utility function and budget line is:

$$a^* = \frac{\alpha}{\alpha + \beta} \frac{W}{p_a} =$$

$$b^* = \frac{\beta}{\alpha + \beta} \frac{W}{p_b}$$

For the given data, the demand function for each of the two goods is:

$$p_a = \frac{3}{a}$$

$$p_b = \frac{1}{b}$$

The demand curves are downward-sloping, as expected for a Cobb-Douglas function, and are depicted in Fig. 2.18.

2.

The equation for the Engel curve is given by Eq. 2.52. Applying to good b leads to:

$$W = \frac{\alpha + \beta}{\beta} p_b b = 4b$$

which is an upward-sloping straight line crossing the origin of the axis.

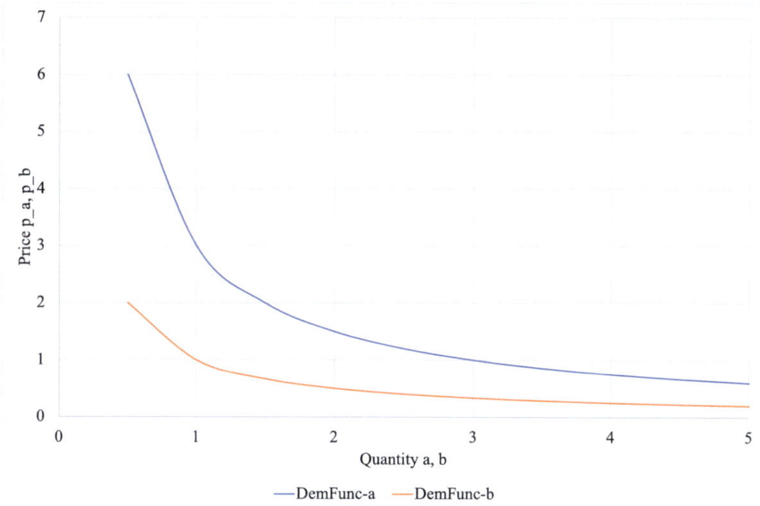

Fig. 2.18 Demand functions for Example 2.9

3.

For the price of 1 and income of 4, the optimal quantity of b is 1 as can be easily verified in Fig. 2.18. Replacing known values, we have:

$$b^* = \frac{\beta}{\alpha + \beta} \frac{W}{p_b} = \frac{1}{p_b}$$

The price elasticity of demand (Eq. 2.54) applied to good b is:

$$\varepsilon_p = \frac{\partial b}{\partial p_b} \frac{p_b}{b} = -\frac{1}{p_b^2} \frac{p_b}{b} = -1$$

When the price increases 1%, the quantity decreases 1%. This is not a Giffen good because the price elasticity of demand is negative. In a Giffen good, the quantity increases as the price increases.

4.

Given the problem's data, the relationship between income and quantity of good b is:

$$b = \frac{W}{4}$$

The income elasticity of demand (Eq. 2.56) applied to good b is:

$$\varepsilon_W = \frac{\partial b}{\partial W} \frac{W}{b} = \frac{1}{4} \frac{4}{1} = 1$$

This means that for an increase of 1% in the income, an increase of 1% in the quantity of good b is observed. This is a normal good ($\varepsilon_W > 0$), more specifically a necessity good.

5.

The quantity of good b does not depend on the price of good a, because the consumer uses a fixed proportion of the income in purchasing good b. So, the cross-price elasticity of demand is zero, and the goods are not substitutes, nor complements. ◂

Example 2.10

The preferences of a consumer are driven by the utility function $U(x, y) = \ln x + y$.

1. Determine the optimal consumption basket and the derived utility, considering $p_x = 2$, $p_y = 1$, $W = 4$. Show evidence of the obtained results by plotting illustrative graphics.
2. Derive the demand function for each of the two goods x and y. Plot the respective graphics. In the sequence, consider income and prices as defined before whenever appropriate.
3. At the optimal quantity of good y, compute the price elasticity of demand, the income elasticity of demand, and the cross-price elasticity of demand. Classify good y according to the results obtained.

◂

Solution

1.

At this stage, it should be easy for the reader to figure out that the equations to be solved are:

$$\frac{\frac{1}{x}}{1} = \frac{2}{1}$$

$$2x + y = 4$$

which leads to the optimal bundle being:

$$x^* = 0.5;\ y^* = 3$$

and the utility is:

$$U = \ln(0.5) + 3 = 2.3069$$

The results are plotted in Fig. 2.19. Note that $y = 2.3069 - \ln x$ and $y = 4 - 2x$.

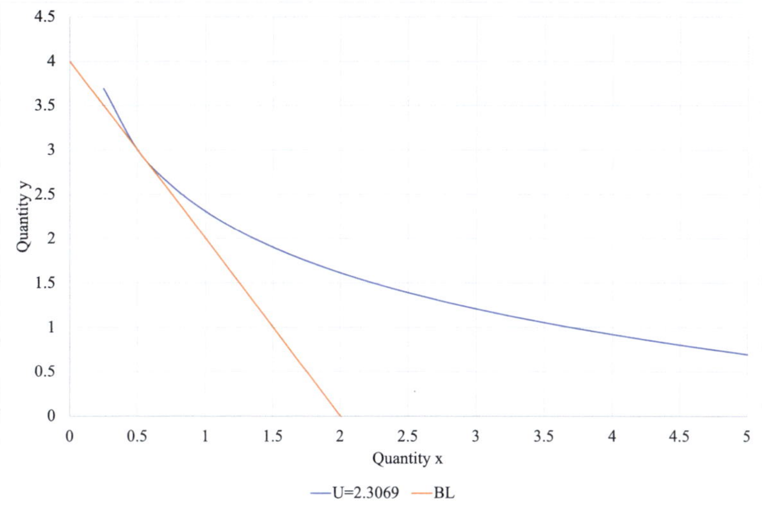

Fig. 2.19 Indifference curve and budget line for Example 2.10

2.
From the relationship between the marginal utilities ratio and the price ratio, we conclude:

$$p_x = \frac{p_y}{x} = \frac{1}{x}$$

The budget line equation leads to:

$$p_y = \frac{W - p_x x}{y} = \frac{W - 1}{y} = \frac{3}{y}$$

The corresponding plottings are in Fig. 2.20.

3.

$$\varepsilon_p = \frac{\partial y}{\partial p_y}\frac{p_y}{y} = -\frac{3}{p_y^2}\frac{p_y}{y} = -\frac{3}{1}\frac{1}{3} = -1$$

$$\varepsilon_W = \frac{\partial y}{\partial W}\frac{W}{y} = 1\frac{4}{3} = \frac{4}{3}$$

To confirm this result, the reader is invited to demonstrate that:

$$y = \frac{W - 1}{p_y} = \frac{4 + 1\% \times 4 - 1}{1} = 3.04 = 3 + \frac{4}{3}\% \times 3 = 3.04$$

$$\varepsilon_{cp} = \frac{\partial y}{\partial p_x}\frac{p_x}{y} = 0$$

This is a normal good, more precisely, a luxury good. ◄

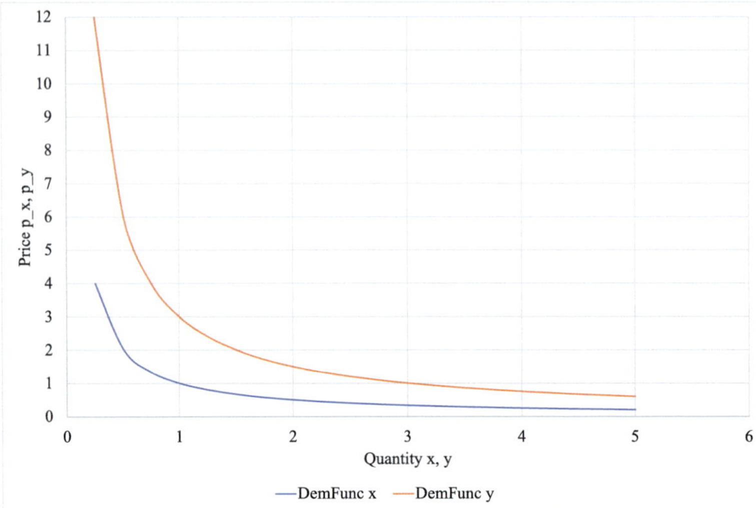

Fig. 2.20 Demand functions for Example 2.10

2.8 Conclusions

In this chapter, we addressed the theory of consumers in the power systems context. As demonstrated, this is a vital component of the power systems industry, as it describes the behaviour of the important economic agent, the consumers. Throughout this chapter, we have explored key concepts including the utility function, the budget line, the optimisation problem using the Lagrangian method, typical utility functions, and the demand function, all of which are highly relevant to understanding consumer behaviour and decision-making in power systems.

The utility function lies at the heart of consumer theory, representing the preferences and satisfaction of consumers regarding different combinations of goods and services. By quantifying utility, energy engineers can gain insights into consumers' preferences and design strategies to maximize consumer satisfaction and well-being. Key concepts such as total utility, marginal utility, indifference curves, and marginal rate of substitution were tackled in this section.

The budget line serves as a critical constraint that consumers face in power systems economics. It illustrates the combinations of goods and services that consumers can afford given their income and the prevailing prices. By analysing the budget line, it is possible to understand the trade-offs consumers make and devise strategies to optimise resource allocation and pricing decisions.

The optimisation problem tackled through the Lagrangian method, is essential in consumer theory. By maximising utility subject to the constraints imposed by the budget line, the optimal consumption bundle that maximises consumer satisfaction within their budgetary limits can be determined. This approach enables stakeholders to make informed decisions about resource allocation and pricing strategies to meet consumer needs efficiently. Further in this section, the physical meaning of the Lagrange multiplier in the context of the theory of consumers, changes in the budget line, and substitution and income effects were explained.

Typical utility functions employed in power systems economics, such as the Cobb-Douglas utility function, CES utility function, and quasi-linear function, capture different aspects of consumer preferences and behaviour. These functions allow modelling and analysing various scenarios, considering factors such as relative goods importance and elasticity of substitution between them.

The demand function, derived from the optimisation problem, provides valuable insights into the relationship between price and quantity demanded by consumers. Understanding the demand function empowers energy engineers to assess the responsiveness of consumers to price changes, predict consumer behaviour, and make informed decisions regarding power systems operation, pricing, and market dynamics. The concepts of Marshallian demand curve and Engel curve were approached, as well as the important definition of the elasticity of demand were introduced in this section.

By integrating the theory of consumers into the engineering of power systems economics, energy engineers can optimise resource allocation, design effective pricing strategies, and enhance consumer welfare. This knowledge contributes to the efficient and sustainable operation of power systems, ensuring the delivery of reliable and affordable electricity to meet the ever-evolving needs of consumers in a dynamic energy landscape.

2.9 Proposed Exercises

Problem 2.1

A consumer is characterized by the following expression for the utility function where a_1 and a_2 are the consumer goods. Note that the utility function gives an order for the preferences.

$$U(a_1, a_2) = 100 a_1^{0.3} a_2^{0.5}$$

The consumer goods a_1 and a_2 have the unitary prices given by:

$$p_1 = 2; \; p_2 = 4$$

The available consumer budget is $W = 90$.

1. In the plane (a_1, a_2), draw the indifference curves for utility equal to 1000 and to 700. What do these curves mean?
2. What are the conditions for utility maximisation given a specific available budget? Calculate for the parameters given in the problem and for a utility equal to 875.

3. What happens if the available budget increases to 100, while keeping the unitary costs constant? What if the price of good a_2 decreases to $p_2 = 3$, keeping the budget constant?

Key Problem 2.1

1.

The utility function of the problem is:

$$U(a_1, a_2) = 100a_1^{0.3}a_2^{0.5}$$

This function represents the combination of goods quantities (a_1, a_2) that correspond to the same utility, U, to the consumer. For each particular utility, we obtain a different indifference curve, showing that the consumer is able to get the same utility from the consumption of different bundles of goods.

For $U = 1000$, we obtain:

$$1000 = 100a_1^{0.3}a_2^{0.5}$$

$$a_2 = \left(\frac{10}{a_1^{0.3}}\right)^2$$

In Fig. 2.21, we show the indifference curves for $U_1 = 1000$, $U_2 = 700$ and $U_3 = 875.3957$. We also show the budget line for an income $W = 90$. The budget line shows all combinations of goods that are available to the consumer given his/her

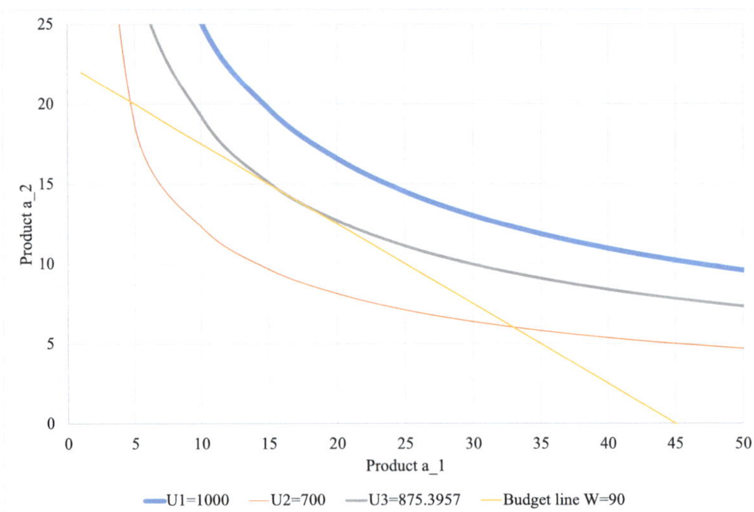

Fig. 2.21 Indifference curves and budget line for Problem 2.1

money income, W and the prices of the goods he/she purchases (p_1, p_2). The budget line is given by:

$$W = p_1 a_1 + p_2 a_2 \overset{\text{yields}}{\rightarrow} a_2 = \frac{W}{p_2} - \frac{p_1}{p_2} a_1$$

We can see in the figure that the budget line intercepts the indifference curve for $U_3 = 875.3957$ at a tangency point. This point defines the conditions (combination of goods quantities) for maximum utility given the budget line corresponding to an income $W = 90$.

The budget line does not intercept the indifference curve for $U_1 = 1000$. This means that, for the given income, it is not possible to obtain such a high utility. On the other side, the budget line intercepts the indifference curve for $U_2 = 700$ in two points. These two points define the two possible combinations of goods quantities that exhaust the income. However, the same utility can be obtained with a lower income (and a different combination of goods).

2.

Problem formulation:

$$\max \left[100 a_1^{0.3} a_2^{0.5} \right]$$

$$\text{s.t. } 2a_1 + 4a_2 - 90 = 0$$

Solution method (Lagrangian method):

$$Lag = 100 a_1^{0.3} a_2^{0.5} - \lambda(2a_1 + 4a_2 - 90)$$

Optimal conditions:

$$\frac{\partial Lag}{\partial a_1} = 100 a_2^{0.5} 0.3 a_1^{-0.7} - 2\lambda = 0 \quad (1)$$

$$\frac{\partial Lag}{\partial a_2} = 100 a_1^{0.3} 0.5 a_2^{-0.5} - 4\lambda = 0 \quad (2)$$

$$\frac{\partial Lag}{\partial \lambda} - 90 - 2a_1 - 4a_2 = 0 \quad (3)$$

Most common solution procedure: divide (1) by (2) and replace it in (3).

$$\frac{1}{2} = \frac{a_2^{0.5} 0.3 a_1^{\;0.7}}{a_1^{0.3} 0.5 a_2^{-0.5}} \overset{\text{yields}}{\rightarrow} \frac{1}{2} = \frac{3a_2}{5a_1} \overset{\text{yields}}{\rightarrow} a_1 = \frac{6}{5} a_2$$

$$\frac{12a_2}{5} + 4a_2 = 90 \overset{\text{yields}}{\rightarrow} a_2^* = \frac{5}{32} 90 = 14.0625$$

$$a_1^* = 16.875$$

$$100 a_2^{0.5} 0.3 a_1^{-0.7} = 2\lambda \overset{\text{yields}}{\rightarrow} \lambda = 7.7813$$

a_1^*, a_2^* are the quantities of the goods that maximise the consumer's utility, given an income $W = 90$. The maximum utility is:

$$U^* = 100a_1^{0.3}a_2^{0.5} = 875.3957$$

$\lambda = \frac{\partial U}{\partial W}$ is the marginal utility of wealth, i.e., the change in utility when the income is increased in one unit. Recall that, in general, the λ parameter is the derivative of the objective function with respect to the restriction function.

Let us relate the utility with the income.

$$W = 2a_1 + 4a_2 = 2\left(\frac{6}{5}a_2\right) + 4a_2 = \frac{32}{5}a_2 \overset{\text{yields}}{\rightarrow} a_2 = \frac{5}{32}W$$

$$U = 100a_1^{0.3}a_2^{0.5} = 100\left(\frac{6}{5}a_2\right)^{0.3}a_2^{0.5} = 100\left(\frac{3}{16}W\right)^{0.3}\left(\frac{5}{32}W\right)^{0.5} = 100\left(\frac{3}{16}\right)^{0.3}\left(\frac{5}{32}\right)^{0.5}W^{0.8}$$

$$\frac{\partial U}{\partial W} = 100\left(\frac{3}{16}\right)^{0.3}\left(\frac{5}{32}\right)^{0.5}0.8W^{-0.2}$$

$$W = 90 \overset{\text{yields}}{\rightarrow} \frac{\partial U}{\partial W} = 7.7813 = \lambda$$

The relationship between both the utility and the marginal utility of wealth with income is depicted in Fig. 2.22. The utility increases with the income, which is quite obvious. On the contrary, the marginal utility of wealth decreases with the income increase: as the income becomes higher and higher, the utility gain when the income increases one unit is decreasing. Eventually, the marginal utility of wealth becomes negative, meaning that one loses utility from the consumption of goods if the income is too high.

3.

Let us recall that:

$$a_2 = \frac{W}{p_2} - \frac{p_1}{p_2}a_1$$

If the budget increases to $W = 100$, both the x- and y-intercept of the budget line increase, but the slope remains constant. This means that one can buy products that deliver a higher utility (this can be clearly seen in the previous Fig. 2.22).

If the budget remains constant, but the unitary price of good 2 decreases to $p_2' = 3$, both the intercept and the slope will increase. Note, however, that the intercept with the x-axis ($\frac{W}{p_1}$) will remain constant.

$$0 = a_2 = \frac{W}{p_2} - \frac{p_1}{p_2}a_1 \overset{\text{yields}}{\rightarrow} W = p_1a_1 \overset{\text{yields}}{\rightarrow} a_1 = \frac{W}{p_1}$$

The intercept with the x-axis is independent of p_2.

Once again, one can purchase products that deliver a higher utility.

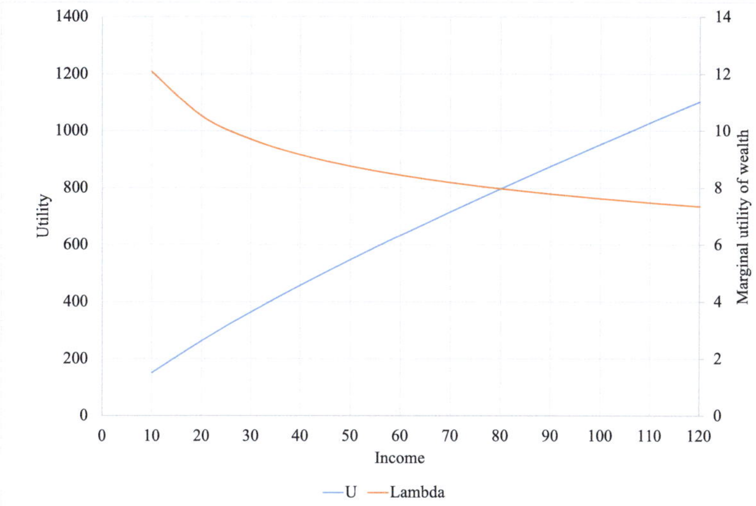

Fig. 2.22 Utility and marginal utility of wealth as a function of income. Problem 2.1

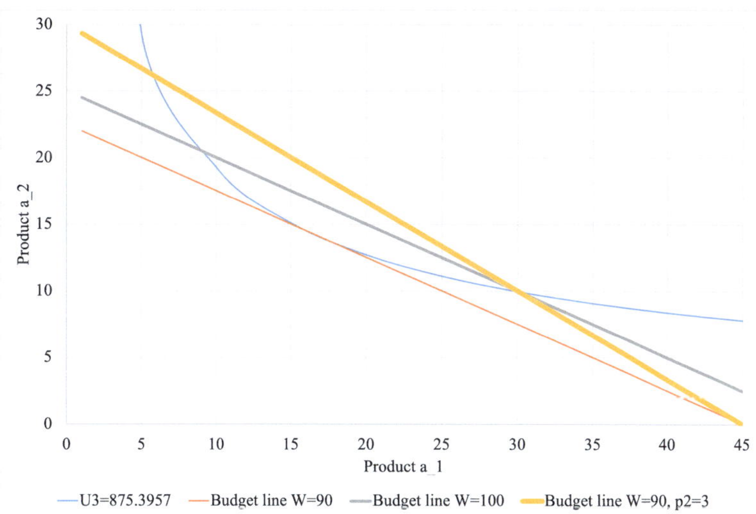

Fig. 2.23 Different budget lines. Problem 2.1

Both situations are depicted in Fig. 2.23.

At this time, we could formulate the dual problem as *Compute the minimum income that allows obtaining a utility equal to 875.3957.*

The problem is formulated as follows:

$$\min\ [2a_1 + 4a_2]$$

$$\text{s.t. } 100a_1^{0.3}a_2^{0.5} - 875.3957 = 0$$

The reader is invited to solve this problem. Of course, the optimal goods 1 and 2 quantities are identical as before and the minimum income is $W = 90$. The λ parameter is $\lambda = \frac{\partial W}{\partial U} = 0.1285$ and represents how many units of income are needed to obtain one additional utility unit.

Problem 2.2

A consumer is characterised by the utility function $U = xy+x+y$. The prices for goods x and y are 2 and 4, respectively and the income the consumer has available to spend in purchasing the goods is 18.

1. Find the optimal consumption bundle, the respective utility, and the marginal utility of wealth. Plot the graphics to show you found the correct solution.
2. Figure out the demand function for good x and plot the graphics.
3. For good x, compute the price elasticity of demand, the income elasticity of demand, and the cross-price elasticity of demand at the optimal point. Classify this good.

Key Problem 2.2

1.

Based on the Lagrangian, it is possible to write:

$$\frac{y+1}{x+1} = \frac{2}{4} \overset{\text{yields}}{\rightarrow} y = 0.5x - 0.5$$

Replacing in the budget equation $2x + 4y = 18$, one obtains:

$$x^* = 5;\ y^* = 2$$

At this optimal point (consumption bundle), the utility gained is:

$$U(5, 2) = 17$$

Figure 2.24 displays the related graphics. We note that $y = \frac{17-x}{x+1}$ and $y = 4.5 - 0.5x$.

To compute the Lagrange multiplier, we use:

$$\lambda = \frac{MU_x}{p_x} = \frac{y+1}{2} = 1.5$$

i.e., when the income is raised in one unit, the utility increases by 1.5 units. The reader is invited to verify that if the wealth is $W' = 19$, then $x'^* = 5.25$, $y'^* = 2.125$ and $U' = 18.5$.

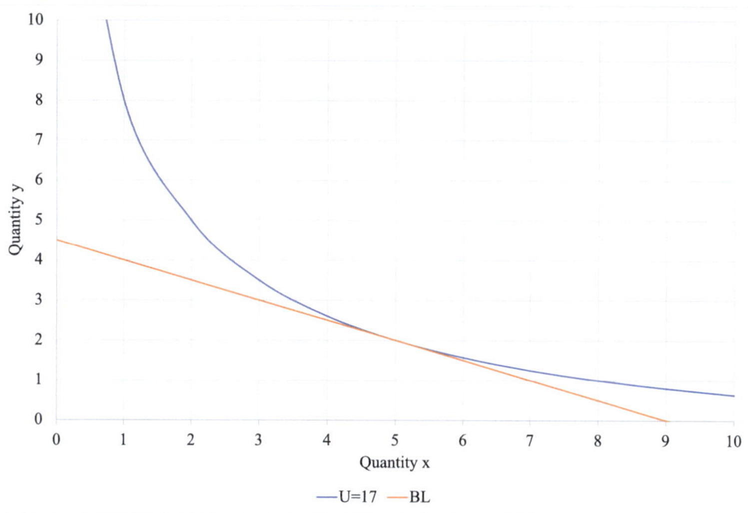

Fig. 2.24 Indifference curve and budget line. Problem 2.2

2.

The demand function for good x is $p_x = f(x)$, with p_y and W held constant. To this aim, we pick up the fundamental equation and express all variables as a function of x.

$$p_x(x+1) = p_y(y+1) \overset{\text{yields}}{\rightarrow} y = \frac{p_x}{p_y}(x+1) - 1$$

Then, replace in the budget equation and rearrange. After some manipulation, we obtain:

$$p_x = \frac{W + p_y}{2x+1} \overset{\text{yields}}{\rightarrow} p_x = \frac{22}{2x+1}$$

The result of the downward-sloping curve is shown in Fig. 2.25.

3.

Writing x as a function of p_x leads to:

$$x = \frac{W + p_y}{2p_x} - 0.5 = \frac{11}{p_x} - 0.5$$

The price elasticity of demand is:

$$\varepsilon_p = \frac{\partial x}{\partial p_x}\frac{p_x}{x} = -\frac{11}{p_x^2}\frac{p_x}{x} = -\frac{11}{4}\frac{2}{5} = -1.1$$

To confirm this result, the reader is invited to demonstrate that:

$$x = \frac{11}{p_x} - 0.5 = \frac{11}{2 + 1\% \times 2} - 0.5 = 4.945 = 5 - 1.1\% \times 5 = 4.945$$

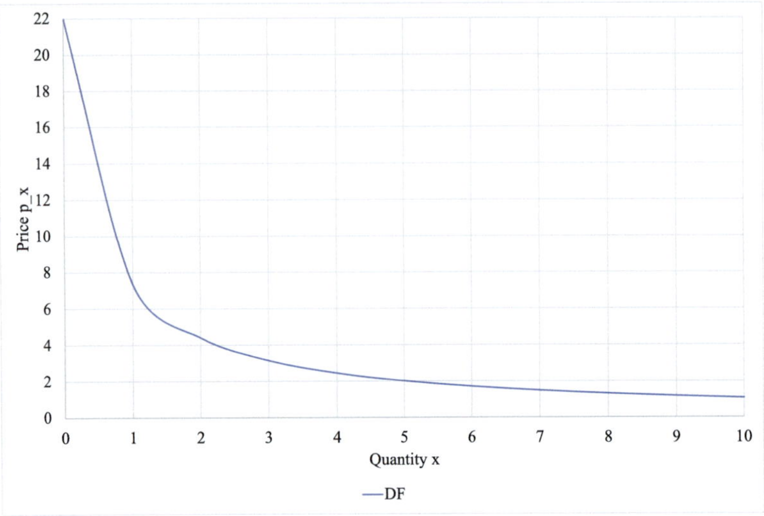

Fig. 2.25 Demand function for good x. Problem 2.2

The income elasticity of demand is given by:

$$\varepsilon_W = \frac{\partial x}{\partial W}\frac{W}{x} = \frac{1}{2p_x}\frac{W}{x} = \frac{1}{4}\frac{18}{5} = 0.9$$

Finally, the cross-price elasticity is:

$$\varepsilon_{cp} = \frac{\partial x}{\partial p_y}\frac{p_y}{x} = \frac{1}{2p_x}\frac{p_y}{x} = \frac{1}{4}\frac{4}{5} = 0.2$$

From the results obtained, one can conclude that this is a normal good, more specifically, a necessity good. The two goods are substitutes.

Problem 2.3

Assume a consumer is driven by the utility function $U = \alpha \ln x_1 + (1-\alpha)\ln x_2$. The prices for goods x_1 and x_2 are p_1 and p_2, respectively, and the available income is W.

1. Find the general equation of the optimal consumption basket and the utility that can be derived.
2. Show that the marginal utility of wealth is the reciprocal of the income.

Key Problem 2.3

1.

The first equation one may derive is:

$$\frac{\alpha p_2}{x_1} = \frac{(1-\alpha)p_1}{x_2} \overset{\text{yields}}{\rightarrow} x_1 = \frac{\alpha p_2}{(1-\alpha)p_1}x_2$$

Plugging in the budget equation, $p_1 x_1 + p_2 x_2 = W$, allows to obtain:

$$x_2^* = \frac{(1-\alpha)}{p_2} W$$

$$x_1^* = \frac{\alpha}{p_1} W$$

The function:

$$U(p_1, p_2, W) = \alpha \ln \left[\frac{\alpha}{p_1} W \right] + (1-\alpha) \ln \left[\frac{(1-\alpha)}{p_2} W \right]$$

is called the *Indirect Utility Function*.

2.

The marginal utility of wealth is:

$$\lambda = \frac{MU_{x_1}}{p_1} = \frac{\alpha}{x_1 p_1}$$

Replacing the optimal quantity of x_1 leads to:

$$\lambda = \frac{1}{W}$$

The same result would be achieved if we recall that the marginal utility of wealth is:

$$\lambda = \frac{\partial U}{\partial W} = \alpha \frac{1}{W} + (1-\alpha) \frac{1}{W} = \frac{1}{W}$$

We observe that:

$$\frac{d \ln u(x)}{dx} = \frac{1}{u(x)} \frac{du}{dx}$$

3 The Single Market Behaviour

Abstract

The single market behaviour is a crucial aspect of understanding the dynamics of power systems economics. It encompasses the market equilibrium, market adjustment, and the concept of dynamic equilibrium, which can be analysed either through the Marshall and Walras methods. Market equilibrium refers to a state where the demand for and supply of a product or service are balanced. In the context of power systems economics, it represents the point at which the quantity of electricity demanded matches the quantity supplied. Market adjustment occurs when there is a disparity between demand and supply, leading to price and quantity changes that stimulate market participants to adjust their behaviour accordingly. Dynamic equilibrium, as envisioned by Alfred Marshall, suggests that markets are constantly evolving and adjusting over time. It recognises the influence of factors such as technology advancements, consumer preferences, and government regulations on market behaviour. Marshall's method focuses on the short-run dynamics of market equilibrium and considers the quantity as the adjustment variable. On the other hand, Walras' approach offers a comprehensive understanding of the complexities involved in the single market behaviour by considering the price as the adjustment variable. The study of single market behaviour in power systems economics is crucial for comprehending market equilibrium, market adjustment, and dynamic equilibrium.

3.1 Introduction

In the realm of economics, the behaviour of a single market is influenced by various factors, including supply and demand dynamics, price mechanisms, and market equilibrium.

R. Castro, *Engineering of Power Systems Economics*,
https://doi.org/10.1007/978-3-031-55251-9_3

Market equilibrium refers to a state in which the quantity demanded by buyers matches the quantity supplied by sellers, resulting in a stable market price. At equilibrium, there is no inherent tendency for prices or quantities to change. This balance is achieved when the forces of demand and supply reach a point of agreement. The demand curve represents the willingness of buyers to purchase a product at different prices, while the supply curve illustrates the quantity of the product that suppliers are willing to provide at various prices. The intersection of these curves determines the equilibrium price and quantity.

Markets are not static entities; they constantly undergo adjustments in response to changes in demand, supply, or other factors. When there is a disruption in the market, such as a shift in consumer preferences, changes in production costs, or the introduction of new technologies, the existing equilibrium is disturbed. In such cases, the market moves towards a new equilibrium through the process of market adjustment.

During market adjustment, the forces of demand and supply work to eliminate the existing imbalance. If demand increases, meaning buyers are willing to purchase more of a product at each price level, there will be upward pressure on prices. As prices rise, suppliers are incentivised to increase production, leading to an expansion of the quantity supplied. This adjustment continues until a new equilibrium is reached, with a higher price and quantity. Similarly, if demand decreases, the reverse process occurs, resulting in a lower equilibrium price and quantity.

Dynamic equilibrium recognises that markets are subject to continuous changes, and equilibrium is not a static state but rather an ongoing process. Factors such as technological advancements, shifts in consumer preferences, and changes in input costs continually impact supply and demand conditions. As a result, markets experience shifts in equilibrium over time.

In this chapter, we are going to analyse the interactions between producers and consumers in a single market. Producers and consumers may be called *market agents*. The market is defined as a platform that allows for an exchange of products, usually the exchange of a product by currency.

Before we begin, we recall that the market is ruled by the selling bids from the producers and the buying offers from the consumers. Both offers are drawn in the plane (q, p), where q is the quantity of the product and p is its unitary price. It goes without saying that all the other factors affecting supply and demand remain constant. We usually use the Latin phrase, *ceteris paribus*, which roughly means *all other things remain the same*.

The selling bids—supply curves—are based on the producers' cost functions. These are functions $p = f_1(q)$, where p is the unitary price and q is the quantity. Depending on the returns to scale, the cost function may be constant, upward-sloping, or downward-sloping. The buying offers—demand curves—are based on the consumers' demand functions, $p = f_2(q)$. For the majority of the products, the demand function is downward-sloping (for Giffen goods it is not). In classical Economics textbooks, it is usual to consider the cost function as upward-sloping. We stress that this is not universal, in fact, it implies considering decreasing returns to scale.

In the previous chapters, we have introduced the producers and the consumers as market agents, i.e., firms that operate in the market, by selling and buying a product (electricity, in the power system framework), respectively. The objective of this chapter is to analyse how these agents interact in a market. To this aim, we begin by addressing how aggregated supply and demand curves, aggregating several producers and consumers, respectively, are built to include the effect of the individual curves of each producer and consumer. Then, we expose the groundings of the *market equilibrium*, by delving into *demand curves* and *supply curves*, *surplus* and *shortage*, and *price ceiling* and *price floor*. The *market adjustment*, which follows market disturbances, is next addressed, by recurring to the concept of *state variable*, which can be price or quantity, according to the different lines of thought. The control theory is used to introduce the *dynamic equilibrium*, explaining the two methodologies available in the literature. At the end of the chapter, some exercises are offered to apply the knowledge just taught.

3.2 Aggregation of Market Agents

Usually, a market is composed of several producers and several consumers. We have learned how to derive the individual cost function (demand function) of a single producer (consumer). In a market, a single supply curve and a single demand curve exist. The supply curve represents the aggregation of all the individual supply curves and the demand curve represents the aggregation of all individual demand curves.

The market supply (demand) curve is derived from the individual curves, as follows. For a given price, p, the quantities, q, offered by each producer (consumer) are summed up, thus obtaining the global quantity for the given price.

The implementation of this procedure is not straightforward, as the individual supply (demand) functions are represented as $p_i = f_i(q_i)$. So, to obtain the aggregated supply (demand) functions, the following steps are necessary:

1. Obtain the individual inverse supply (demand) functions, $q_i = g_i(p_i)$.
2. Sum the quantities for the same price.
3. Obtain the aggregated inverse supply (demand) function, $q = g_{agg}(p)$.
4. Revert to the usual representation as $p = f_{agg}(q)$.

As a matter of fact, this is not as simple as it seems because there are kinks that must be considered. The next examples shed some light on these issues.

Example 3.1

Lewis and George are two friends who show identical preferences for music applications. The individual demand for music streaming applications is:

$$p = 15 - q$$

Compute the aggregated demand function. ◄

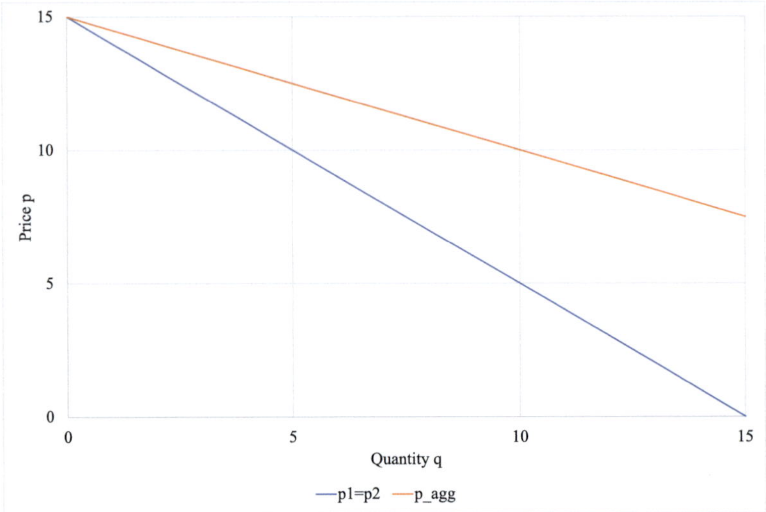

Fig. 3.1 Demand functions for Example 3.1

Solution

This is a simple example, to begin with.

$$p_1 = 15 - q_1 \overset{\text{yields}}{\rightarrow} q_1 = 15 - p_1$$

$$p_2 = 15 - q_2 \overset{\text{yields}}{\rightarrow} q_2 = 15 - p_2$$

$$q = q_1 + q_2 = 2q_1 = 30 - (p_1 + p_2)$$

$$q = 30 - 2p; \text{ for } p \leq 15 \overset{\text{yields}}{\rightarrow} p = 15 - \frac{q}{2}$$

Illustrative graphics are shown in Fig. 3.1. ◀

Example 3.2

The supply functions of two firms, A and B, are given by:

$$p_A = 0.4 + 0.6q_A$$

$$p_B = 0.6 + 0.4q_B$$

Determine the aggregated supply function. ◀

Solution

The situation gets more complicated when the functions are different for each market agent.

From the given data, one finds the inverse supply functions as:

$$q_A = \frac{p_A - 0.4}{0.6}\ ; \text{for}\ \ p \geq 0.4$$

$$q_B = \frac{p_B - 0.6}{0.4}\ ; \text{for}\ \ p \geq 0.6$$

We can see that $p = 0.4$ and $p = 0.6$ are kinks. A first attempt to get the aggregated inverse supply function is:

$$q = q_A + q_B = \frac{p - 0.52}{0.24}$$

However, this function does not represent the aggregated inverse supply function for every price, because, if $0.4 < p \leq 0.6$, only firm A supplies. Let us look at an example: suppose $p = 0.5$; at this price level, firm A supplies $1/6 = 0.1(6)$ units, and firm B does not supply (no negative quantities are admissible). So, the aggregated supply is $0.1(6)$ and not $-0.08(3) = 0$ as given by the aggregated supply function. We need to focus on the prices where the supply goes to zero. Therefore, the aggregated inverse supply function is:

$$q = \begin{cases} 0, & \text{for } p \leq 0.4 \\ \frac{p-0.4}{0.6} = q_A, & \text{for } 0.4 < p \leq 0.6 \\ \frac{p-0.52}{0.24}, & \text{for } p > 0.6 \end{cases}$$

Take a look at Fig. 3.2 where the aggregated inverse supply function is represented and the kinks are visible.

This is not the way we want the supply function to be. This is the inverse supply function, $q = g(p)$, and we want the supply function, $p = f(q)$. Based on the inverse supply function, we can write for the supply function.

$$p = \begin{cases} 0.4 + 0.6q = p_A, & \text{for } 0 \leq q \leq 0.(3) \\ 0.52 + 0.24q, & \text{for } q > 0.(3) \end{cases}$$

The illustrative graphics may be found in Fig. 3.3 wherein the kinks are again visible. ◀

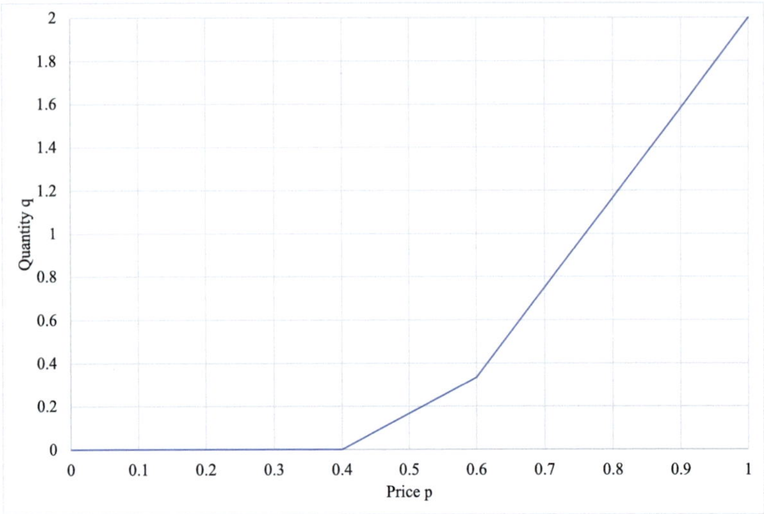

Fig. 3.2 Aggregated inverse supply function for Example 3.2

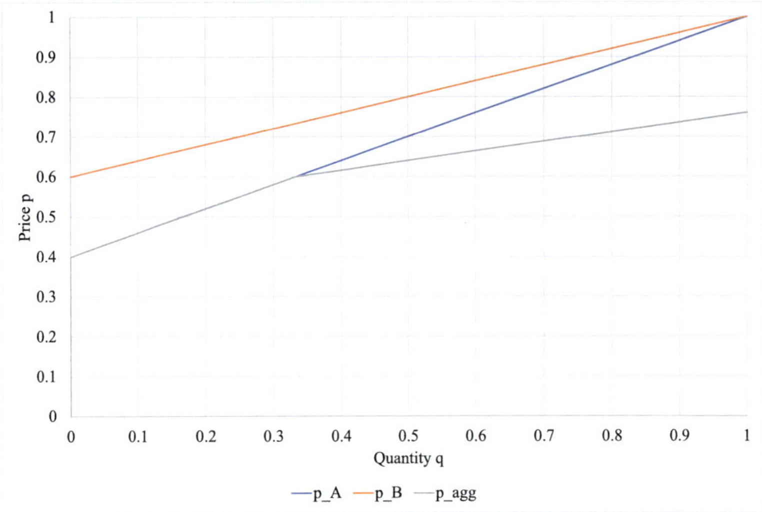

Fig. 3.3 Supply functions for Example 3.2

3.3 Market Equilibrium

There is a theory, called *General Equilibrium Theory*, that explains the behaviour of supply, demand, and prices in a market by seeking to prove that the interaction of supply and demand will result in an overall equilibrium. This equilibrium consists of obtaining an equilibrium price, also known as *market clearing price*, as a result of quantity demanded and quantity supplied being equal. The commonly demanded and supplied quantity is called equilibrium quantity. In other words, at the price at which these two quantities are identical, i.e., at the price at which the quantity demanded equals the quantity supplied, the market is in equilibrium. This means that, at the clearing price, the sellers sell exactly the quantity they want to sell at this price, and the buyers buy exactly the quantity they want to buy at this very same price.

It is now clear that, if we plot the supply curve and the demand curve in the same graphics, the equilibrium point is the interception point of the two curves. On the y-axis, we read the clearing price, and on the x-axis, we find the equilibrium quantity. We can say that the equilibrium point is the quantity-price pair at which both buyers and sellers are satisfied: the market is clear.

Example 3.3

Suppose the supply and demand curves for pencils are given, respectively, by:

$$p_S = 1 + 0.5q$$

$$p_D = 10 - q$$

Determine the equilibrium price and quantity. ◀

Solution

$$1 + 0.5q = 10 - q \overset{\text{yields}}{\rightarrow} q^* = 6$$

To find the equilibrium price, p^*, we replace the equilibrium quantity in any of the two functions, for instance:

$$p^* = p_D(6) = 4$$

The situation is pictured in Fig. 3.4. We recall that this is the classical example in Economics textbooks. However, if the slopes of the supply and demand curves were not the classical ones, the market equilibrium would be found in the same way. ◀

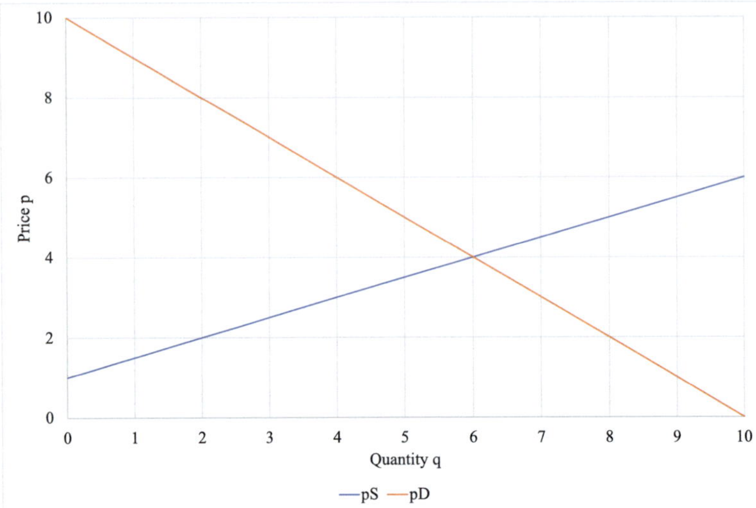

Fig. 3.4 Determination of the equilibrium point. Example 3.3

3.3.1 Demand and Supply Curve Shifting

The market is not static, disequilibrium happens. For instance, suppose the consumers of pencils of Example 3.3 are willing to buy more pencils for the same price. Hence, the demand curve shifts right, as seen in Fig. 3.5. The consequence of the consumers' willingness to buy more pencils is a new market equilibrium at a quantity of 8 and a price of 5. With respect to the initial situation, both the price and quantity increase, when demand increases.

Several factors can affect demand making the demand curve shift. Examples of these factors are changes in:

- Consumer preferences.
- Income of consumers.
- Prices of other consumer goods (substitutes and complements).
- Number of consumers in the market.
- Expectations about the future.

Now, let us suppose the cost of raw materials has increased and the sellers are offering the same quantity at a higher price. The supply curve shifts left, as seen in Fig. 3.6. The consequence of the producers' willingness to sell at higher prices is a new market equilibrium at quantity 4 and price 6. We note that the quantity decreases and the price increases when supply decreases (suppliers are offering a lower quantity for the same price).

The main factors affecting the supply curve (making it to shift) are:

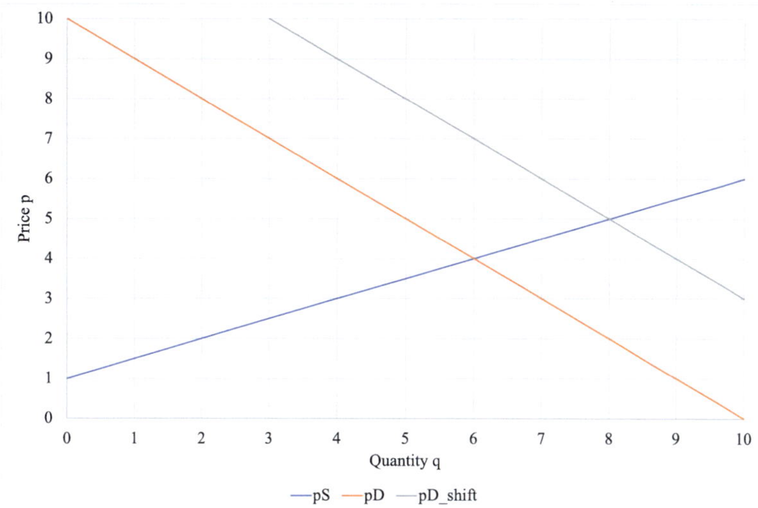

Fig. 3.5 Shift right in the demand

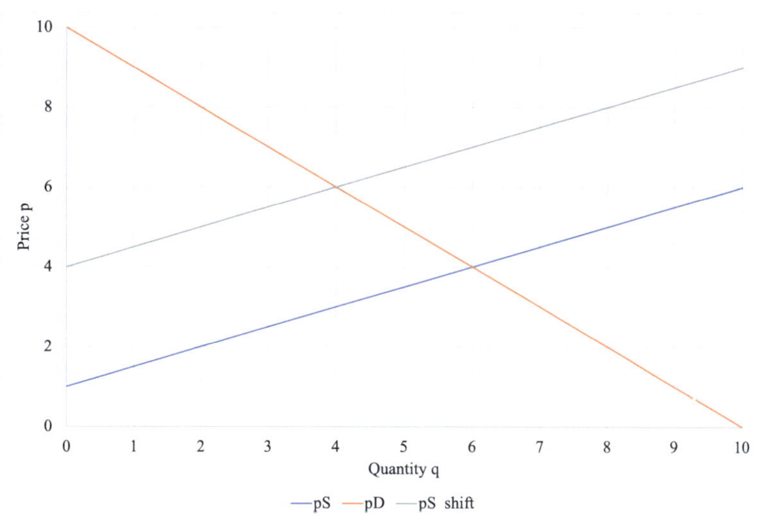

Fig. 3.6 Shift left in the supply

- Prices of inputs.
- Technology.
- Number of sellers in the market.
- Economic environment (taxes, government regulations, ...)
- Future prices and economic conditions.

And what if, the supply decreases and the demand increases in the proportions seen before? The situation is displayed in Fig. 3.7. We observe that the new price is higher (7) but the quantity was kept constant (6), with respect to the initial situation. This outcome is solely determined by the amplitude of the changes in supply and demand. Should the amplitude of the changes be different a different result would be obtained. This allows us to conclude that when both the supply and demand shift, it is not possible to determine beforehand what will happen to the quantity and price. It depends on the amplitude of the shifts.

The effects on prices and quantities are summarised in Table 3.1, taken from https://www.khanacademy.org/economics-finance-domain/ap-macroeconomics.

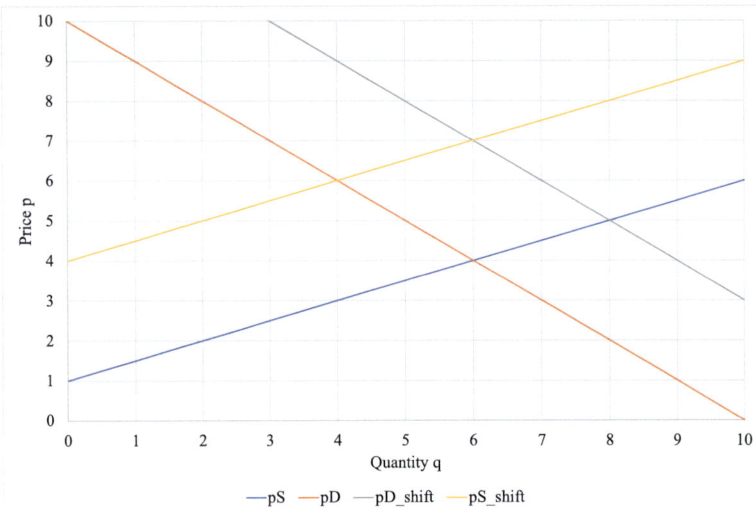

Fig. 3.7 Shift left in the supply and shift right in the demand

Table 3.1 Summary of equilibrium changes. Legend: ind. = indeterminate

	p up	p down	p ind.	q up	q down	q ind.
Supply increases (shifts right)		x		x		
Supply decreases (shifts left)	x				x	
Demand increases (shifts right)	x			x		
Demand decreases (shifts left)		x			x	
Demand increases, supply increases			x	x		
Demand increases, supply decreases	x					x
Demand decreases, supply increases		x				x
Demand decreases, supply decreases			x		x	

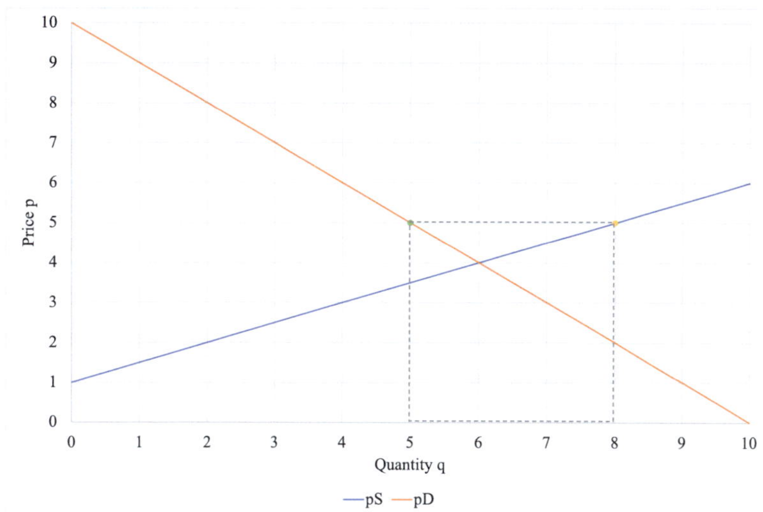

Fig. 3.8 Surplus

3.3.2 Surplus and Shortage

Let us consider again our example with the market for pencils. Suppose that the producer has a lot of excess pencil inventory that he/she cannot sell. This creates a *surplus*, i.e., the quantity supplied is greater than the quantity demanded. In the case of Fig. 3.8, the quantity supplied is 8 at a price of 5; at this very same price, the quantity demanded is 5. Most likely, the producer will put the inventory on sale. The market is going to react, by adapting price and quantity (demand increases and supply decreases), until equilibrium is reached. The conclusion is that surplus drives prices down.

To illustrate the inverse situation, suppose pencils are always out of stock. This is a *shortage* typical situation because the quantity demanded is greater than the quantity supplied. Figure 3.9 shows a shortage for a price of 3: at this price, the quantity demanded is 7 but only 4 is supplied. The producer will raise the price, to increase profits. The market will react, by changing price and quantity (demand decreases and supply increases), until equilibrium is reached. Conclusion: shortage drives the price up.

3.3.3 Consumer Surplus and Producer Surplus

The demand curve traces consumers' willingness to pay for different quantities. *Consumer surplus* is defined as the difference between what consumers are willing to pay (based on their preferences) and what they actually pay. On the other side, the supply curve shows the quantity that firms are willing to supply at each price. *Producer surplus* is the difference between what the producer is paid, and what the

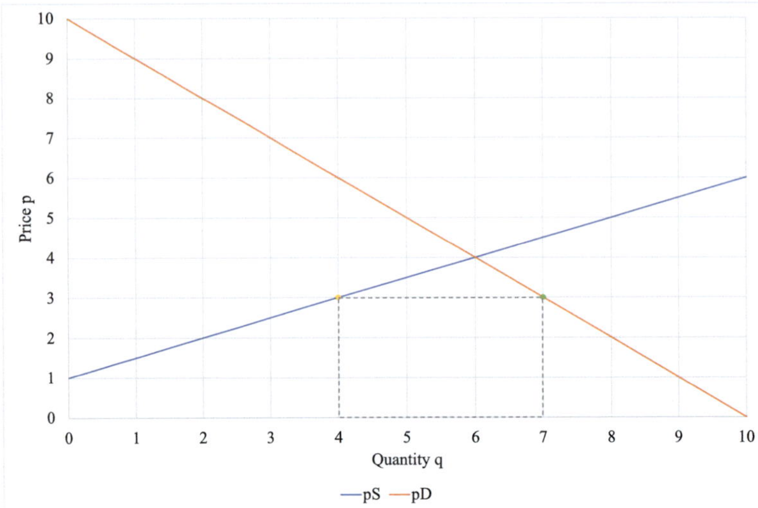

Fig. 3.9 Shortage

producer is ready to sell (based on their costs). In other words, consumer surplus can be measured by calculating the difference between the maximum willingness to pay and the actual price for each consumer and then summing those differences. In the same way, producer surplus can be measured by calculating the difference between the minimum acceptable price and the actual price for each unit sold and then summing those differences.

In the example we are following, the consumer is willing to pay 9 for each unit of the good, but he/she will pay only 4 per unit. So, the surplus is 9 – 4. This must be integrated for each (quantity, price) pair, therefore, the consumer surplus is the triangular area below the demand curve and limited by the market price horizontal line. Following similar reasoning, the producer surplus is the triangular area above the supply curve until the market price horizontal line. The situation is represented in Fig. 3.10, where the consumer surplus is the orange area and the producer surplus is the blue one. Putting some numbers on it, the consumer surplus is $CS_1 = 0.5 \times 6 \times (10 - 4) = 18$ and the producers surplus is $PS_1 = 0.5 \times 6 \times (4 - 1) = 9$.

Consumer surplus always increases as the price of a good falls and decreases as the price of a good rises. With the producer surplus, the situation is reversed.

The sum of consumer surplus and producer surplus is called the *social surplus*, or *economic surplus*. The total surplus is larger at the equilibrium quantity and price than it will be at any other quantity and price. In this example, $ES_1 = 18 + 9 = 27$.

Example 3.4

In a single market, the demand function is given by: $p_D = 200 - 0.02q^2$, and the supply function is $p_S = 100 + q$, where p is price and q is quantity. At the

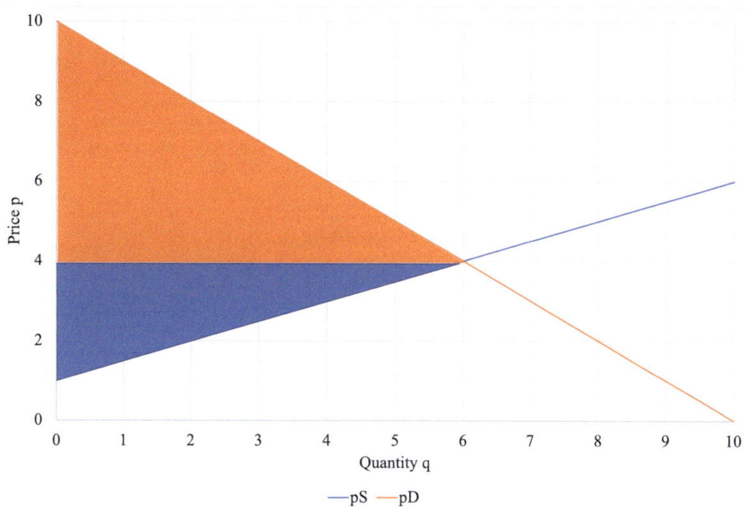

Fig. 3.10 Consumer surplus (orange) and producer surplus (blue) at market price

market equilibrium, compute the consumer surplus, the producer surplus, and the economic surplus (taken from https://www.usu.edu/math/koebbe). ◀

Solution

First, it is needed to find the market equilibrium. This is achieved by solving the next equation for q.

$$200 - 0.02q^2 = 100 + q \overset{\text{yields}}{\rightarrow} 0.02q^2 + q - 100 = 0$$

Using the quadratic formula, the solution is:

$$q_1 = 50 \vee q_2 = -100$$

The solution to be retained is $q^* = 50$, as the quantity cannot be negative. The corresponding price is $p^* = p_S(q^*) = 150$.
The consumer surplus is:

$$CS = \int_0^{q^*} (p_D(q) - p^*)\, dq$$

$$CS = \int_0^{50} (200 - 0.02q^2 - 150)\, dq = \left(50q - \frac{0.02}{3}q^3\right)\Big|_0^{50} = 1666.67$$

The producer surplus is:

$$PS = \int_0^{q^*} (p^* - pS(q))\,dq$$

$$PS = \int_0^{50} (150 - 100 - q)\,dq$$

We could solve the integral using integration rules, but, in this case, there is no need to do that because this is the area of a rectangular triangle.

$$PS = \frac{50 \times 50}{2} = 1250$$

The economic surplus is:

$$ES = CS + PS = 2916.67$$

◀

3.3.4 Price Ceiling and Price Floor

The market equilibrium can be disturbed by external actions. Examples of such disturbances are the imposition of price ceilings or price floors by the government. *Price ceiling* is the establishment of a maximum price, below the market price, with the objective of protecting the consumers from high market prices. A typical example of a price ceiling is the rental market, where the government imposes a limit on the price landlords can charge the tenants. The situation is represented in Fig. 3.11 where a price ceiling of 3, below the market price of 4, was imposed.

It can be seen that the consequence of the price ceiling is a shortage. Before the price ceiling, 6 units were traded in the market. At the price ceiling of 3, only 4 units are supplied, but the demand is 7, therefore, creating a shortage, i.e., some buyers will not be able to buy, due to this shortage.

And what about the impact on producer surplus and consumer surplus? Fig. 3.12 sheds some light on it. Producers are obliged to sell at 3, so the producers' surplus is the blue area. Remember that the producer's surplus is the area above the supply curve till the selling price. In regards to the consumers, they will buy at 3 instead of 4, so the consumers' surplus is the orange area. We recall that consumer surplus is everything below the demand curve and the buying price.

Putting some numbers on it, we have $PS_2 = 0.5 \times 4 \times (3 - 1) = 4$ (blue area). We can see that $PS_2 < PS_1$. $CS_2 = 0.5 \times 4 \times (10 - 6) + 4 \times (6 - 3) = 20$, hence, $CS_2 > CS_1$ (orange area). The economic surplus is $ES_2 = 4 + 20 = 24$. As $ES_1 > ES_2$, there is a loss of the market's efficiency, which is measured by the *deadweight loss* (grey area), $DWL = ES_1 - ES_2 = 27 - 24 = 3$.

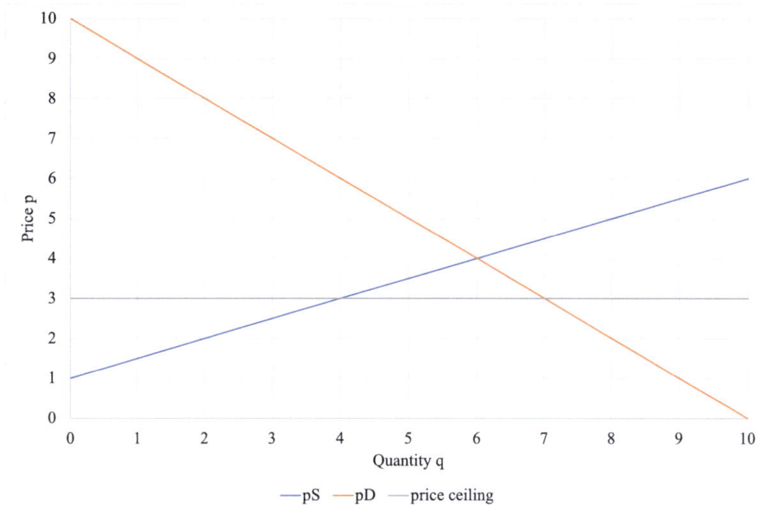

Fig. 3.11 Market equilibrium and price ceiling

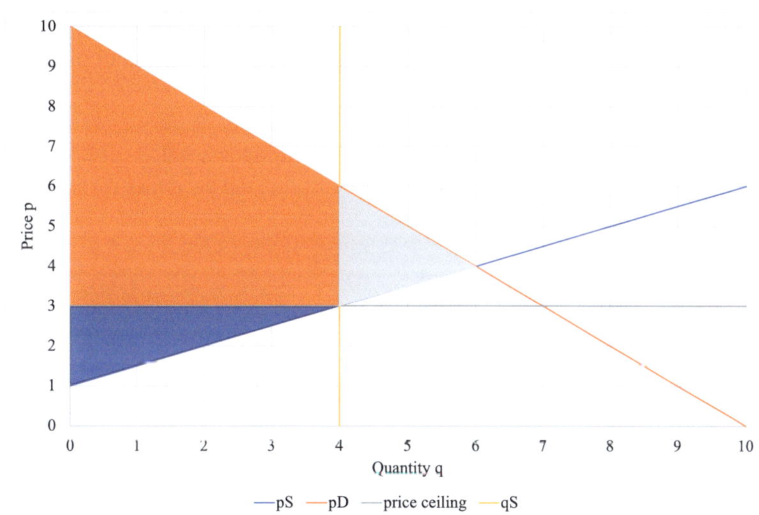

Fig. 3.12 Price ceiling: consumer surplus (orange), producer surplus (blue), and deadweight loss (grey)

A similar analysis can be made to the price floor. *Price floor* is the establishment of a minimum price for a good or service to be purchased, in order to protect the producers from a considered low market price. A typical example of a price floor is a minimum wage policy. This is aimed at protecting the producers of labour (firms are

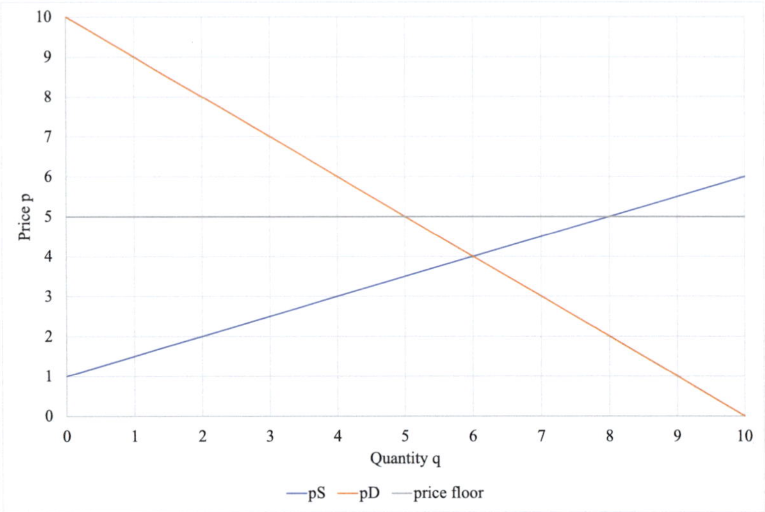

Fig. 3.13 Market equilibrium and price floor

the consumers of labour), the workers, from low wages. The condition is depicted in Fig. 3.13 where a price floor of 5, above the market price of 4, was imposed by the government.

It is visible that, at a price of 5, 8 units are supplied, but only 5 are demanded. This corresponds to a surplus, meaning that not all producers will be able to sell their products. Figure 3.14 illustrates the consumer surplus, the producer surplus, and the deadweight loss, for a price floor situation.

Regarding Fig. 3.14, we may write. $PS_3 = 0.5 \times 5 \times (3.5 - 1) + 5 \times (5 - 3.5) = 13.75$ and $PS_3 > PS_1$. $CS_3 = 0.5 \times 5 \times (10 - 5) = 12.5$. $ES_3 = 13.75 + 12.5 = 26.25$ and $CS_3 < CS_1$. $DWL = ES_1 - ES_3 = 0.75$. As expected, the consumers lose surplus, because they are obliged to pay above market price. The opposite happens with the producers, who gain surplus; they receive more than the market price. Both price ceiling and price floor cause deadweight loss, which is visible by a transfer of surplus from one player to the other.

Detailed information about price controls can be found in
https://pressbooks.bccampus.ca/uvicecon103/chapter/4-6-price-controls/

Example 3.5

This is a typical example that can be found in several textbooks. The inverse demand function for corn is $q_D = 10 - 2p$ and the inverse supply function is $q_S = 3p - 5$. The government is concerned that the market price of corn is too low and implements a price floor policy to protect farmers, by setting the minimum selling price at $p_S = 4$.

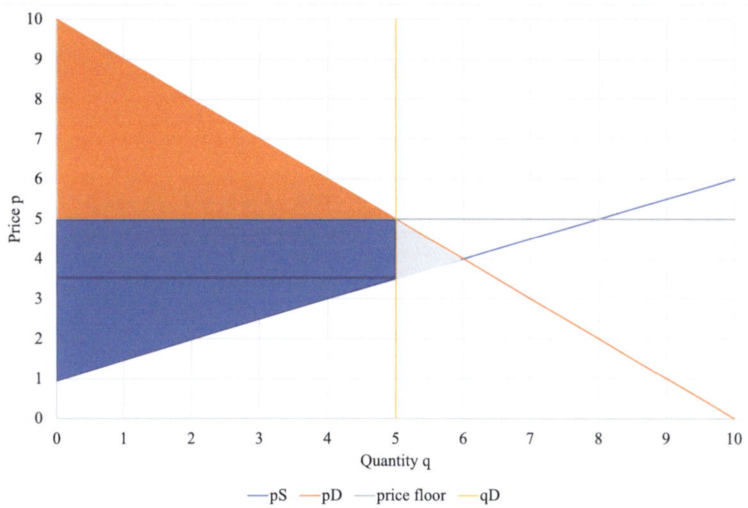

Fig. 3.14 Price floor: consumer surplus (orange), producer surplus (blue), and deadweight loss (grey)

1. Calculate the original equilibrium price and quantity.
2. How much is the consumer surplus, producer surplus, and economic surplus?
3. After the price floor policy, what is the new quantity of corn traded in the market?
4. Calculate the consumer surplus, producer surplus, and economic surplus after the implementation of the price floor. If it exists, calculate the deadweight loss. Illustrate the computations with appropriate graphics.

◀

Solution

1.

$$q_S = q_D \overset{\text{yields}}{\rightarrow} 10 - 2p = 3p - 5 \overset{\text{yields}}{\rightarrow} p^* = 3$$

$$q^* = q_D(p^*) = 4$$

2.

To compute the surplus, we need the demand and supply functions:

$$p_D = 5 - \frac{q}{2};\ p_S = \frac{5}{3} + \frac{q}{3}$$

$$CS_1 = \frac{1}{2} \times (5 - 3) \times 4 = 4$$

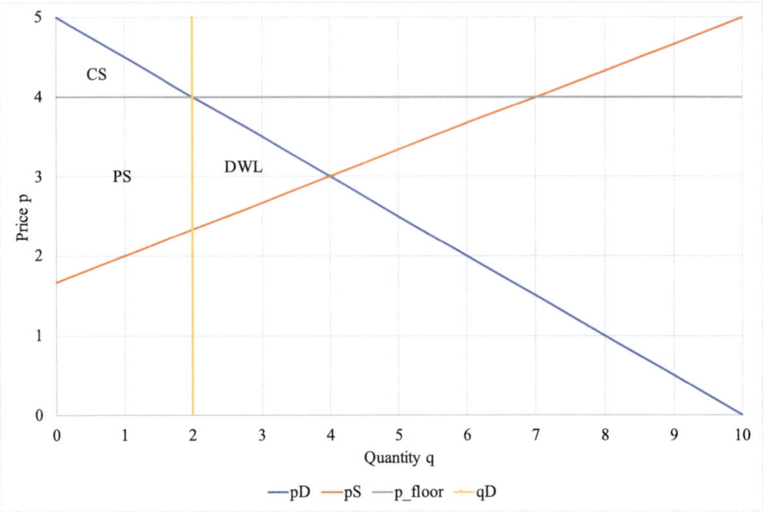

Fig. 3.15 Price floor: consumer surplus (CS), producer surplus (PS), and deadweight loss (DWL). Example 3.5

$$PS_1 = \frac{1}{2} \times (3 - \frac{5}{3}) \times 4 = 2.67$$

$$ES_1 = 6.67$$

3.
As a consequence of the price floor, there will be an excess supply of corn. The quantity sold in the market is determined by the number of persons willing to buy corn at this higher price.

$$4 = 5 - \frac{q_2}{2} \overset{\text{yields}}{\rightarrow} q_2 = 2$$

4.
The situation becomes clearer by looking at Fig. 3.15.

$$CS_2 = \frac{1}{2} \times (5 - 4) \times 2 = 1$$

$$PS_2 = \frac{1}{2} \times (2.33 - 1.67) \times 2 + (4 - 2.33) \times 2 = 4$$

Please note that $p_S(2) = 2.33$ and $p_S(0) = 1.67$.

$$ES_2 = 5$$

As $ES_2 < ES_1$, there is deadweight loss, given by $DWL = ES_1 - ES_2 = 1.67$
◀

3.4 Market Adjustment

Consider a body with mass M subject to two forces, acting in opposite directions. It is well-known that if the net force is zero, then the speed is constant (including at zero) and, therefore, the acceleration is zero. We say that the body is in *static equilibrium*. If, however, the net force is not zero, the speed is not constant and the body is subject to acceleration. We say that the body is in *dynamic equilibrium*. The speed defines the state of the system, therefore, the speed is the *state variable*. The speed is the changing quantity (with time) that makes the body change between states.

3.4.1 State Variable, Price or Quantity?

Starting from here, one could ask, "In the analysis of market dynamics, what is the variable, price or quantity, that must be considered as the state variable, equivalent to the speed in the equilibrium of motion?" The opinion of the Economists is not uniform in answering this question. Economists led by Leon Walras considered the price as the independent variable. On the other side, another set of economists led by Alfred Marshall considered the quantity as the independent variable. In this textbook, we will devote more attention to the last option, considering the quantity as the state variable. However, we will approach also the first option with the main objective of highlighting the differences.

In other words, the problem could be stated as follows. When there is a disturbance that moves the market away from the equilibrium, what is the adjustment variable that rules the market behaviour afterward? Price or quantity? As said before, the following analysis will consider quantity as the state variable and, whenever appropriate, references to the price acting as the state variable will be made. Moreover, we will follow traditional economics textbooks that consider the upward supply function and the downward demand function.

3.4.2 Law of Demand, Law of Supply and Quantity Adjustment

In traditional economics, there are two laws applying to market behaviour.

Law of Demand: the empirical observation that when the price of a product falls, people demand larger quantities of it. This observation is based on the diminishing marginal utility. As we know, this law is not general, as it does not apply to Giffen goods, for instance.

Law of Supply: the empirical observation that when the price of a product rises, firms offer more of it for sale. This is related to the firms' objective of maximising profits. As we know, this law is not general, as increasing returns to scale supply functions do not follow this law.

We define *Quantity Adjustment* as the process by which a market surplus leads to a cut-back in the quantity supplied or a market shortage causes an increase in supplied quantity. If for a given quantity of the product, q, the unitary price that the consumers are willing to pay, p_D, is greater (lesser) than the unitary price offered by the producers, p_S, then, the producers will be willing to increase (decrease) the quantity of the product over time. A similar rationale may be applied to consumers. This can be easily understood by looking at the graphical representation of the demand and supply functions.

3.5 Dynamic Equilibrium

In the previous section, we made an analogy with the physics of the motion of a body. The respective governing equation is:

$$M\frac{d\dot{x}}{dt} = F_I - F_R \tag{3.1}$$

where M is the mass, $\dot{x}$ is the speed and F_I, F_R are the two forces acting on the body. t stands for time.

3.5.1 Marshall Method—Quantity Adjustment

The problem we want to analyse may be formulated as follows. When the market is disturbed and moves away from the equilibrium, how does quantity change in time after the disturbance? Does the system return to the initial equilibrium? Or does it reach another equilibrium? Or does it reach any equilibrium at all? The answers to these questions can be found below. To begin with, let us recall that:

$$\begin{aligned} p_D(q) > p_S(q) &\implies \frac{dq}{dt} > 0 \\ p_D(q) < p_S(q) &\implies \frac{dq}{dt} < 0 \end{aligned} \tag{3.2}$$

Based on this assumption and making an analogy with physics, we will write the following equation, after Marshall's methodology:

$$I\frac{dq}{dt} = p_D(q) - p_S(q) \tag{3.3}$$

This equation describes the dynamic equilibrium of a single market, establishing a continuous model over time. I is a constant that represents the inertia of the market, in a similar way to the physical system. Following Marshall's methodology, the quantity of the product is the state variable of the system, which is going to adjust till a new equilibrium is found if it exists.

Now, we need to solve Eq. 3.3 and find $q(t)$. Before we proceed, we are going to make a simplification: we will assume the supply and demand curves are linear, which is a valid approximation within the vicinity of a given point. This is a simplification that is often considered in Economics textbooks. Therefore, from now on, we assume:

$$\begin{aligned} p_D(q) &= a - bq \\ p_S(q) &= c + dq \end{aligned} \tag{3.4}$$

Concerning the demand function, it is assumed $a > 0$; $b > 0$, which represents the typical downward behaviour of demand functions, excluding a few exceptions (Giffen goods, for instance), as already mentioned. In regards to the supply function, no doubt that $c > 0$, but $d > 0$, for decreasing returns to scale—normally considered in traditional Economics textbooks—and $d < 0$, for increasing returns to scale.

Based on the considered simplification, Eq. 3.3 can be rewritten as:

$$I\frac{dq}{dt} + (b + d)q = a - c \tag{3.5}$$

This is a differential equation whose solution is:

$$q = q_n + q_f \tag{3.6}$$

where q_n and q_f are the natural and forced responses of the system, respectively.

The natural response, q_n, is the solution to the homogeneous equation:

$$I\frac{dq}{dt} + (b + d)\, q = 0 \tag{3.7}$$

The functional form of q_n is:

$$q_n = Ke^{st} \tag{3.8}$$

where K and s are constants to be determined.

The value of s can be found by substituting the functional form in the homogeneous differential equation:

$$\begin{aligned} I\frac{d(Ke^{st})}{dt} + (b + d)\,(Ke^{st}) &= 0 \\ IKse^{st} + (b + d)\left(Ke^{st}\right) &= 0 \\ Kse^{st}\,(Is + b + d) &= 0 \\ Is &= -(b + d) \\ s &= -\frac{b + d}{I} \end{aligned} \tag{3.9}$$

We will deal with the K parameter later. With regards to the forced solution, we note that our differential equation is of the type:

$$f(t) = a - c \tag{3.10}$$

The functional form of q_f is:

$$q_f = A \tag{3.11}$$

where A is a constant to be determined.

Substituting for q_f in our differential equation, we get:

$$\begin{aligned} I\frac{dA}{dt} + (b+d)\,A &= a - c \\ A &= \frac{a-c}{b+d} \end{aligned} \tag{3.12}$$

We can now write:

$$q(t) = q_n + q_f = Ke^{-\frac{b+d}{I}t} + \frac{a-c}{b+d} \tag{3.13}$$

We use the initial condition $q(0) = q_0$ to find K. The initial condition is regarded as the observed quantity before the disturbance.

$$\begin{aligned} q(0) = q_0 &= K + \frac{a-c}{b+d} \\ K &= q_0 - \frac{a-c}{b+d} \end{aligned} \tag{3.14}$$

The final solution is:

$$q(t) = \frac{a-c}{b+d} + \left(q_0 - \frac{a-c}{b+d}\right)e^{-\frac{b+d}{I}t} \tag{3.15}$$

What is the meaning of this equation? It shows the time evolution of the quantity starting from an initial point, q_0. Is an equilibrium point always reached? Not really. We say the system is stable if an equilibrium point is reached and unstable if not.

Stability is assured if $(b+d) \geq 0$, given that $I > 0$—I is a positive constant, normally taken as 1, unless otherwise stated. In this condition, the system tends to an equilibrium point (static equilibrium) and is said to be stable. The equilibrium quantity is the forced response of the system:

$$q^* = \frac{a-c}{b+d} \tag{3.16}$$

If $(b+d) < 0$, the system grows to infinity and the system is said to be unstable, no equilibrium is attained. It is noted that, for the examples in traditional Economics, equilibrium is always reached. The condition $b+d \geq 0$ is fulfilled when the aggregate demand curve has a negative slope, $b > 0$, and the aggregate supply curve represents non-increasing returns to scale, $d \geq 0$.

Example 3.6

Let us recycle the example of the pencils market. The supply and demand curves for pencils are given, respectively, by:

$$p_S = 1 + 0.5q$$

$$p_D = 10 - q$$

Suppose the market is initially operating at a quantity equal to 5. Equate the time evolution of the quantity and plot the respective graphics. ◀

Solution

Given the data, we have:

$$a = 10;\ b = 1;\ c = 1;\ d = 0.5$$

$b + d = 1.5 \geq 0$, therefore the market is stable and an equilibrium point exists (we already knew that).

The equilibrium quantity is:

$$q^* = \frac{10 - 1}{1 + 0.5} = 6$$

The time evolution of the quantity from the initial quantity, $q_0 = 5$ is:

$$q(t) = 6 - 1e^{-1.5t}$$

The second term goes to 0 as t goes to infinity. That is why $q^* = 6$.

The plotting is in Fig. 3.16, for the time evolution of the quantity.

We may also plot the time evolution of the two prices, as they are dependent only on the quantity (see Fig. 3.17).

As the system is stable, equilibrium is reached. ◀

Now, let us analyse the non-traditional (from the Economics textbooks standpoint) supply curve displaying increasing returns to scale (downward-sloping). To facilitate, we are going to write the supply function in the form:

$$p_S(q) = c - b'q;\ b' > 0;\ b' = -d \tag{3.17}$$

Stability is determined by the sign of $b - b'$. If it is positive, the system is stable, if it is negative, the system is unstable. $b - b' > 0$, if $b > b'$; $b - b' < 0$, if $b < b'$.

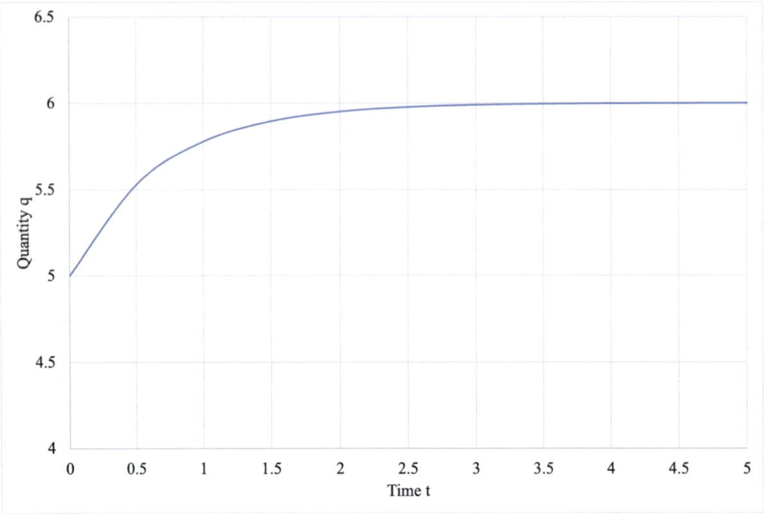

Fig. 3.16 Quantity over time. Example 3.6

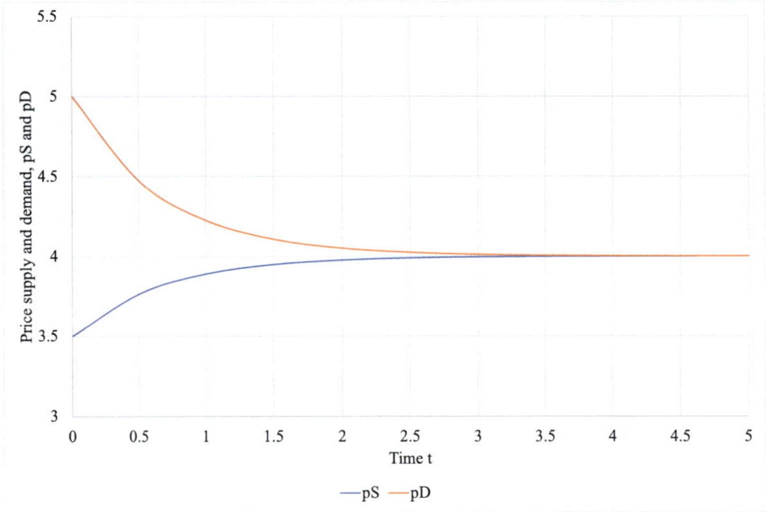

Fig. 3.17 Supply price and demand price over time. Example 3.6

The problem that we face, is better illustrated with an example. Let us come again to the pencils market example that we have been following. Suppose the demand curve remains unchanged,

$$p_D(q) = 10 - q$$

and the supply function changes to:

$$p'_S(q) = 8 - 0.6q$$

In this case, $b = 1$ and $b' = 0.6$, therefore, $b - b' > 0$ and the market reaches the equilibrium whose quantity is:

$$q'^* = \frac{10 - 8}{1 - 0.6} = 5.$$

The equilibrium price is:

$$p'^* = p_D(5) = 5$$

Now, let us suppose the supply curve changes again, this time to:

$$p''_S = 12 - 1.25q$$

Then, $b' = 1.25$ and $b - b' = -0.25 < 0$. According to the quantity adjustment methodology (Marshall), the market is unstable and no equilibrium is reached. However, we know that this is not true. If we make $p_D(q) = p''_S(q)$, we obtain $q''^* = 8$ and $p''^* = 10 - 8 = 2$. The conclusion we draw is that the quantity adjustment methodology is unable to find the equilibrium point, for the case where the downward supply curve is steeper than the demand curve. This opens space for addressing the price adjustment methodology (Walras methodology) that we have introduced before.

3.5.2 Walras Method—Price Adjustment

Using price adjustment, the equation that describes the dynamic equilibrium of a single market is:

$$I'\frac{dp}{dt} = q_D(p) - q_S(p) \tag{3.18}$$

We highlight that the product's price is the state variable of the system. Considering that the supply and demand curves may be linearised, we can write:

$$\begin{aligned} p_D(q) = a - bq &\overset{\text{yields}}{\rightarrow} q_D(p) = \frac{a}{b} - \frac{1}{b}p \\ p_S(q) = c + dq &\overset{\text{yields}}{\rightarrow} q_S(p) = -\frac{c}{d} + \frac{1}{d}p \end{aligned} \tag{3.19}$$

and replacing the above, we sequentially obtain:

$$\begin{aligned} I'\frac{dp}{dt} &= \frac{a}{b} - \frac{1}{b}p + \frac{c}{d} - \frac{1}{d}p \\ I'\frac{dp}{dt} &= \left(-\frac{1}{b} - \frac{1}{d}\right)p + \left(\frac{a}{b} + \frac{c}{d}\right) \\ I'\frac{dp}{dt} &= \frac{ad + bc}{\text{bd}} - \left(\frac{b + d}{bd}\right)p \\ I'\frac{dp}{dt} + \left(\frac{b + d}{bd}\right)p &= \left(\frac{ad + bc}{bd}\right) \end{aligned} \tag{3.20}$$

At this point, we recall, as seen in the quantity adjustment method, that the solution of the equation:

$$I\frac{dq}{dt} + (b + d)q = a - c \tag{3.21}$$

is:

$$q\,(t) = \frac{a-c}{b+d} + \left(q_0 - \frac{a-c}{b+d}\right)e^{-\frac{b+d}{I}t} \tag{3.22}$$

By similarity, we can write:

$$p\,(t) = \frac{ad+bc}{b+d} + \left(p_0 - \frac{ad+bc}{b+d}\right)e^{-\left(\frac{b+d}{bd}\frac{1}{I'}\right)t} \tag{3.23}$$

The system is stable—an equilibrium point is found –, according to Walras method (price adjustment), if:

$$\frac{b+d}{bd} = -\frac{b-b'}{bb'} \geq 0 \tag{3.24}$$

and unstable—no equilibrium point is found –, if:

$$\frac{b+d}{bd} = -\frac{b-b'}{bb'} < 0 \tag{3.25}$$

We observe that I' is a positive constant, generally taken as 1 when no more information is given. We recall that, for downward-sloping supply functions, it is more convenient to consider $b' = -d;\ b' > 0$. Moreover, $p_0 = p(0)$ is the initial price.

The equilibrium price is the forced response of the system:

$$p^* = \frac{ad+bc}{b+d} = \frac{bc-ab'}{b-b'} \tag{3.26}$$

Now, let us retake the example with the downward-sloping supply functions we were following. The first case study is given by:

$$p_D(q) = a - bq = 10 - q$$

$$p'_S(q) = c - b'q = 8 - 0.6q$$

We have seen that using Marshall's methodology, the system was stable because $b - b' = 1 - 0.6 = 0.4 > 0$. To solve for Walras stability, we compute:

$$-\frac{b-b'}{bb'} = -\frac{1-0.6}{1\times 0.6} = -0.67 < 0.$$

We conclude that price adjustment is unable to find an equilibrium point for this market, the price goes to infinity. But quantity adjustment does. The market is stable, according to Marshall, and unstable, according to Walras.

The second case study is given by:

$$p_D(q) = a - bq = 10 - q$$

$$p''_S(q) = c - b'q = 12 - 1.25q$$

For this market, quantity adjustment was unable to find the equilibrium because:

$$b - b' = 1 - 1.25 = -0.25 < 0$$

Using price adjustment, we find that:

$$-\frac{b - b'}{bb'} = -\frac{1 - 1.25}{1 \times 1.25} = 0.2 > 0$$

So, the market is stable, price adjustment methodology is able to find the equilibrium price at:

$$p''^{*} = \frac{1 \times 12 - 10 \times 1.25}{1 - 1.25} = 2$$

and the equilibrium quantity at:

$$q''^{*} = q_D\left(p^{*}\right) = \frac{a}{b} - \frac{1}{b}p^{*} = 8$$

3.5.3 Summary of Market Stability

A summary of market stability according to quantity adjustment (Marshall) and price adjustment (Walras) is shown in Table 3.2.

Table 3.2 Summary of market stability according to quantity adjustment (Marshall) and price adjustment (Walras)

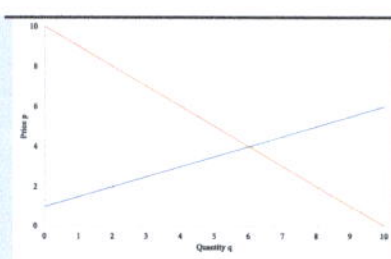	$b > 0; d > 0$	Marshal: stable Walras: stable
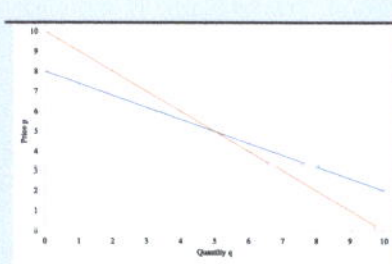	$b > 0; b' < b$	Marshal: stable Walras: unstable
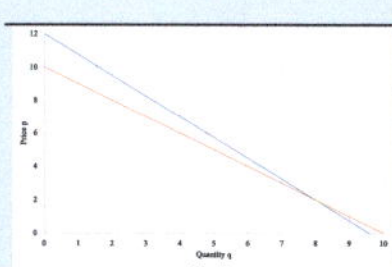	$b > 0; b' > b$	Marshal: unstable Walras: stable

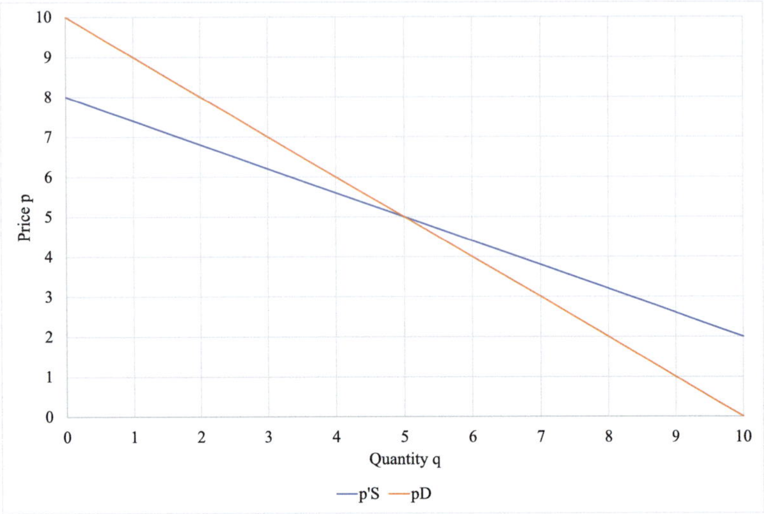

Fig. 3.18 Supply and demand functions. Supply function: $p'_S = 8 - 0.6q$. Example 3.7 – 1

Example 3.7

Again, we are going to use the example of the pencils market. The demand curve for pencils is:

$$p_D = 10 - q$$

1. Using quantity adjustment, plot the time evolution of the quantity, when: (i) $p'_S = 8 - 0.6q$; (ii) $p''_S = 12 - 1.25q$, for an initial quantity equal to 4.
2. Using price adjustment, plot the time evolution of the price, when: (i) $p'_S = 8 - 0.6q$; (ii) $p''_S = 12 - 1.25q$, for an initial price equal to 4.

◀

Solution

1.

The supply and demand functions are plotted in Figs. 3.18 and 3.19. We can see that $p'^* = 5$ and $q'^* = 5$, regarding $p'S(q)$ (Fig. 3.18) and $p''^* = 2$ and $q''^* = 8$, regarding $p''S(q)$ (Fig. 3.19).

For the supply functions $p'_S = 8 - 0.6q$ and $p''_S = 12 - 1.25q$, the time evolution of $q(t)$ is, respectively:

$$q'(t) = 5 - e^{-0.4t}$$

$$q''(t) = 8 - 4e^{0.25t}$$

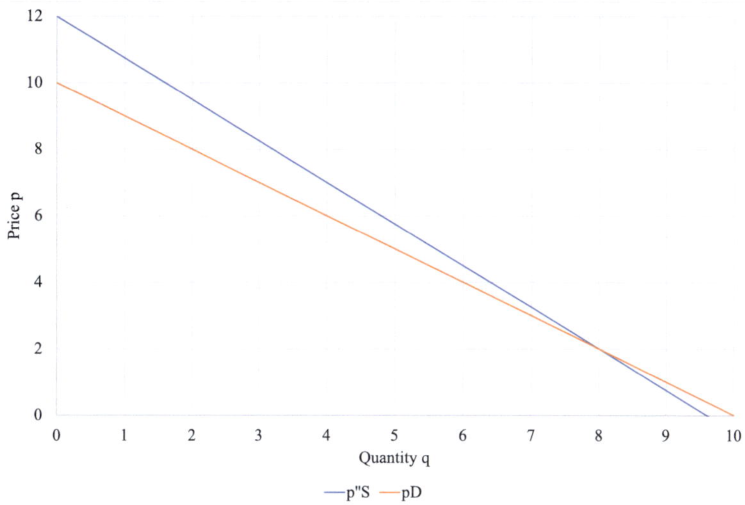

Fig. 3.19 Supply and demand functions. Supply function: $p''_S = 12 - 1.25q$. Example 3.7 – 1

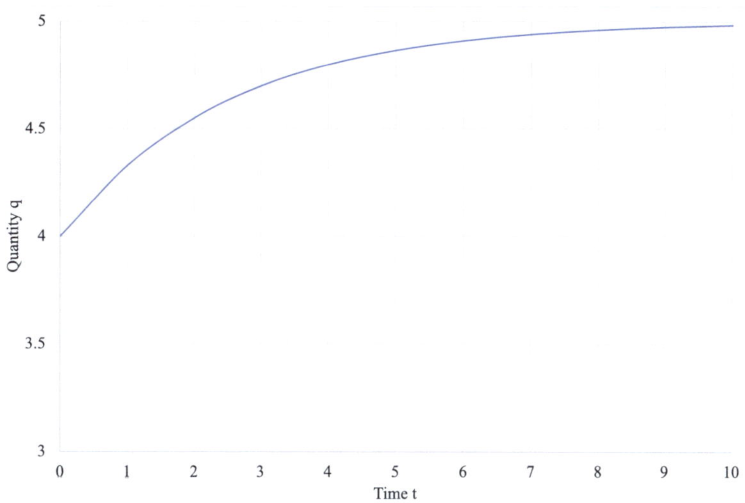

Fig. 3.20 Time evolution of the quantity. Supply function: $p'_S = 8 - 0.6q$. Example 3.7 – 1

The corresponding graphics are depicted in Figs. 3.20 and 3.21. We conclude that the quantity adjustment method is able to converge to the optimal quantity, in the case the downward supply function is less steep than the demand function (Fig. 3.20)—stable market. Conversely, when the supply function is steeper than the demand function, the quantity adjustment method fails to find the optimal quantity (Fig. 3.21)—unstable market.

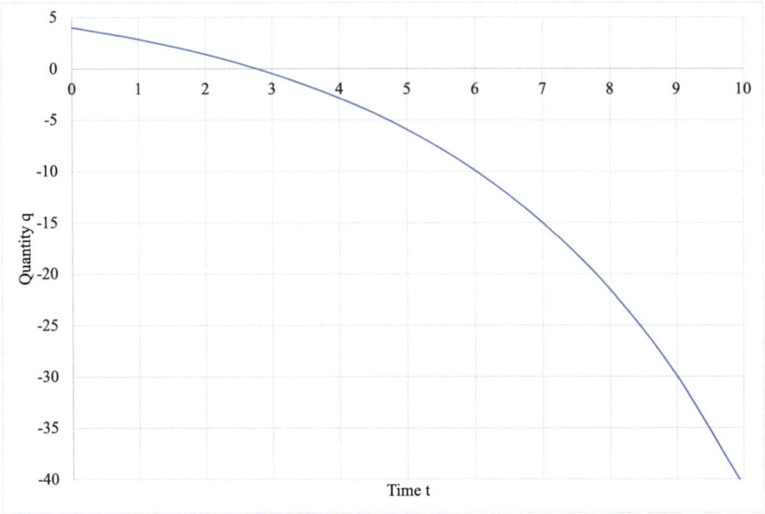

Fig. 3.21 Time evolution of the quantity. Supply function: $p''_S = 12 - 1.25q$. Example 3.7 – 1

2.
Now, we are using the price adjustment method. For the supply functions $p'_S = 8 - 0.6q$ and $p''_S = 12 - 1.25q$, the time evolution of $p(t)$ is, respectively:

$$p'(t) = 5 - e^{0.67t}$$

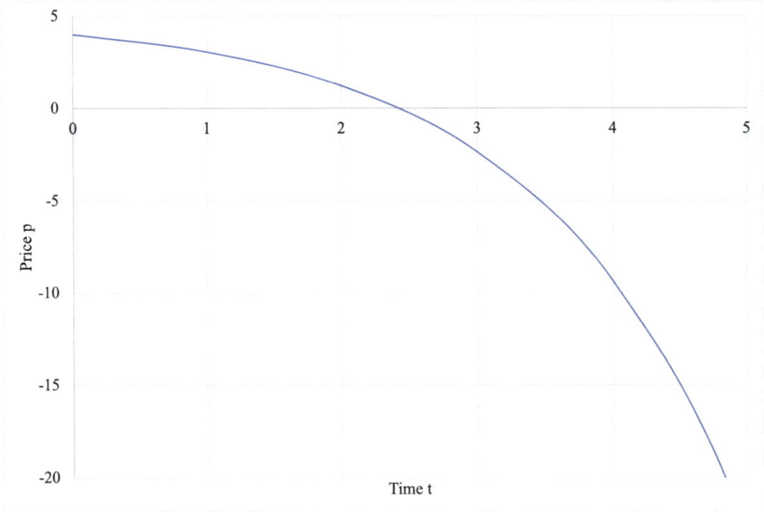

Fig. 3.22 Time evolution of the price. Supply function: $p'_S = 8 - 0.6q$. Example 3.7 – 2

$$p''(t) = 2 + 2e^{-0.2t}$$

The related graphics are depicted in Figs. 3.22 and 3.23. With the price adjustment method, the situation is reversed. When the downward supply function is less steep than the demand function, the equilibrium price is not found and the price goes to infinity (Fig. 3.22)—unstable market. On the other side, when the supply function is steeper than the demand function, the equilibrium price is reached and the market is stable (Fig. 3.23). ◄

3.6 A Note on Interaction Between Markets

The interaction between markets is a fundamental aspect of economics, as it shapes the dynamics of resource allocation, prices, and overall economic activity. Markets do not operate in isolation; they are interconnected and influence each other through various channels. Understanding the interaction between markets is crucial for policymakers, businesses, and individuals to make informed decisions and navigate the complexities of the global economy.

One way markets interact is through the concept of complementarity. Complementary markets are those where the demand or supply of goods or services in one market depends on the demand or supply in another market. For example, the market for smartphones and the market for mobile apps are complementary, as the demand for smartphones drives the demand for apps, and vice versa. Changes in one market

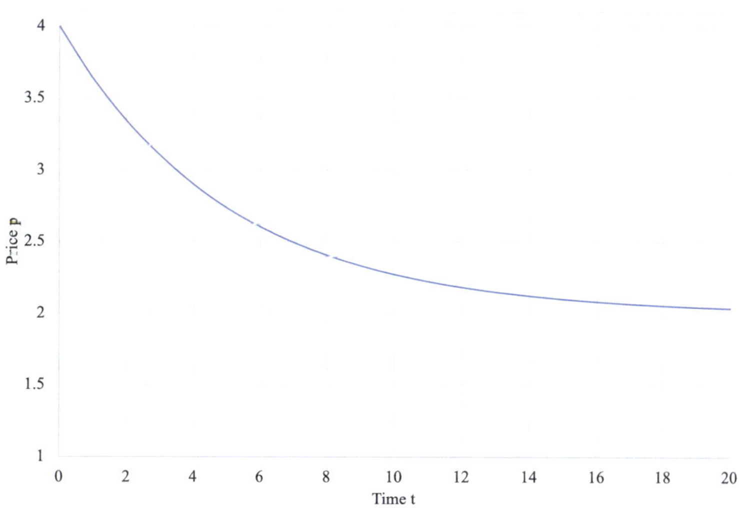

Fig. 3.23 Time evolution of the price. Supply function: $p''_S = 12 - 1.25q$. Example 3.7 – 2

can have ripple effects on the other, leading to adjustments in prices, quantities, and investments.

Another form of market interaction is substitution. Substitutable markets are those where consumers can replace one good or service with another based on relative prices and preferences. For instance, if the price of beef increases significantly, consumers may opt for chicken as a substitute. This substitution effect impacts both the beef and chicken markets, leading to changes in prices and quantities demanded.

Market interactions also occur through the transmission of information. Price signals in one market can convey valuable information to participants in related markets. For instance, changes in oil prices can impact various markets, such as transportation, manufacturing, and energy. These price signals influence decision-making, production levels, and investment patterns across multiple sectors.

Furthermore, the interaction between markets is influenced by factors such as government policies, regulations, and global economic conditions. Trade policies, fiscal and monetary measures, and macroeconomic trends can have widespread effects on interconnected markets. Financial markets, including stock exchanges and currency markets, also play a crucial role in transmitting information and impacting various sectors of the economy.

In conclusion, the interaction between markets is a dynamic and complex phenomenon. Complementarity, substitution, and the transmission of information are some of the mechanisms through which markets interact. Understanding these interactions is essential for policymakers and economic actors to anticipate and respond effectively to changes, promote efficiency, and ensure stability in the interconnected global economy.

3.7 Conclusions

In the framework of power systems economics, understanding the behaviour of the single market is of utmost importance. This chapter has explored key concepts such as market equilibrium, market adjustment, and dynamic equilibrium, with a focus on the methods proposed by Marshall and Walras.

The chapter begins with a discussion of the methodologies to aggregate the supply and demand curves from the producers and the consumers, respectively. This is relevant because the market is operated considering aggregated supply and demand curves.

Market equilibrium represents a state of balance between the demand for and supply of electricity within a single market. It is the point where the quantity of electricity sought by consumers matches the quantity supplied by producers. Achieving market equilibrium is essential for ensuring stability and efficiency in the power systems sector. Topics such as surplus and shortage, consumer surplus and producer surplus, and price ceiling and price floor are addressed.

Market adjustment is the process through which a single market responds to changes in demand or supply. When there is an imbalance between these two factors, prices and quantities fluctuate, prompting market participants to adapt their behaviours accordingly. This mechanism enables the market to self-correct and restore equilibrium. A discussion on the choice of price or quantity as a state variable is offered and the well-known laws of supply and demand are discussed in the context of quantity adjustment.

Dynamic equilibrium, as elucidated by Alfred Marshall, acknowledges that markets are not static but evolve over time. It takes into account technological advancements, shifts in consumer preferences, and regulatory changes that influence the behaviour of the power systems market. Marshall's approach provides valuable insights into the short-term dynamics of market equilibrium, enabling analysts to understand and predict fluctuations in a single market. Marshall assumes the quantity as being the adjustment variable.

By studying Walras' approach, engineers and economists can gain a deeper understanding of the complexities involved in single-market behaviour and devise effective strategies to ensure long-term stability and optimal resource allocation. The adjustment variable in the Walras method is the price.

The chapter ends with a brief note on the different types of interactions between markets.

By studying the single market behaviour through the lens of engineering power systems economics, policymakers and industry professionals gain insights that enable them to make informed decisions. Strategies can be devised to ensure the stability, efficiency, and sustainability of the power systems sector. It is essential to monitor market equilibrium, facilitate market adjustment mechanisms, and understand the behaviour of dynamic equilibrium to promote a robust and resilient single market for electricity.

3.8 Proposed Exercises

Problem 3.1

Max and Sergio are two friends who are fond of gadgets. Max inverse demand curve for gadgets is $q_M = 10 - 2p$ and Sergio's is $q_S = 5 - 0.5p$.

1. Find the two friends' aggregated demand for gadgets.
2. What is the consumption of gadgets when the price is 4? And if it is 6?

Key Problem 3.1

1.

The inverse demand functions are:

$$q_M = 10 - 2p; \text{ for } p \leq 5$$

$$q_S = 5 - 0.5p; \text{ for } p \leq 10$$

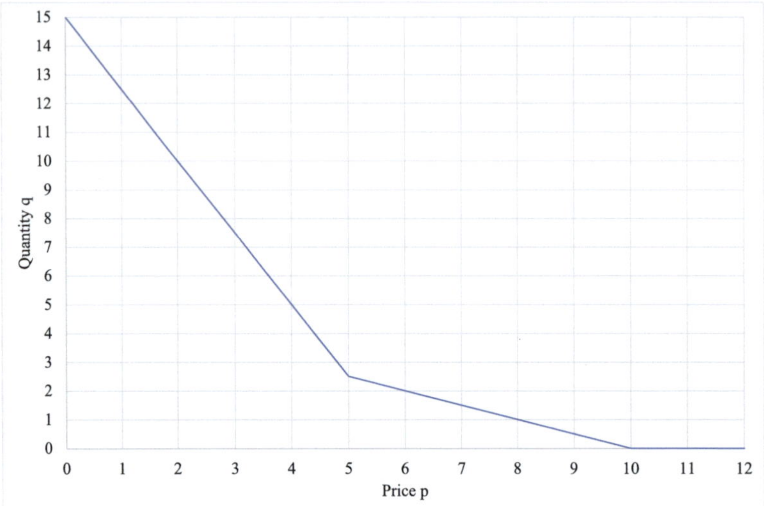

Fig. 3.24 Inverse demand function. Problem 3.1

It follows that $p = 5$ and $p = 10$ are kinks. For $5 \leq p \leq 10$, only Sergio buys gadgets. When $p < 5$, both buy gadgets. So, the aggregated inverse demand function is:

$$q = \begin{cases} 15 - 2.5p, & \text{for } p \leq 5 \\ 5 - 0.5p, & \text{for } 5 < p \leq 10 \\ 0, & \text{for } p > 10 \end{cases}$$

Note that it is required to sum the inverse demand functions and not the demand functions.

The graphics of the inverse demand function are represented in Fig. 3.24.

Based on Fig. 3.24 it is straightforward to conclude that the demand function is:

$$p = \begin{cases} 10 - 2q, & \text{for } q < 2.5 \\ 6 - 0.4q, & \text{for } 2.5 \leq q \leq 15 \end{cases}$$

2.

From Fig. 3.24, one concludes that $q(4) = 5$ (both consume gadgets) and $q(6) = 2$ (only Sergio consumes).

Problem 3.2

Consider two markets ruled by the following demand and supply functions:

$$p_D(q) = 1.4 - 0.4q$$

$$p_{S1}(q) = 0.4 + 0.6q$$

and:

$$p_D(q) = 1.4 - 0.4q$$

$$p_{S2}(q) = 1.2 - 0.2q$$

1. Explain how the supply and demand curves of the markets are obtained.
2. What are the returns to scale presented by the producers in the two markets? Justify.
3. Considering quantity adjustment, establish and solve the equations of evolution of the two markets. Consider $I = 1$ and $q_0 = 0.5$.

Key Problem 3.2

1.

The aggregated market supply and demand curves are obtained as follows: for a given price, p, the quantities, q, offered by each producer are summed up, thus obtaining the global quantity for the given price.

2.

The supplier with an upward supply function shows DRS because the unitary price increases with the quantity's increase. The supplier with a downward supply function presents IRS because the unitary price decreases with the quantity's increase.

3.

The equation that describes the dynamic equilibrium of a single market is:

$$I\frac{dq}{dt} = p_D(q) - p_S(q)$$

We consider that the quantity of the product is the state variable of the system. Considering that the supply and demand curves may be linearised, which is valid in the vicinity of a given point, we can write:

$$I\frac{dq}{dt} = a - bq - c - dq$$

$$I\frac{dq}{dt} + (b + d)\,q = a - c$$

The solution of this differential equation is:

$$q(t) = \frac{a - c}{b + d} + \left(q_0 - \frac{a - c}{b + d}\right) e^{-\frac{b+d}{I}t}$$

Stability is assured if $(b + d) \geq 0$, given that $I > 0$. I is a positive constant. In this condition, the system tends to an equilibrium point (static equilibrium), and the system is said to be stable. If $(b + d) < 0$, the system grows to infinity and the system is said to be unstable.

Let us now apply this theory to the cases approached in the problem.

We have: $a = 1.4, \quad b = 0.4$, for the demand function.

We have two supply functions:

$c_1 = 0.4,\ d_1 = 0.6,\ c_2 = 1.2,\ d_2 = -0.2.$

Both systems are stable, because:

$b + d_i > 0,\ i = 1,\ 2$

The variation of the quantity in time is:

$$q_1(t) = 1 - 0.5e^{-t}$$

$$q_2(t) = 1 - 0.5e^{-0.2t}$$

Figure 3.25 shows the plotting of the two functions, which is known as quantity adjustment.

The equilibrium quantity is 1 in both cases.

$$p_D(q) = p_{Si}(q),\ i = 1,\ 2$$

$$a + bq = c_i + d_i q$$

$$q^* = \frac{a - c_i}{b + d_i} = 1$$

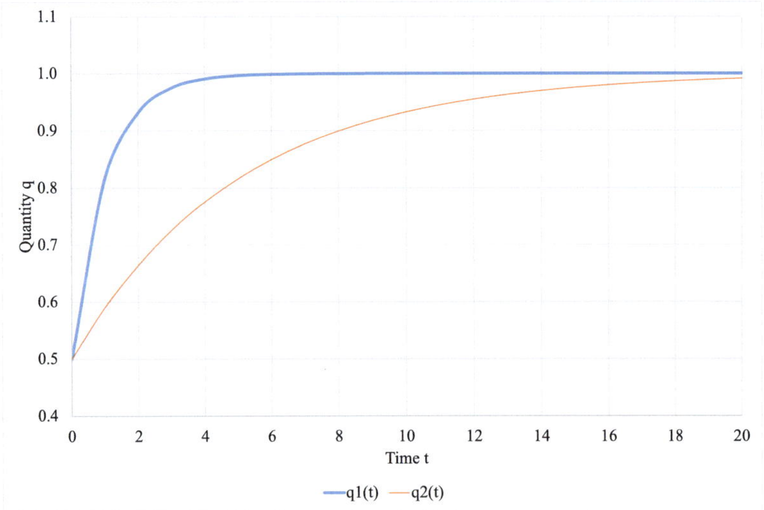

Fig. 3.25 Time evolution of quantity. Problem 3.2

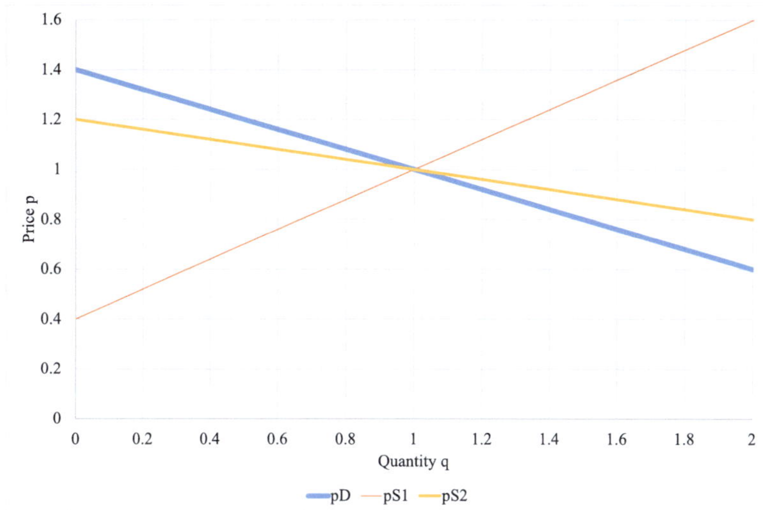

Fig. 3.26 Market equilibrium. Problem 3.2

Figure 3.26 shows the market equilibrium at $q^* = 1$ and $p^* = 1$.

We can also plot the adjustment of the supply and demand prices, for the two cases (see Figs. 3.27 and 3.28), respectively:

$$p_{Di}(t) = a - bq_i(t), \ i = 1, \ 2$$

$$p_{Si}(t) = c + dq_i(t), \ i = 1, \ 2$$

Problem 3.3

The interaction through the market of consumers and producers of a given product is ruled by:

Market demand: $p_D(q) = 1.4 - 0.4q$

Market supply: $p_S(q) = 1.4 - 0.44q$

1. With the price adjustment, establish and solve the equation governing the market's behaviour. Consider $I' = 1$ and $p_0 = 1.2$. Is the system stable or unstable? Justify.
2. What is the result obtained for the selling price of the product on the market? What is the quantity of the product? Comment.

Key Problem 3.3

1.

The equation that describes the dynamic equilibrium of a single market, using price adjustment, is:

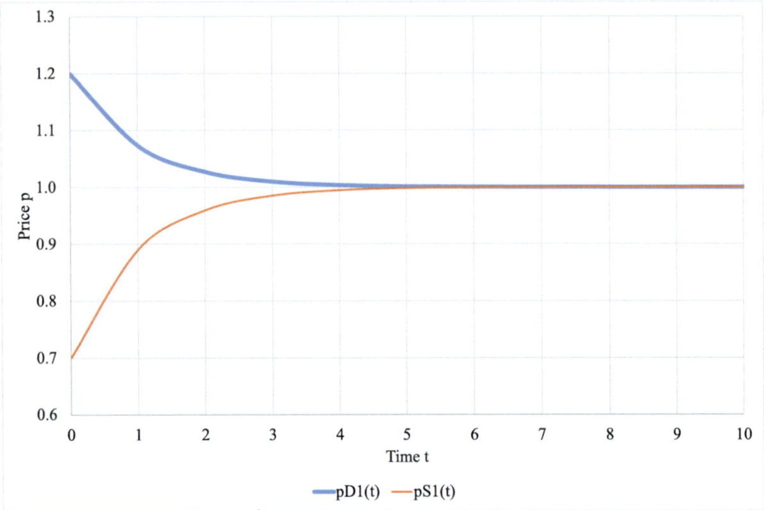

Fig. 3.27 Time evolution of price for $p_{S1}(q) = 0.4 + 0.6q$. Problem 3.2

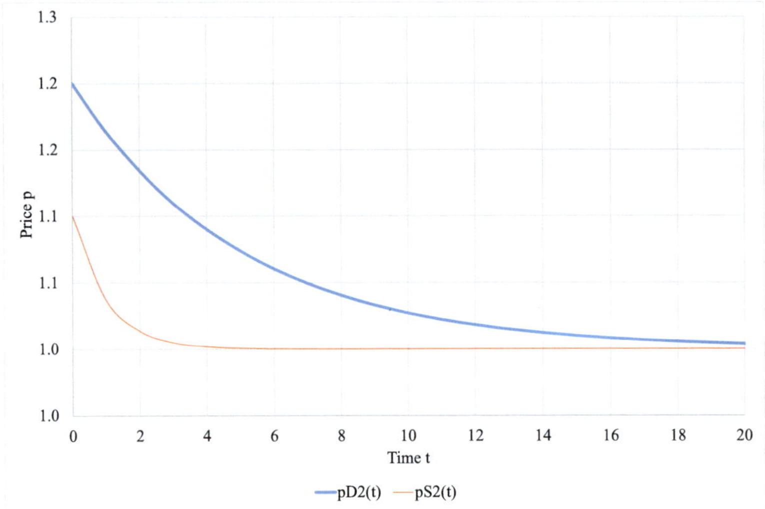

Fig. 3.28 Time evolution of price for $p_{S2}(q) = 1.2 - 0.2q$. Problem 3.2

$$I'\frac{dp}{dt} = q_D(p) - q_S(p)$$

We highlight that the product's price is the state variable of the system. Considering that the supply and demand curves may be linearised, we can write:

$$p_D(q) = a - bq \overset{\text{yields}}{\rightarrow} q_D(p) = \frac{a}{b} - \frac{1}{b}p$$

$$ps(q) = c + dq \overset{\text{yields}}{\rightarrow} qs(p) = -\frac{c}{d} + \frac{1}{d}p$$

and replacing the above, we obtain:

$$I'\frac{dp}{dt} + \left(\frac{b+d}{bd}\right)p = \left(\frac{ad+bc}{bd}\right)$$

At this point, we recall that the solution of the equation:

$$I\frac{dq}{dt} + (b+d)q = a - c$$

is:

$$q(t) = \frac{a-c}{b+d} + \left(q_0 - \frac{a-c}{b+d}\right)e^{-\frac{b+d}{I}t}$$

By similarity, we can write:

$$p(t) = \frac{ad+bc}{b+d} + \left(p_0 - \frac{ad+bc}{b+d}\right)e^{-\left(\frac{b+d}{bd}\frac{1}{I'}\right)t}$$

The system is stable, according to Walras, if:

$$\frac{b+d}{bd} \geq 0$$

and unstable, if:

$$\frac{b+d}{bd} < 0$$

Applying to the conditions of our problem, we have $b = 0.4$ and $d = -0.44$. Therefore:

$$\frac{b+d}{bd} = 0.2273 > 0$$

The system is stable according to Walras's methodology.

The time evolution of the adjustment price is (considering $p(0) = p_0 = 1.2$):

$$p(t) = 1.4 - 0.2e^{-0.2273t}$$

which is represented in Fig. 3.29.

This model is able to predict the equilibrium price at $p^* = 1.4$.

The time evolution of the demand and supply quantities is:

$$q_D(t) = \frac{a}{b} - \frac{1}{b}p(t)$$
$$q_S(t) = -\frac{c}{d} + \frac{1}{d}p(t)$$

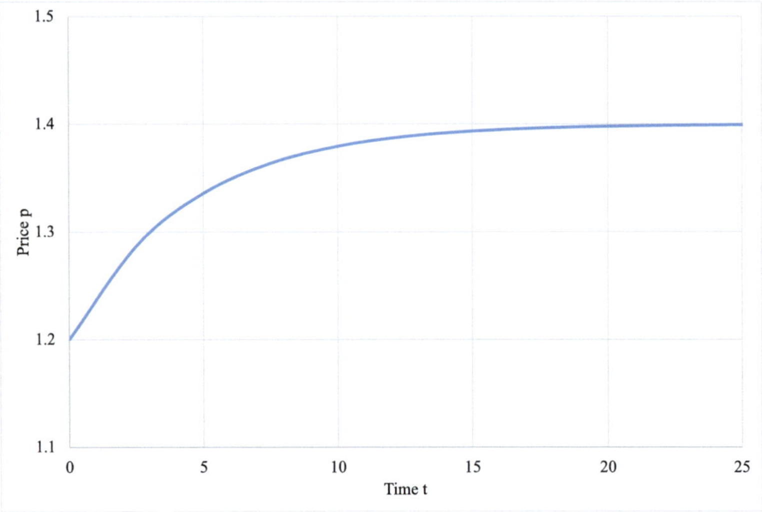

Fig. 3.29 Price adjustment. Problem 3.3

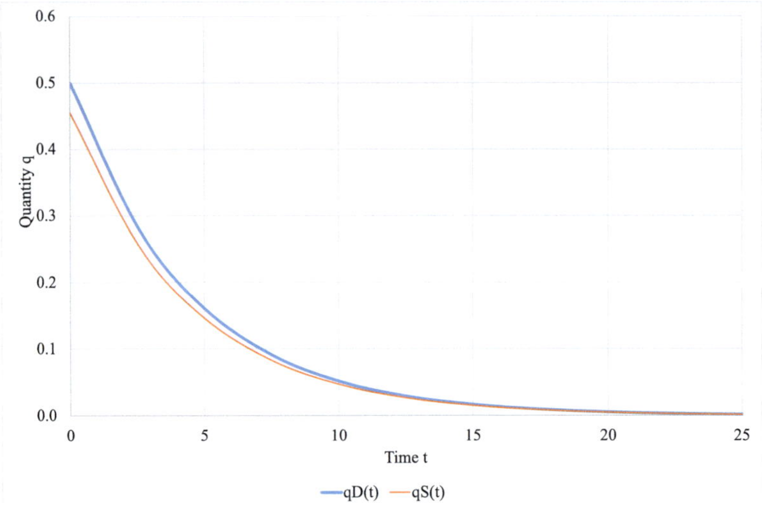

Fig. 3.30 Time evolution of the quantities. Problem 3.3

and is pictured in Fig. 3.30.

It can be observed that the quantity goes towards zero, which is the equilibrium quantity. Also, the equilibrium price is correctly predicted, as we have concluded

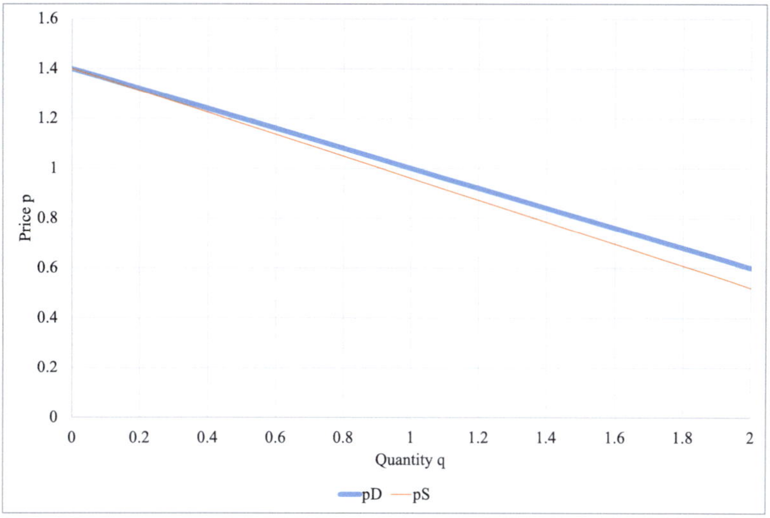

Fig. 3.31 Market equilibrium. Problem 3.3

before. As the equilibrium quantity is 0, in practical terms, does not allow for a market to be established, as we will discuss in the next answer.

2.

If we plot the supply and demand functions, we can see that the equilibrium quantity is zero and the equilibrium price is 1.4, as we show in Fig. 3.31.

The model we used (Walras methodology—price adjustment) can predict the equilibrium point. Nevertheless, as the equilibrium quantity is zero, there is no way to establish a market in such conditions.

4 Market Structures

Abstract

The chapter on market structures in the context of power systems provides a detailed examination of the diverse market forms that characterise the energy sector. Developed and modern power systems, essential for today's and future societies, exhibit a range of market structures, each with distinct characteristics and implications. This chapter delves into five primary market structures: perfect competition, monopoly, monopolistic competition, oligopoly, and monopsony. Perfect competition serves as an ideal benchmark, where numerous producers and consumers operate in a competitive environment, ideally promoting efficiency and low prices. In contrast, a monopoly represents a market dominated by a single supplier, which most probably leads to higher prices and reduced innovation. The monopolistic competition combines elements of both perfect competition and monopoly, with numerous producers offering slightly differentiated products. Oligopoly, somehow prevalent in the energy sector, involves a small number of powerful firms, resulting in complex dynamics, potential collusion, and strategic behaviour. Furthermore, the chapter explores the less-discussed concept of monopsony, where a single buyer holds significant influence over the market, impacting the pricing and availability of power. Understanding these market structures in the context of power systems is essential for policymakers, regulators, and industry stakeholders. It informs decisions regarding market design, competition, and regulation, ensuring that energy markets efficiently deliver reliable and affordable power to consumers while fostering innovation and sustainability.

4.1 Introduction

In the world of economics, market structures define the nature and characteristics of a market environment. These structures play a crucial role in determining the behaviour of firms and the overall outcomes of market interactions.

R. Castro, *Engineering of Power Systems Economics*,
https://doi.org/10.1007/978-3-031-55251-9_4

The market structure is the term usually used to define the type of market in which sellers supply goods or services and consumers buy these goods or services. The determinants of the market structure can be:

- Number of firms.
- Barriers to entry and exit.
- Perfect or imperfect information.
- Nature of the product.
- Price levels.
- Demand curve.

Depending on these determinants, different types of market structures are usually considered:

- Perfect competition—many sellers of a homogeneous (identical) product with perfect substitutes.
- Monopolistic competition—many sellers of a slightly differentiated product.
- Oligopoly—few sellers of a homogeneous or a differentiated product.
- Monopoly—a single seller of a homogeneous product for which there are no substitutes.

The perfect competition represents an idealised market structure, characterised by a large number of buyers and sellers who are price takers. In this scenario, no individual firm has the ability to influence the market price. The products sold in perfect competition are homogeneous, meaning they are identical across all sellers. Entry and exit barriers are low, facilitating easy market entry for new firms. Examples of perfect competition can be found in agricultural markets, such as the trading of staple crops.

Monopolistic competition shares similarities with perfect competition but differs in one significant aspect: product differentiation. In this market structure, numerous firms exist, each offering a slightly differentiated product. These variations can be based on branding, quality, design, or other factors that create a perceived difference among products. Consequently, firms possess a limited degree of market power, allowing them to have some influence over the prices of their products. Monopolistic competition can be observed in industries such as restaurants, clothing, and personal care products.

An oligopoly represents a market structure characterised by a small number of interdependent firms dominating the market. Due to the limited number of players, the actions of one firm can have significant effects on its competitors. Oligopolistic markets often display high barriers to entry, making it challenging for new firms to enter and compete. Strategic decision-making becomes crucial, as firms must consider the potential reactions of their rivals. Examples of oligopolies can be found in industries such as automobile manufacturing, telecommunications, and airline companies.

A monopoly refers to a market structure with a single dominant firm that has exclusive control over the supply of a particular product or service. As the sole provider, the monopolist possesses substantial market power, enabling it to set prices and output levels without significant competition. Monopolies may arise due to barriers to entry, including legal restrictions, patents, or economies of scale. Although monopolies can lead to efficiency in some cases, they are often regulated to prevent abuse of market power. Historically, examples of monopolies include national postal services or public utilities.

These four market structures represent a generic and abstract characterisation of a type of real market. These somewhat abstract models intend to determine some, but not all, details of a specific real-world market where buyers and sellers actually meet to trade goods and services.

Market structure is usually defined in a continuous line where at the starting end stays the perfect competition, and on the other end the monopoly, the monopolistic competition, and the oligopoly stay in between. Table 4.1 summarises the main characteristics of each market structure's determinants.

Recommended reading include https://www.studysmarter.us/explanations/microeconomics/imperfect-competition/market-structures/

A market structure similar to the monopoly but somehow reversed is the monopsony. A monopsony is a market structure in which a single buyer exists.

In this chapter, all these market structures, some of which may be found in the power system industry, are deepened and further developed. We begin by presenting the *perfect competition*, defining how the *optimal quantity and price* are obtained, and, furthermore, establishing the *operating conditions* for a perfectly competitive market to function. Then, we move to the other end of the competition line and go into the *monopoly*. Again, how to obtain the *optimal quantity and price* is explained, and the concept of *deadweight loss* is introduced as a measure of the market's inefficiency. This section ends with an approach to *regulated monopoly*, which is the modern form of monopoly, emphasising that the power system's natural monopolies often use this market structure. The next section is about the *monopolistic competition*, which is somehow a combination of the two previous market structures. Here, we deal with the *short run and long run optimisation* of this type of market. Next,

Table 4.1 General characteristics of market structures

	Perfect competition	Monopolistic competition	Oligopoly	Monopoly
No. of firms	Very large	Large	Few	One
Entry and exit	Free	Relatively easy	Some barriers	Restricted
Information	Perfect	Imperfect	Imperfect	Imperfect
Product	Homogeneous	Slightly different	Homogeneous	Unique
Price levels	Marginal cost (MC)	Higher than MC	Higher than MC	Higher than MC
Demand curve	Perfectly elastic	Elastic	Less elastic	Market

it is the *oligopoly* turn, where different models are introduced, namely, the *game theory*, *Cournot model* and *Bertrand model*. Moreover, the important concept of *Nash equilibrium* is dealt with. Finally, the *monopsony*, a single-buyer market, dual of the monopoly, is introduced to take a look at the consumers' side as well. The monopsonist *optimal quantity and price* is established, together with the *deadweight loss* and the new concept of *input elasticity of supply*. The chapter concludes with some exercises regarding the concepts just introduced.

4.2 Perfect Competition

A market operating under perfect competition is a theoretical market structure that is difficult to find in the real world. However, it is helpful to explain how demand and supply affect prices, revenues, and costs. Certain raw materials and agricultural goods, the stock exchange, are examples of markets with some contact points with perfect competition.

A perfectly competitive market is characterised by:

- Many independent agents—producers and consumers.
- Each agent is small relative to the whole market.
- The product is homogeneous (identical).
- The agents have complete or perfect information (in the past, present, and future) about the product being traded and, therefore, are able to make rational decisions.
- Easy entry and exit (no barriers to entry and exit).

The perfectly competitive firm is said to be a price-taker because it takes the market price as given and has no control over it. Why is this? The firm would lose all its business if it tried to charge a higher price because customers could go elsewhere to buy the same product for less. Moreover, since the firm is very small, it can sell as much as it wants at the market price. So, there is no reason to charge a lower price.

A perfectly competitive firm is free to sell any quantity of the product without changing the market price. Therefore, the challenge is not deciding what price to charge consumers, but rather what quantity to produce at the pre-set market price established by the interaction of demand and supply.

4.2.1 Optimal Quantity and Price

Profit is defined as the difference between total revenue and total cost. The total revenue is quantity-dependent and is given by:

$$R(q) = pq \tag{4.1}$$

where p is the (unitary) price and q is the product's quantity.

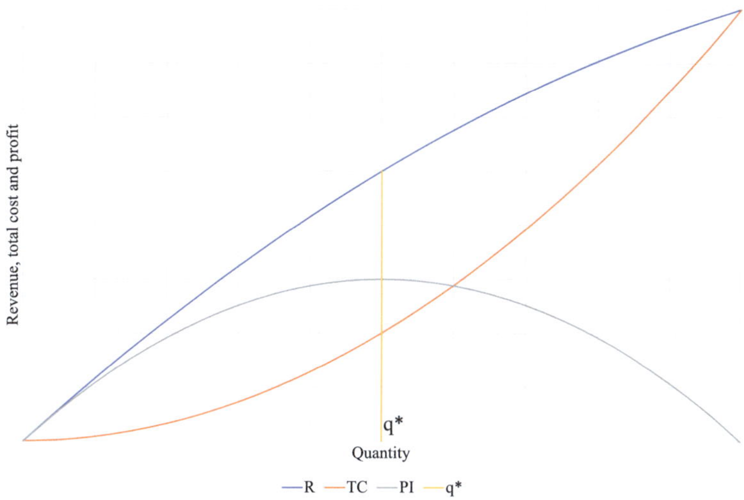

Fig. 4.1 Revenue, total cost and profit

The total cost, as we have already seen, depends on the output quantity. In general, the total cost has two parcels: fixed cost (FC) and variable cost (VC).

$$TC(q) = FC + VC(q) \tag{4.2}$$

So, the profit, Π, is:

$$\Pi(q) = R(q) - TC(q) \tag{4.3}$$

Figure 4.1 illustrates a typical situation of profit maximisation. The blue curve represents the revenue, the orange is the total cost and the grey is the profit. At quantity, q^*, profit is maximum, i.e., the difference between revenue and cost is maximum. Firms select the output (quantity) at the point where the profit is maximum.

It is interesting to note that profits are maximised at the point where the *Marginal Revenue*, i.e., the slope of the revenue curve, and the *Marginal Cost*, i.e., the slope of the total cost curve, are equal.

$$\max \pi \overset{\text{yields}}{\rightarrow} \frac{\partial \Pi}{\partial q} = \frac{\partial R}{\partial q} - \frac{\partial TC}{\partial q} = 0 \overset{\text{yields}}{\rightarrow} MR = MC \tag{4.4}$$

Marginal revenue is the change in revenue resulting from a one-unit increase in output (quantity). Mathematically, we can write (AR is the average revenue and MR is the marginal revenue):

$$MR = \frac{\partial R}{\partial q} = p = AR = \frac{R}{q} = p \overset{\text{yields}}{\rightarrow} MR = AR = p \tag{4.5}$$

We recall that the price is constant and imposed by the market, the firms have no control over it. Marginal Cost is, as already seen, the additional cost of producing an additional unit of output.

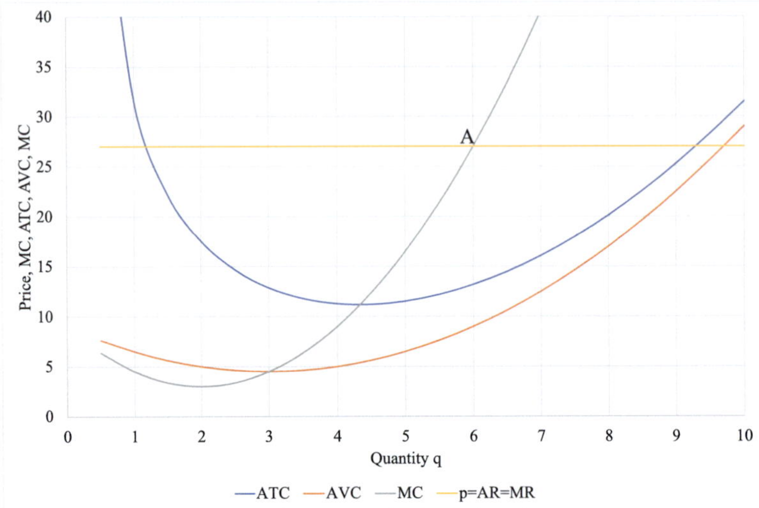

Fig. 4.2 Profit maximisation in a perfectly competitive market

Let us put some numbers on the equation to facilitate. Suppose a firm with a total cost given by: $TC = 0.5q^3 - 3q^2 + 9q + 25$. We know that:

- Average Total Cost: $ATC = TC/q = 0.5q^2 - 3q + 9 + 25/q$
- Average Variable Cost: $AVC = VC/q = 0.5q^2 - 3q + 9$
- Marginal Cost: $MC = \partial TC/\partial q = 1.5q^2 - 6q + 9$

Figure 4.2 shows the plotting of said three curves plus the straight line representing $p = AR = MR = 27$. At the optimal quantity $q^* = 6$ (point A), profit is maximum because $MC = MR$. This is the output firms choose to produce. At any other quantity, either it is with $MR > MC$ or $MR < MC$, profit is not maximum. At a quantity less than q^*, $MR > MC$, so, more profit can be gained by increasing output: this means that the additional revenue from selling one more is greater than the cost of making one more. At a quantity greater than q^*, $MR < MC$, too much output is being sold, and profit can increase by decreasing output: this means that it costs more to make one more than it will bring in when it is sold.

At the optimal quantity, the (maximum) profit is:

$$\Pi(q^*) = (p - ATC(q^*))q^* \tag{4.6}$$

This is the maximum area represented in Fig. 4.3 (red area–$(27 - 13.167) \times 6 = 83$). At any other quantity, the area is smaller (blue and green areas).

If the price is reduced to 5, keeping the same output would make the firm produce at a loss, given by the grey area represented in Fig. 4.4. This is because $p < ATC$.

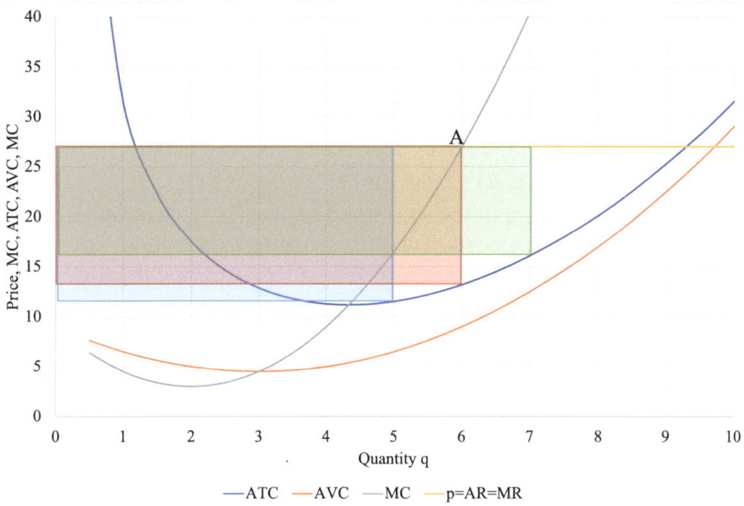

Fig. 4.3 Profit at the optimal quantity

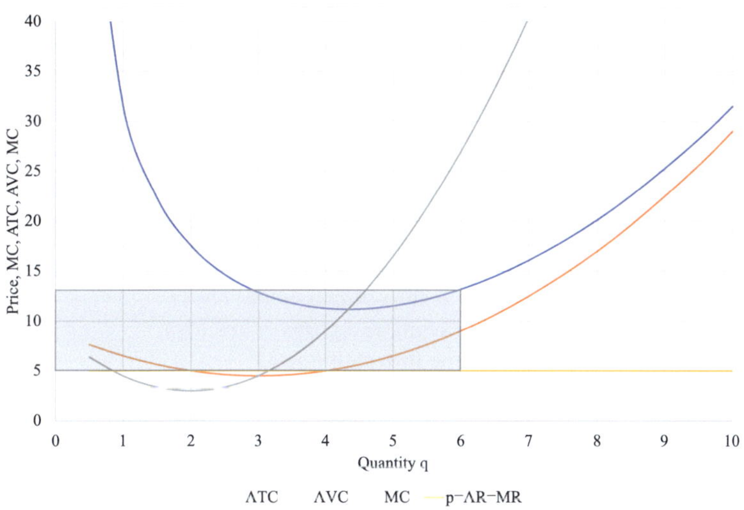

Fig. 4.4 Production at loss

4.2.2 Operating Conditions and Thresholds

At this point, it makes sense to address the question: "Why should a firm produce at a loss in the short run"? The firm might think the price will increase in the near future. Or it might happen that shutting down and starting up again could be costly. At the end of the day, the firm has two options: either continue producing in the short run or shut down temporarily. The firm has to compare which option is more advantageous.

To help the firms in the decision process, the following rules are established:

- If the price is higher than the ATC, $p > ATC$, the firm makes a profit.
- If the price is between the AVC and the ATC, i.e., $AVC < p < ATC$, the firm should continue producing in the short run because it can cover all of its variable costs and some of its fixed costs. The firm is operating at a loss.
- If the price is lower than the ATC, i.e., $p < AVC < ATC$, the firm should shut down as it cannot cover its variable costs or any of its fixed costs.

Therefore, we can summarise as follows:

- Operating conditions: $p = MC \wedge p \geq AVC$.
- Shut down threshold: min AVC.
- Profitability threshold: min ATC.

A firm should continue to produce as long as the price is above the AVC. When the price falls below AVC, then the firm should minimise its losses by shutting down. Why? If the price is below AVC, the firm cannot even recover the variable costs (cannot pay the workers, for instance), so the best choice is to shut down.

Example 4.1

Let us retake the firm with the total cost function given by: $TC = 0.5q^3 - 3q^2 + 9q + 25$.

1. For a price equal to 9, what is the optimal output level?
2. At this price, is the firm operating at a profit or at a loss?
3. At what price level should the firm shut down?
4. Find the minimum price that assures the firm's profitability.

◀

Solution

1.

For the given total cost, we have:

- Average Total Cost: $ATC = TC/q = 0.5q^2 - 3q + 9 + 25/q$
- Average Variable Cost: $AVC = VC/q = 0.5q^2 - 3q + 9$
- Marginal Cost: $MC = \partial TC/\partial q = 1.5q^2 - 6q + 9$

The optimal output is given by:

$$p = MC \overset{\text{yields}}{\rightarrow} 1.5q^2 - 6q = 0 \overset{\text{yields}}{\rightarrow} q^* = 4$$

as the other solution, $q = 0$, is unfeasible.

2.

$$ATC(4) = 11.25; \; AVC(4) = 5$$

Therefore, the firm is producing at a loss because $p < ATC$, but the firm should continue producing in the short run as $AVC < p < ATC$. The loss is:

$$L = (p - ATC(q^*))q^* = -9$$

We highlight that the rule $p = MC$ applies, even though the firm is producing at a loss. This guarantees minimum losses. At this price, the revenue is $R = p^*q^* = 36$ and the total cost is $TC = ATC(q^*)q^* = 45$.

3.

The firm should shut down, at the price corresponding to min AVC.

$$\min AVC \overset{\text{yields}}{\rightarrow} \frac{\partial AVC}{\partial q} = 0 \overset{\text{yields}}{\rightarrow} q - 3 = 0 \overset{\text{yields}}{\rightarrow} q = 3$$

For this quantity, the price is:

$$p = AVC(3) = MC(3) = 4.5$$

At this price level, the loss is the fixed cost. At any lower market price, the firm's loss would be greater than its fixed costs. In this case, the firm can avoid losses that are greater than its fixed cost by shutting down.

4.

The condition for minimum profitability is:

$$\min ATC \overset{\text{yields}}{\rightarrow} \frac{\partial ATC}{\partial q} = 0 \overset{\text{yields}}{\rightarrow} q^3 - 3q^2 - 25 = 0 \overset{\text{yields}}{\rightarrow} q = 4.3321$$

as the imaginary solutions are unfeasible. A cubic equations solver can be found in https://www.calculatorsoup.com/calculators/algebra/cubicequation.php.
The corresponding price is:

$$p = ATC(4.3321) = 11.1581$$

◄

4.2.3 Supply Function

An inverse supply curve shows the quantity supplied given the price, ceteris paribus. If the price in the market equals p_1 then the firm will produce q_1, where $p = MC$. If the price rises to p_2, then the firm produces q_2, where the new price equals the MC curve. Notice in both cases that the quantity is taken from the MC curve, which is equal to the price. Thus, the MC curve is the firm's inverse supply curve. This holds true until the price reaches the AVC. For prices lower than the minimum AVC, the firm shuts down in the short run and the quantity supplied is zero. Therefore, the inverse supply curve is:

$$q_S(p) = \begin{cases} MC(p), & \text{if } p \geq p_{sd} \\ 0, & \text{if } p < p_{sd} \end{cases} \tag{4.7}$$

where p_{sd} is the shut down price.

We recall that we know the function $MC(q)$, not the function $MC(p)$. To obtain $MC(p)$, we make $p = MC(q)$ and solve for p, thus obtaining $MC(p)$. To find the shutdown price, p_{sd}, we compute the quantity corresponding to min AVC, i.e., we make $\partial AVC/\partial q = 0$ and solve for q. Let us call this quantity q_{sd}. Then:

$$p_{sd} = AVC(q_{sd}) \tag{4.8}$$

Using the inverse supply function is the most intuitive way of showing that the MC curve plays a major role in the process. However, we are more acquainted with the supply curve, $p_S(q)$. Now, it should be straightforward to understand that:

$$p_S(q) = \begin{cases} MC(q), & \text{if } q \geq q_{sd} \\ 0, & \text{if } q < q_{sd} \end{cases} \tag{4.9}$$

In Fig. 4.5, we represent the supply function of the producer in Example 4.1.

We have already learned in the previous chapter how to obtain the aggregated supply curve from the individual supply curves, in the usual case where there are several producers in the market. The market supply curve is the horizontal sum of all the firm's individual supply curves. The simplest case is when all the firms show identical supply curves. The market supply curve is obtained by multiplying the number of firms by the individual supply curve. This is true for both the supply curve and the inverse supply curve.

Example 4.2

Let us retake again the firm with the total cost function given by: $TC = 0.5q^3 - 3q^2 + 9q + 25$. The marginal cost is $MC = \partial TC/\partial q = 1.5q^2 - 6q + 9$. Derive the inverse supply function of the firm and plot it. ◄

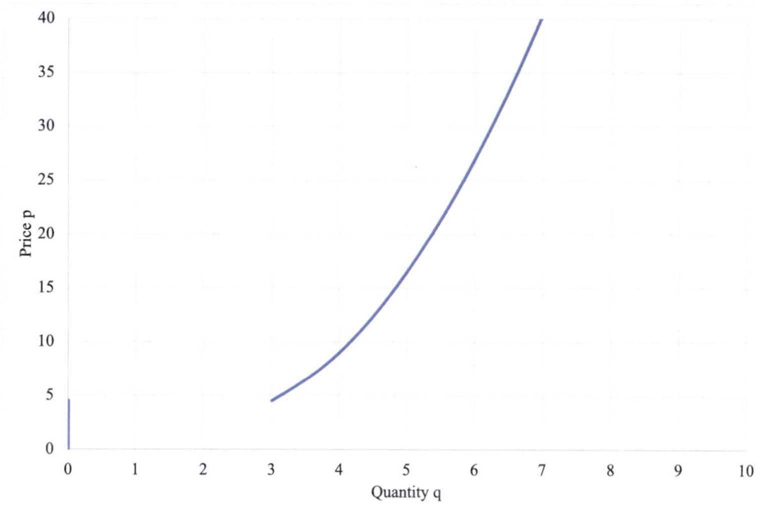

Fig. 4.5 Supply function. Example 4.1

Solution

Let us take the equation of the market in perfect competition:

$$p = MC = 1.5q^2 - 6q + 9$$

and solve it for p.
The resulting equation is:

$$1.5q^2 - 6q + (9 - p) = 0$$

Disregarding negative quantities, the solution is:

$$q_S(p) = \begin{cases} \frac{6+\sqrt{6p-18}}{3}, & \text{if } p \geq p_{sd} = 4.5 \\ 0, & \text{if } p < p_{sd} = 4.5 \end{cases}$$

We recall, from the previous example, that q_{sd} is determined from $\partial AVC/\partial q = 0$, and $p_{sd} = AVC(q_{sd})$.
The result is plotted in Fig. 4.6. ◀

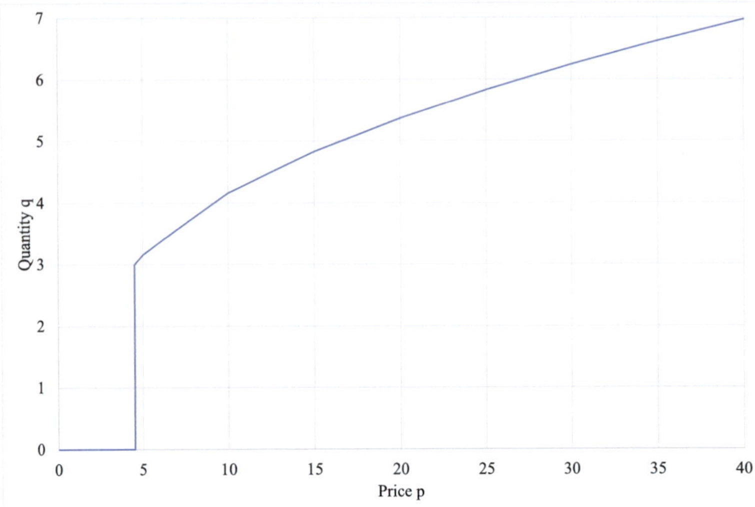

Fig. 4.6 Inverse supply function. Example 4.2

4.3 Monopoly

A *Monopoly* is a market type characterised by the existence of a single seller and many buyers. Entry to this market is restricted. The monopolist is not a price-taker, indeed it is a price-maker, as we will see further in this section. In this type of market, there is a lack of competition, and substitutes for the product are unavailable as there is only one producer. Examples of monopolies are markets with a product subject to a patent, e.g., a new drug, sole ownership of a resource, e.g., a toll bridge, and utility companies. Classical monopolies are undesirable market structures. When they exist, governments regulate their profits, preventing unfair profits at the expense of the customers.

Companies become monopolies by controlling the entire supply chain, from production to sales through vertical integration, or buying competing companies in the market through horizontal integration, becoming the sole producer.

One possible justification for monopolies is economies of scale. If there are significant economies of scale, a monopoly can benefit from lower average costs. This can lead to lower prices for consumers. If it does not, governments need to regulate monopolies to take advantage of economies of scale, without raising the prices.

Two types of monopolies can be devised:

- *Unregulated monopoly*, which is the traditional monopoly where the monopolist has no external controls over its activity.
- *Regulated monopoly*, which are the modern forms of monopolies where there is a regulation entity that regulates the activity of the monopolist.

4.3.1 Optimal Quantity and Price

While a competitive firm is a price-taker, as seen before, a monopoly firm is a price-maker. In a competitive market, the firm cannot directly influence price because if the price-taking competitive firm tries to raise the price, it will simply lose sales to other firms in the market. For the monopolist, there are no other firms to worry about. Therefore, the monopolist fixes the price that is more convenient to serve its profit maximisation purposes. The ability of the monopolist to raise prices above the competitive level is known as *market power*.

The objective of the monopolist is to maximise profit, i.e.:

$$\max\,(R(q) - TC(q)) \tag{4.10}$$

where $R(q)$ is the quantity-dependent revenue and $TC(q)$ is the quantity-dependent total cost.

Equation 4.10 leads to:

$$\frac{dR(q)}{dq} = \frac{dTC(q)}{dq} \tag{4.11}$$

i.e., marginal revenue (MR) equal to marginal cost (MC).

The revenue is given by:

$$R(q) = p_D(q)q \tag{4.12}$$

where $p_D(q)$ is the market demand curve.

The demand curve of the monopolist is the downward-sloping market demand curve because it is the only seller; this curve is used for the monopolist to set the price on it.

The marginal revenue is:

$$MR = \frac{dR(q)}{dq} = p_D(q) + q\frac{dp_D(q)}{dq} \tag{4.13}$$

This is in contrast with the perfectly competitive market where $p = MR$.

If we make the usual approximation of the linearity of the demand curve, $p_D(q) = a - bq$, and consider also linearity of the total cost function, $TC(q) = FC + cq$, the marginal revenue becomes:

$$MR = a - 2bq \tag{4.14}$$

and the marginal cost is:

$$MC = c \tag{4.15}$$

The maximum profit condition, $MR = MC$, leads to the optimal quantity (from the monopolist standpoint), q^*:

$$a - 2bq = c \overset{\text{yields}}{\rightarrow} q^* = \frac{a-c}{2b} \tag{4.16}$$

To establish the selling price, the monopolist resorts to the market demand curve (which is its own demand curve), where it sets the price at the highest price consumers are willing to pay for that quantity:

$$p^* = a - bq^* = a - b\left(\frac{a-c}{2b}\right) = \frac{a+c}{2} \tag{4.17}$$

A monopoly does not have a supply curve, because there is only one price and quantity at which it will maximise profits given its costs and market demand, as we have just seen.

Example 4.3

Consider a monopoly where the market demand curve is given by $p_D(q) = 800 - 2q$. The monopolist faces a total cost function given by $TC(q) = 10{,}000 + 100q$. Compute the quantity and price that maximise the monopolist's profit and the maximum profit achieved. ◀

Solution

$$R(q) = p_D(q)q = (800 - 2q)q \overset{\text{yields}}{\rightarrow} MR = 800 - 4q$$

$$MC = 100$$

$$MR = MC \overset{\text{yields}}{\rightarrow} q^* = 175$$

$$p^* = p_D(q^*) = 450$$

The profit is:

$$\Pi = p^*q^* - TC(q^*) = 51{,}250$$

Figure 4.7 shows the relevant lines. Point A is the monopolist's operating point, i.e., the point where its profits are maximum.
In Fig. 4.8, we display the revenue, $(R(q) = p_D(q)q)$, the total cost, $(TC(q) = FC + cq)$ and the profit $(\Pi(q) = R(q) - TC(q))$.
For the quantity, $q^* = 175$, the profit is maximum (51,250). ◀

4.3.2 Deadweight Loss

The monopoly causes a deadweight loss. By charging higher prices, the monopoly gets more surplus and consumers get less surplus. Monopoly also reduces the size of the economic surplus, by producing less than the socially efficient quantity and

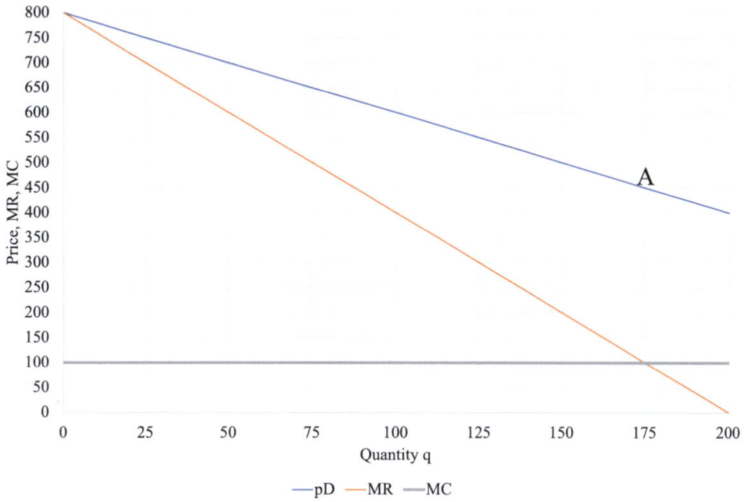

Fig. 4.7 Demand curve, marginal revenue, and marginal cost. Monopolist's operating point. Example 4.3

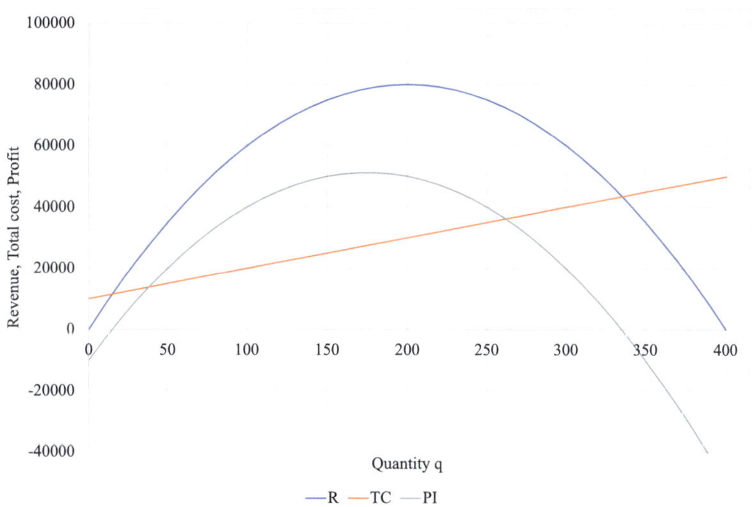

Fig. 4.8 Revenue, total cost, and profit. Example 4.3

causing a deadweight loss. Figure 4.9 depicts the situation, taking the data from the example we have been following.

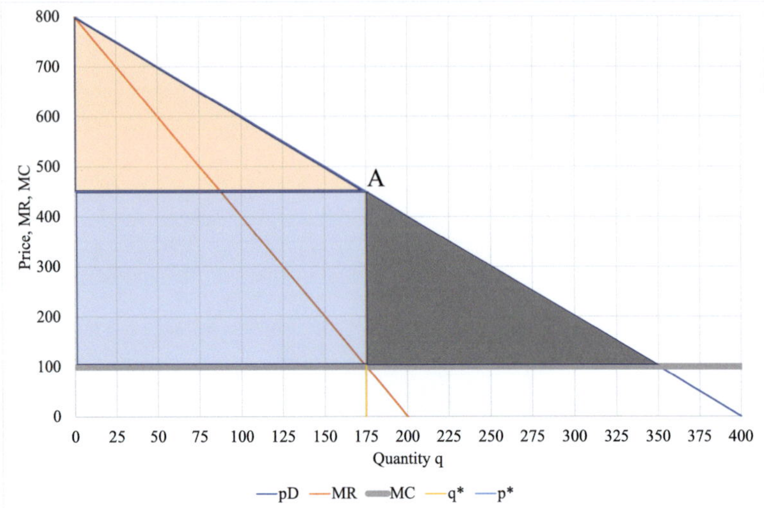

Fig. 4.9 Consumer surplus (orange), producer surplus (blue), and deadweight loss (grey)

4.3.3 Price Markup

At this point, we would like to introduce a useful relationship between the monopolist's marginal cost (MC) and the selling price at which it effectively sells the product, p^*.

We have seen that Eq. 4.13:

$$MR = p_D(q) + q\frac{dp_D(q)}{dq} \tag{4.18}$$

On the other side, we have seen in Chap. 2 that the price elasticity of demand (ratio of relative changes in quantity and price) is:

$$\varepsilon_p(q) = \frac{dq}{dp_D(q)}\frac{p_D(q)}{q} \tag{4.19}$$

Therefore, Eq. 4.18 can be written as:

$$MR = p_D(q)\left(1 + \frac{1}{\varepsilon_p(q)}\right) \tag{4.20}$$

In the monopolist's optimal point, we have:

$$MR = MC \overset{\text{yields}}{\rightarrow} p_D(q)\left(1 + \frac{1}{\varepsilon_p(q)}\right) = MC \tag{4.21}$$

Hence, at the optimal point, we can write:

$$p^* = p_D(q^*) = \frac{1}{1 + \frac{1}{\varepsilon_p(q)}} MC \tag{4.22}$$

The factor:

$$f = \frac{1}{1 + \frac{1}{\varepsilon_p(q)}} \tag{4.23}$$

measures how much the monopolist's selling price is higher than its marginal cost, at the maximum profit point. This is often known as *price markup*.

Example 4.4

Retake the data given in the previous example. Using the price elasticity of demand concept, show that the monopolist's selling price is $450/100 = 4.5$ times higher than its marginal cost. ◀

Solution

The price elasticity of demand at the monopolist's optimal point is:

$$\varepsilon_p(q^*) = \frac{dq}{dp_D(q)} \frac{p^*}{q^*} = -\frac{1}{2}\frac{450}{175} = -1.2857$$

The price elasticity of demand is -1.2857 meaning that when the price of the product increases by 1%, the consumers are prepared to decrease the consumption level by 1.2857%.
The factor is:

$$f = \frac{1}{1 + \frac{1}{\varepsilon_p(q)}} = 4.5$$

◀

4.3.4 Price Discrimination

If the right conditions are met, the monopolist may implement a *price discrimination* scheme. By price discrimination, we mean the practice of charging different prices for the same product. Economic theory distinguishes three types of price discrimination.

First-degree price discrimination involves charging each client the maximum price he/she is willing to pay for a certain quantity of the good or service (reflected by the demand curve). By implementing this type of price discrimination, the monopolist captures all the surplus at the expense of the consumer surplus which becomes zero. Doctors and lawyers may practice this type of price discrimination.

Second-degree price discrimination occurs when the monopolist knows there are several groups of consumers with different willingness to pay and charges each group based on the quantity consumed. An example is when airlines lower the prices for frequent travellers or bulk purchase discounts. In this type of price discrimination, the high-demand group still retains some consumer surplus.

Third-degree price discrimination happens when the monopolist identifies groups of consumers with some identifiable characteristic such as age, time of purchase, or location, and charges different prices based on their price elasticity of demand. The more inelastic the demand, the higher the price. Examples may be found in airlines that discriminate the price based on the time of purchase, movie theatres that discriminate based on age, and gas stations that charge a higher price on the highways.

Another form of price discrimination is the *two-part tariff*. Two-part tariffs relate to the imposition of an upfront fixed fee, which must be paid to consume any amount of good, and then a variable fee based on usage. Examples are golf clubs or health clubs that charge a membership fee plus a per-use fee. Normally the per-use fee is the marginal cost and the membership fee is set with the objective of capturing all the consumer surplus.

Detailed information about price discrimination in monopolies is available at https://nmiller.web.illinois.edu/documents/notes/notes9.pdf and https://sites.bu.edu/manove-ec101/files/2017/11/VarianHalPriceDiscrimination1989.pdf

4.3.5 Regulated Monopoly

So far, we have presented the theory of the monopoly as seen by traditional Economics, i.e., an unregulated monopoly. Modernly, most of the existing monopolies are regulated by an independent energy regulatory agency. Examples of regulated monopolies are the electrical transmission and distribution grids which are called *natural monopolies*. By natural monopolies, we mean assets that historically were developed as single infrastructures without competition, due to the extremely high investments required to set up these grids. This makes the fixed cost to be very high, and the marginal cost very low, therefore, implying decreasing average cost curves (increasing returns to scale). This means a single firm can produce the entire market quantity at a lower cost than several firms.

From the highest market efficiency standpoint, the optimal quantity to sell is where the demand function meets marginal cost. However, for natural monopolies, marginal cost is below average cost and the monopolist will not be able to cover its costs. In order for the monopolist to cover its production costs, it must be allowed to charge a higher price than the marginal cost. However, as it increases price above marginal cost, the quantity drops below the efficient quantity, and there is a corresponding deadweight loss. The objective of regulation is to allow the monopolist to cover its costs and, at the same time, keep prices (and deadweight loss) low. The price that does this is called the *Ramsey price*.

As we will see later in this book, the regulator aims at protecting the consumers, who eventually pay the bill, by limiting the earnings of the monopolist to a reasonable level. Most of the time, a reasonable profit level is to impose the selling price equal to the average cost. We highlight that the average cost has an implicit capital rate of return on its calculation, as we will see later in this book. The earnings of the monopolist are measured by this capital rate of return.

The condition applying to a regulated monopoly is to set the selling price at the point where the market demand curve meets the average cost curve, i.e.:

$$ATC(q) = p_D(q) \tag{4.24}$$

Considering a linear total cost curve, $TC(q) = FC + cq$, and a linear downward-sloping market demand function, $p_D(q) = a - bq$, Eq. 4.24 becomes:

$$\frac{FC}{q'} + c = a - bq' \overset{\text{yields}}{\rightarrow} bq'^2 + (c-a)q' + FC = 0 \tag{4.25}$$

There are two solutions to this quadratic equation (q' is the solution, we do not call it q^* because it is not optimal from the monopolist point-of-view). One out of two is a quantity close to zero and should be disregarded, the other solution is the one to be retained. More details on this on the following example.

Example 4.5

Retake the data given in the example we are following: $p_D(q) = 800 - 2q$; $TC(q) = 10{,}000 + 100q$. Compute the quantity and price imposed by the regulator to the regulated monopoly and the profit achieved. Compare this with the unregulated monopoly. ◀

Solution

The equation that rules the regulated monopoly is:

$$\frac{FC}{q'} + c = a - bq' \overset{\text{yields}}{\rightarrow} 2q'^2 - 700q' + 10{,}000 = 0$$

The solution of the quadratic equation is:

$$q'_1 = 335.0781 \vee q'_2 = 14.9219$$

The solution to be retained is q'_1 (corresponding to the B point in Fig. 4.10). q'_2 is the solution close to the zero quantity, which evidently is not the solution we are looking for.

The corresponding price is $p' = p_D(q') = 129.8438$.

Figure 4.10 shows a graphical representation of this example.

Comparing Figs. 4.7 and 4.10, the regulator is imposing the regulated monopolist to sell a higher quantity (335.0781 versus 175) at a lower unitary price (129.8438 versus 450) as compared to the unregulated monopolist.

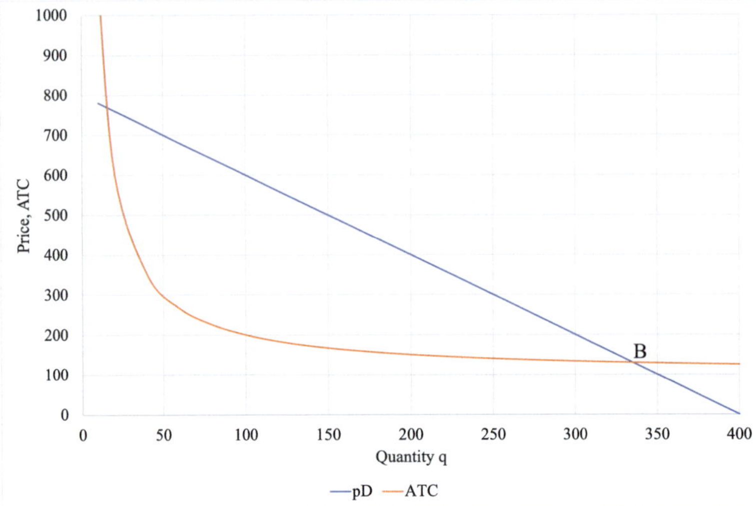

Fig. 4.10 Demand curve and average total cost. Regulated monopolist's operating point. Example 4.5

The profit of the regulated monopoly is obviously zero:

$$\Pi = p'q' - ATC(q')q' = 0$$

However, as previously mentioned, we note that the average cost of the monopolist already has a capital rate of return implicitly included. So, the profit is zero beyond the capital rate of return fixed by the regulator. ◄

4.4 Monopolistic Competition

The theory of *monopolistic competition* was first introduced in 1933 by Edward Chamberlin of Harvard University and Joan Robinson of Cambridge University.

Monopolistic competition is a market structure involving several firms competing against each other (but that do not take into account rivals' reactions), a product that is somehow similar but can be differentiated by the consumer, and free entry and exit in the market. A differentiated product is a product that consumers perceive as distinctive in some way. On the basis of this differentiation, firms spend large amounts of money on marketing and advertisement of their product to attract more and more customers.

Examples of industries that could be somehow regarded as monopolistic competition are restaurants, hotels, and clothing stores.

4.4.1 Demand Curve

The demand curve faced by a firm in a perfectly competitive market is flat (perfectly elastic) because each firm can sell any quantity it wants at a price it cannot control. The demand curve faced by a monopolist is the downward-sloping market demand curve because there are no substitutes, the product is unique. The demand curve faced by a competitive monopolist is also downward sloping but is flatter than the one faced by the monopolist to reflect product differentiation. If the competitive monopolist raises the price, the consumers may buy a similar (not equal) product from another competing firm. This means the demand curve of a competitive monopolist is not the market demand curve but is a demand curve based on its own product differentiation and on the number of competitors.

4.4.2 Short Run and Long Run Optimal Quantities and Prices

In the short run, the competitive monopolist will behave like a monopolist. It will determine the quantity at the point where its marginal revenue equals the marginal cost and read the price at the demand curve. This will make the competitive monopolist make the maximum profit, called *super normal profit*.

As firms are making super-normal profits, other firms are tempted to enter the market. Remember there are no barriers to entry. The entry of more firms into the market will make the demand faced by the competitive monopolist decrease (shift left) for a given price. The marginal revenue curve will also shift left and the interception point will define the optimal quantity at a lower value. In the long run, the price will be fixed at the tangency point between the demand curve and the average total cost of the firm. The profit is then zero (but remember there is an implicit rate of return in the average total cost). In this case, the firm is making what is called *normal profits*. Even though the firm is earning normal profits, there is still a deadweight loss because the price is above the marginal cost.

To determine the long-run optimal quantity, one needs to solve the following system of equations.

$$\begin{cases} p_D(q) = ATC(q) \\ MR(q) = MC(q) \end{cases} \tag{4.26}$$

The solution of this system of equations is an optimal quantity below the quantity that corresponds to the minimum ATC and a price above MC. The competitive monopoly is inefficient because neither productive nor allocative efficiency is achieved. ATC is not at its minimum in the long run (productive inefficiency) and the price is greater than marginal cost (allocative efficiency).

Information about monopolistic competition can be found at https://pressbooks.bccampus.ca/uvicecon103/chapter/8-3-monopolistic-competition/

Example 4.6

Let us retake the data from the example of the monopoly. The demand function faced by the monopolist competitor is $p_D(q) = 800 - 2q$ and the cost function is $TC(q) = 10{,}000 + 100q$. Find the optimal quantity and price in the short run and in the long run. Assume that in the long run, the demand function shifts left keeping the same slope. Plot illustrative graphics. ◀

Solution

In the short run, the monopolist competitor acts like a monopolist and fixes the quantity at the point where the marginal revenue is equal to the marginal cost. The price is read from the demand curve.

$$MR(q) = MC(q) \overset{\text{yields}}{\rightarrow} 800 - 4q = 100 \overset{\text{yields}}{\rightarrow} q_{SR}^* = 175$$

$$p_{SR}^* = p_D(q_{SR}^*) = 800 - 2q_{SR}^* = 450$$

Attracted by these super-normal profits, more firms enter the market because there are no barriers to entry. The consequence is a decline in demand as seen by our monopolist competitor. The long-run demand faced by the monopolist competitor is a shift left of the short-run demand keeping the same slope (we assume this). The long-run demand curve is:

$$p_D^{LR}(q) = a' - 2q$$

The long-run optimal quantity is found where:

$$p_D^{LR}(q) = ATC(q) \overset{\text{yields}}{\rightarrow} a' - 2q = \frac{10{,}000}{q} + 100$$

The other equation we need is:

$$MR_{LR}(q) = MC(q) \overset{\text{yields}}{\rightarrow} a' - 4q = 100$$

Solving for q and a' results in:

$$2q^2 + (100 - a)q + 10{,}000 = 0 \overset{\text{yields}}{\rightarrow} 2q^2 = 10{,}000$$

$$q_{LR}^* = 70.7107 \wedge a' = 382.8427$$

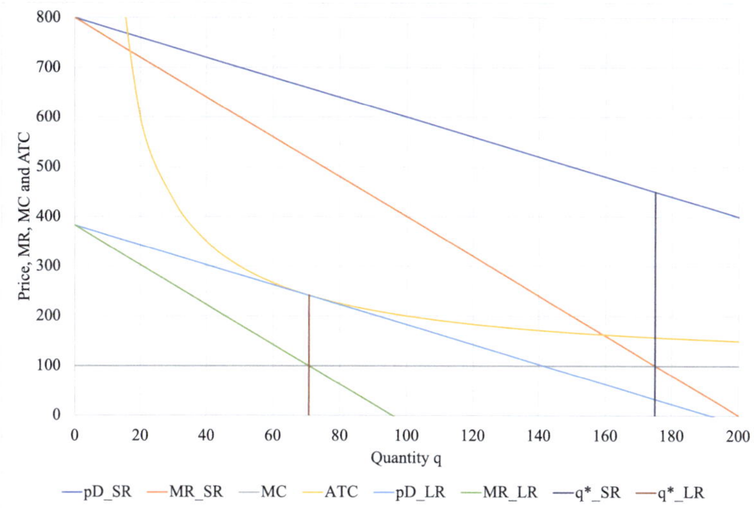

Fig. 4.11 Short-run and long-run operating points of the competitive monopolist. Example 4.6

The price is taken from the long-run demand curve:

$$p_{LR}^{*} = p_{D}^{LR}(q_{LR}^{*}) = 382.8427 - 2q_{LR}^{*} = 241.4213$$

Figure 4.11 depicts both the short-run and long-run operating points of the competitive monopolist. ◀

4.5 Oligopoly

An oligopoly is a market type close to a monopoly but, instead of a single seller, there are a small number of sellers that offer identical or similar products. In a monopoly, there are no competitors, so the monopolist is free to set quantity and price that best fits its objective of maximum profit. In an oligopoly, the situation is completely different because the quantity and price set by one competitor depend on the quantity and price set by the remaining. Each firm must respond to the other's choices, but the others are also responding to the first one choice. In oligopolies, there is always a tension between cooperation and self-interest. This requires strategic thinking.

Operating systems for smartphones and computers provide good examples of potential oligopolies. Other examples are telecommunications operators or aircraft manufacturers.

4.5.1 Principles of Game Theory

There is no single theory of oligopolistic behaviour. The techniques of *game theory* are usually used to solve for the equilibrium of an oligopoly market. Decisions in game theory are based on what the other players in the market are expected to do. All games involve three features: rules, strategies, and payoffs.

A classical example of game theory is the *prisoner's dilemma*. Two prisoners, A and B, are in separate rooms and each is offered the following options:

- If they both confessed, they would both get 5 years in prison.
- If one confessed, the confessor would go free while the other one would get 10 years.
- If neither confessed, then they would each get 2 years.

The intuition of the game is that if the two prisoners "collude" and jointly decide to not confess, they will both go free. But this is not an option because they cannot communicate. So, let us walk through all possible outcomes, given what the other prisoner does.

The options for prisoner A are:

- If B confess, the best option is A to confess too, as a 5 years sentence is better than 10.
- If B does not confess, again the best option for prisoner A is to confess because he/she will go free.

So, the best strategy for prisoner A is to confess, no matter what B does. An exact same reasoning can be made in what concerns prisoner B. The best strategy is the confession of both. We say that the *dominant strategy* of A and B is to confess. A dominant strategy is a strategy that results in the highest payoff to a player regardless of the opponent's action. It is noted that it is possible that a player does not have a dominant strategy because his/her decisions would depend on what the other players do.

4.5.2 Nash Equilibrium

Another important concept in game theory is "Nash Equilibrium", named after the mathematician John Nash, pictured in the movie "A Beautiful Mind". Nash Equilibrium is defined as an outcome where neither firm has the incentive to change its behaviour. In the prisoner's dilemma above, the Nash Equilibrium is, for both prisoners, to confess. If one prisoner confesses and the other does not, this prisoner would serve a 10 years sentence instead of 5. The conclusion is neither prisoner is likely to deviate from the Nash Equilibrium because any change will result in a worse outcome. Again, it is possible to have multiple Nash Equilibrium.

Oligopolies can lead to cartels. A *Cartel* is a group of firms that collude on price and output decisions in an effort to earn monopoly profits. In fact, a cartel acts as a monopoly. Cartels are outlawed in most countries but they are difficult to police.

More information about oligopolies can be found, for instance, in https://pages.charlotte.edu/wp-content/uploads/sites/846/2014/12/bootcampch13out.pdf

Game theory is the modern way of dealing with oligopolies. There are, however, classical models that describe the behaviour of oligopolies. Two classical models are *Cournot model* and *Bertrand model*.

4.5.3 Cournot Model

The Cournot model was introduced by the French economist and mathematician Antoine Augustin Cournot in 1838. In Cournot's model, firms chose output quantity for strategic acting. We will introduce this model by considering a very simple case, in which there are two firms, 1 and 2, (a duopoly) that produce the same product and have the same cost functions, with a constant marginal cost, $MC = c$, and no fixed costs. To further simplify, the demand function the firms face is linear, $p_D = a - bq$, where $q = q_1 + q_2$ is the total quantity produced.

The profit of the company 1 is given by:

$$\Pi_1 = (p_D - c)q_1 = [a - b(q_1 + q_2) - c]\, q_1 \tag{4.27}$$

and, similarly, the profit of the company 2 is:

$$\Pi_2 = (p_D - c)q_2 = [a - b(q_1 + q_2) - c]\, q_2 \tag{4.28}$$

We note that the profit of one company depends on its own output (quantity produced) and on the output of the other.

The objective of each firm is to maximise profit. Therefore:

$$\max \Pi_1 \overset{\text{yields}}{\rightarrow} \frac{\partial \Pi_1}{\partial q} = 0 \overset{\text{yields}}{\rightarrow} a - 2bq_1 - bq_2 - c = 0 \tag{4.29}$$

So, the optimal output of company 1, given the output of company 2, is:

$$q_1^* = \frac{a - c}{2b} - \frac{q_2}{2} \tag{4.30}$$

It is interesting to observe that the maximum profit quantity in a monopoly with a linear demand curve and constant marginal cost is $q = (a - c)/2b$. Here, we have an additional term, $-q_2/2$.

Following the same reasoning, the optimal output of company 2, given the output of company 1, is:

$$q_2^* = \frac{a-c}{2b} - \frac{q_1}{2} \tag{4.31}$$

These are the firms' best response functions, i.e., the profit-maximising output level of one, given the output choice of the other.

To find the Nash equilibrium, we plot the best response functions in the plane (q_1, q_2). This graphical representation is called *reaction functions*. The two reaction functions are straight lines, with equations:

$$\begin{aligned} q_1 &= \frac{a-c}{2b} - \frac{q_2}{2} \overset{\text{yields}}{\rightarrow} q_2 = \frac{a-c}{b} - 2q_1 \\ q_2 &= \frac{a-c}{2b} - \frac{q_1}{2} \end{aligned} \tag{4.32}$$

These are two downward-sloping straight lines. The x-intercept of the first line (firm 1) is $(a-c)/2b$ and the x-intercept of the second (firm 2) is $(a-c)/b$.

As usual, the Nash equilibrium i.e., the quantities both firms are satisfied and with no incentives to change, is the interception of both reaction lines. The equation that needs to be solved is:

$$\frac{a-c}{b} - 2q_1 = \frac{a-c}{2b} - \frac{q_1}{2} \tag{4.33}$$

whose solution is:

$$\begin{aligned} q_1^{eq} &= \frac{a-c}{3b} \\ q_2^{eq} &= q_2(q_1^{eq}) = \frac{a-c}{3b} \end{aligned} \tag{4.34}$$

For an oligopoly with n firms, the optimal quantity each firm should supply is:

$$q_i^{eq} = \frac{a-c}{(n+1)b} \tag{4.35}$$

The total optimal quantity supplied in the Cournot duopoly is:

$$q^* = q_1^{eq} + q_2^{eq} = \frac{2}{3}\frac{a-c}{b} \tag{4.36}$$

It is timely to recall that in perfect competition:

$$q^* = \frac{a-c}{b} \tag{4.37}$$

and in monopoly:

$$q^* = \frac{1}{2}\frac{a-c}{b} \tag{4.38}$$

Cournot optimal quantity is between perfect competition and monopoly.

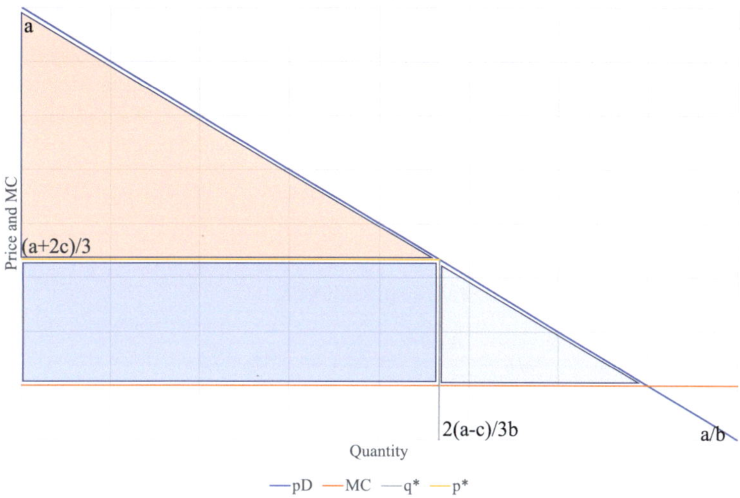

Fig. 4.12 Cournot duopoly optimal quantity and price. Consumer surplus (orange), producer surplus (blue), and deadweight loss (grey)

The price is taken from the demand curve at the optimal quantity.

$$p^* = p_D(q^*) = \frac{a+2c}{3} \tag{4.39}$$

The profit of each firm is:

$$\Pi_1 = \Pi_2 = (p^* - c)q_1^{eq} = \frac{(a-c)^2}{9b} \tag{4.40}$$

Figure 4.12 depicts the Cournot duopoly.
The consumer surplus is:

$$CS = 0.5\left(a - \frac{a+2c}{3}\right)\frac{2(a-c)}{3b} = \frac{2(a-c)^2}{9b} \tag{4.41}$$

The producer surplus is:

$$PS = \left(\frac{a+2c}{3} - c\right)\frac{2(a-c)}{3b} = \frac{2(a-c)^2}{9b} \tag{4.42}$$

The deadweight loss is:

$$DWL = \frac{(a-c)^2}{18b} \tag{4.43}$$

We note that there is a deadweight loss, indicating the oligopoly is not an efficient market.

Example 4.7

Let us retake the data from the example of the monopoly. Find the Nash equilibrium of an oligopoly with two firms, A and B, facing a demand function given by $p_D(q) = 800 - 2q$. The cost function of each firm is $TC(q) = 100q$, in which we have neglected the fixed cost. Use the Cournot model. Plot illustrative graphics. ◀

Solution

The reaction functions of the two companies are:

$$q_B = \frac{a-c}{b} - 2q_A = 350 - 2q_A$$

$$q_B = \frac{a-c}{2b} - \frac{q_A}{2} = 175 - 0.5q_A$$

The Nash equilibrium is:

$$q_A^{eq} = q_B^{eq} = \frac{a-c}{3b} = 116.(6)$$

Details are in Fig. 4.13.

Figure 4.14 allows us to compare the different market types: perfect competition, $q = 350$ and $p = MC = 100$, monopoly, $q = 175$ and $p = 450$, and oligopoly, $q = 233.(3)$ and $p = 333.(3)$. The oligopoly figures are marked in the figure. ◀

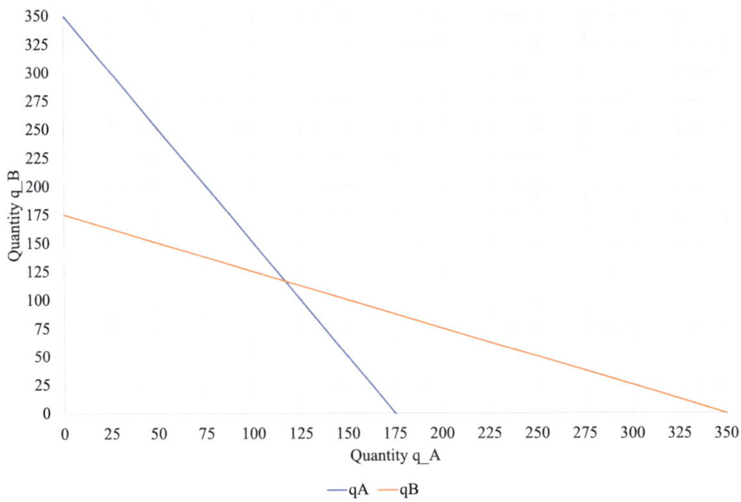

Fig. 4.13 Nash equilibrium using Cournot model. Example 4.7

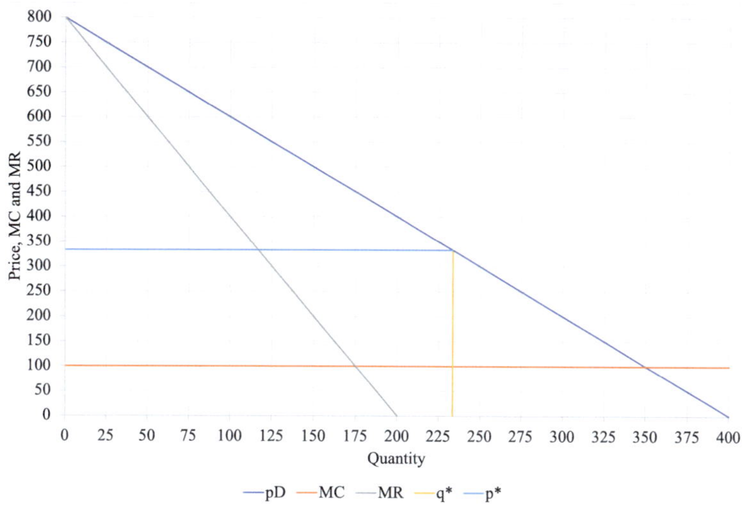

Fig. 4.14 Comparison of perfect competition, monopoly, and oligopoly. Example 4.7

4.5.4 Bertrand Model

In the Cournot model, firms compete on quantity, the strategic decision is the quantity each firm produces. But what if firms compete on price? The answer to this is *Bertrand model*, named after the French mathematician Joseph Bertrand, in 1883.

Let us keep the same assumptions that hold for the Cournot model: two firms, 1 and 2, sell the same product, face a linear inverse demand ($q = a' - b'p$), and have constant marginal cost (c) with no fixed cost. Consumers will choose to buy from the firm offering the lowest price. Under these conditions, the inverse demand function of each company is:

$$q_1 = \begin{cases} a' - b'p_1 & \text{If } p_1 < p_2 \\ \frac{1}{2}(a' - b'p_1) & \text{If } p_1 = p_2 \\ 0 & \text{If } p_1 > p_2 \end{cases} \tag{4.44}$$

$$q_2 = \begin{cases} a' - b'p_2 & \text{If } p_2 < p_1 \\ \frac{1}{2}(a' - b'p_2) & \text{If } p_1 = p_2 \\ 0 & \text{If } p_2 > p_1 \end{cases} \tag{4.45}$$

This means that a firm will get all of the demand if its price is lower than the competitor's and zero if it is higher. If the price is equal, the firms will share the market in equal proportion.

The question now is what is the best response of one firm, given the strategic decision of the other? Let us go through all the options firm 1, for instance, has. Suppose, firm 2 sets the price at $p_2 > c$.

- Firm 1 could respond with $p_1 > p_2$ and they will lose all clients.
- Firm 1 sets the price at $p_1 = p_2$. They will get half of the market and make a profit.
- Firm 1 decides to charge $p_1 = p_2 - \varepsilon$, with ε as small as needed, and get all of the demand. This is the most profitable option.

Now, let us suppose firm 2 sets the price at $p_2 = c$.

- Firm 1 charges at $p_1 = p_2 - \varepsilon$. This is not clever because firm 1 would get all the demand but will make a loss, as it is selling below marginal cost.
- Firm 1 responds with $p_1 > p_2$ and they will get zero clients.
- Firm 1 sets the price at $p_1 = p_2 = c$. They will get half of the demand at a break-even price.

The best response for firm 2 is exactly the same as for firm 1. By setting the price above the marginal cost, one firm is always willing to undercut the other's price, so there is no equilibrium. At $p_1 = p_2 = c$ the firms are giving their best response to each other simultaneously. So the Nash equilibrium is $p_1 = p_2 = c$.

It is surprising that the outcome of Bertrand's model for a duopoly is a perfectly competitive market. This result is so strange and inconsistent with reality, that it is known as *Bertrand paradox*.

The Bertrand paradox can be overcome if some conditions that we have assumed are relaxed. If we assume product differentiation, for instance, geographic differentiation, a consumer near one seller may prefer to buy its products even though they are a little bit more expensive. Production constraints, i.e., one firm is unable to supply all the market, may lead to a different Nash equilibrium. Imperfect information from the buyers about the prices charged by the firm may also disturb the Nash equilibrium we have found.

More information about oligopolies can be found at:
https://open.oregonstate.education/intermediatemicroeconomics/chapter/module-18/
https://www.economics.utoronto.ca/murdockj/eco410/class_02_slides.pdf
http://www.sfu.ca/~shihenl/302/7p%20Oligopoly.pdf

4.6 Monopsony

There is a *monopsony* when there is only one consumer in the market or when the consumers collude.

The behaviour of the monopsonist is similar to that of the monopolist, except that the monopsonist wants to buy at the lowest possible price, whereas the monopolist wants to sell at the highest possible price.

Government procurement, like military hardware, is an example of a monopsony. For instance, there are only 3 firms in the USA that are capable of making a fighter jet. The classic case of monopsony is a "company town," where one employer is the source of all jobs. In general, economists use monopsony in the sense of "monopsony power", which means any situation where the buyer(s) have market power and use that market power to lower the price they pay.

4.6.1 Optimal Quantity and Price

Consider that the single buyer produces an output, whose quantity is y, which must be sold in a competitive market, where he/she does not control the price (p) of the output. Therefore, we have:

$$y = f(x) \tag{4.46}$$

where x is the input's quantity and $f(x)$ is the production function.

The revenue of the monopsonist is:

$$R(x) = py = pf(x) \tag{4.47}$$

The input must be bought in the market following a supply curve, that defines the buying price as a function of the quantity, x. For instance, suppose the input is labour (quantity) and the labour is bought by the firm through wages. The wages follow a supply function, $w(x)$.

The total cost of the monopsonist is given by:

$$TC(x) = w(x)x \tag{4.48}$$

The monopsonist seeks profit maximisation.

$$\max(R(x) - TC(x)) \overset{\text{yields}}{\rightarrow} \frac{dR(x)}{dx} = \frac{dTC(x)}{dx} \overset{\text{yields}}{\rightarrow} MRP = MFC \tag{4.49}$$

Marginal Revenue Product (MRP) is the extra revenue generated when an additional unit of input is used. Since the demand curve shows the buyer's additional willingness to pay for an additional unit, MRP and the demand curve coincide.

Marginal Factor Cost (MFC) is the increment to total costs paid for a factor of production (input) resulting from a one-unit increase in the amount of the factor employed. The average cost is the market price paid for each unit of input, which is determined by the market supply curve. MFC is higher than the average cost because, for instance, to employ one extra worker the firm has to increase the wages of all workers.

When the supply curve is given by a linear function, $w(x) = c + dx$, we have $TC(x) = w(x)x = cx + dx^2$ and $MFC = c + 2dx$.

Let us put some numbers on this to facilitate the explanation. Suppose a monopsonist has a demand curve given by $MRP = 60 - x$ and a supply curve given by $w(x) = 2x + 5$. We have seen that $MFC = 4x + 5$. The situation is depicted in Fig. 4.15.

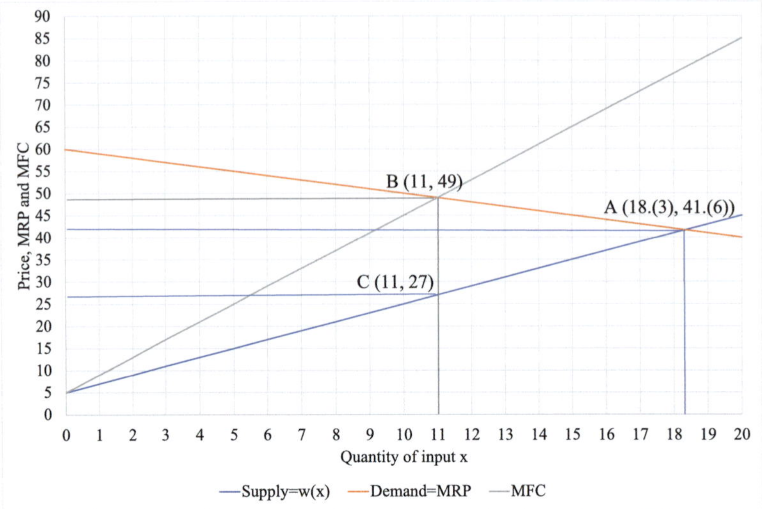

Fig. 4.15 Supply curve, demand curve (MRP) and MFC. The operating point of the monopsonist

If a competitive market was in place, the operating point would be found where the supply curve meets the demand curve (point A), i.e., the monopsonist would buy 18.(3) units of the input at a price of 41.(6) each. But we are under a monopsony environment, so the monopsonist selects the quantity it buys to maximise profits, by setting $MRP = MFC$. The corresponding quantity is 11 and the ideal price should be 49 (point B). Note that the monopsonist is restricting quantity. However, the monopsonist has market power and will pay the price set in the supply curve, which is only 27 (point C).

The monopoly restricts output to drive up the price. The monopoly output is less than the competitive output, and the monopoly price is higher than the competitive price. The monopsony restricts input to drive down the price. The monopsony input is less than the competitive input, and the monopsony price is lower than the competitive price. Both firms are maximising profit by using the market characteristics that they face.

Example 4.8

Suppose that a monopsonist sells a good at a price $p = 10$. To produce this good, the input is labour, whose production function is constant returns to scale, $f(L) = 2L$ (L: Labour). Assume that labour supply to the firm is given by $w(L) = L/50$ (w: wage). Compute the labour quantity and wage that maximises the monopsonist profits. ◀

Solution

The revenue is:

$$R(L) = pf(L) = 20L$$

The marginal revenue product is:

$$MRP = \frac{dR(L)}{dL} = 20$$

The total cost is:

$$TC(L) = w(L)L = \frac{L^2}{50}$$

The marginal factor cost is:

$$MFC = \frac{dTC(L)}{dL} = \frac{L}{25}$$

The maximum profit of the monopsonist is obtained for a labour quantity given by:

$$MRP = MFC \overset{\text{yields}}{\rightarrow} 20 = \frac{L}{25} \overset{\text{yields}}{\rightarrow} L^* = 500$$

The wage is not $MRP(L^*) = 20$, but it is taken from the supply curve, i.e.,

$$w^* = w(L^*) = 10$$

The monopsonist restricts the labour quantity with regard to a competitive market. In the competitive market the quantity would be:

$$w(L) = MRP(L) \overset{\text{yields}}{\rightarrow} \frac{L}{50} = 20 \overset{\text{yields}}{\rightarrow} L' = 1000$$

Furthermore, the monopsonist pays lower wages than would happen in a competitive market. The wage in a competitive market would be:

$$w(L') = 20$$

Illustrative graphics are shown in Fig. 4.16. ◀

4.6.2 Deadweight Loss

Monopsonies, as well as monopolies, are inefficient are markets as they cause a deadweight loss. The consumer surplus (orange), producer surplus (blue), and deadweight loss (grey) can be seen in Fig. 4.17.

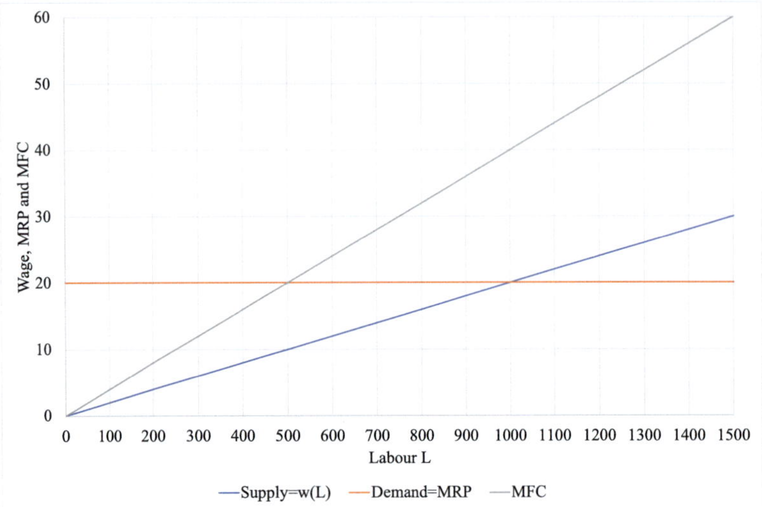

Fig. 4.16 Monopsony optimal operating point. Example 4.8

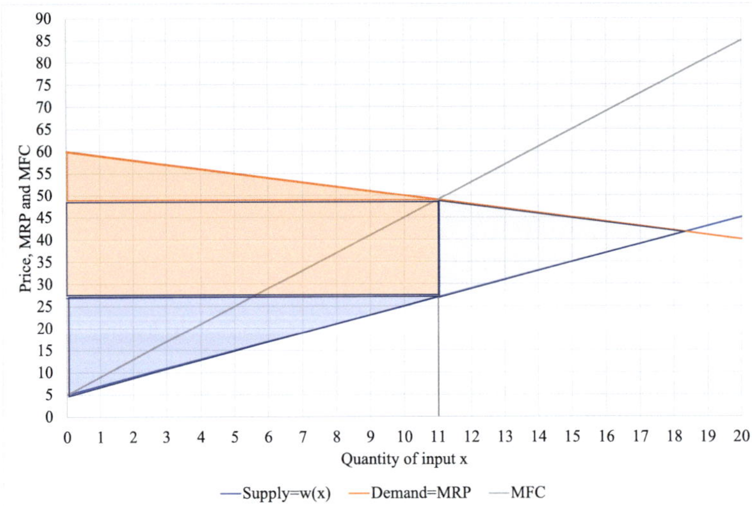

Fig. 4.17 Consumer surplus (orange), producer surplus (blue), and deadweight loss (grey) in a monopsony

4.6.3 Input Elasticity of Supply

As a side note, we observe that:

$$\frac{dTC(x)}{dx} = MFC = w(x)+x\frac{dw(x)}{dx} = w(x)\left(1+\frac{1}{\frac{\frac{dx}{x}}{\frac{dw(x)}{w(x)}}}\right) = w(x)\left(1+\frac{1}{\eta(x)}\right) \tag{4.50}$$

where $\eta(x)$ is called the *input elasticity of supply* defined by:

$$\eta(x) = \frac{dx}{dw(x)}\frac{w(x)}{x} \tag{4.51}$$

Input elasticity of supply measures the percentage change in the quantity of the input supplied caused by a one percent increase in the price of the input. An example would be the change in the supply quantity of cookies caused by a one percent increase in the price of sugar.

From Eq. 4.50, we conclude that the ratio of the price effectively paid to the marginal factor cost:

$$\frac{w(x)}{MFC(x)} = \frac{1}{1+\frac{1}{\eta(x)}} \tag{4.52}$$

where the factor g:

$$g = \frac{1}{1+\frac{1}{\eta(x)}} \tag{4.53}$$

measures how many times less the monopsonist pays the product in relation to the marginal factor cost.

4.7 Conclusions

In conclusion, the examination of various market structures within the context of power systems reveals a complex interplay of economic dynamics, regulatory considerations, and real-world implications. Each market structure, from perfect competition to monopsony, presents its own set of challenges and opportunities.

Perfect competition, characterised by numerous buyers and sellers, achieves the optimal quantity and price at equilibrium. It operates under conditions of price-taking behaviour, where firms have no market power, and resources are allocated efficiently. However, the power sector rarely aligns perfectly with this model, as operational thresholds, grid constraints, and regulatory interventions often distort the ideal competitive landscape.

Monopoly, on the other hand, represents a market dominated by a single supplier. While this may lead to profit maximisation for the monopolist, it results in a deadweight loss to society due to higher prices and reduced output. Regulated monopolies attempt to strike a balance by allowing monopolistic entities to operate under government oversight, aiming to protect consumers while ensuring essential services. An important part of today's power system (namely, transmission and distribution) is operated under a regulated monopoly market structure, in many countries.

Monopolistic competition, with its short-run and long-run optimisation, allows for product differentiation and a degree of market power. Firms can optimise their products and pricing strategies to achieve economic profits in the short run but may face competition that erodes those profits over time.

In the oligopoly market structure, modern game theory models and older Cournot and Bertrand models provide insights into strategic decision-making among a few dominant firms. Nash equilibrium represents a point of stability in which firms make decisions to maximise their own profits, considering the reactions of others. This delicate balance can lead to various outcomes, including collusion or competitive pricing, depending on the strategic choices made.

Lastly, monopsony examines markets with a single buyer's dominance, impacting input prices and quantities. Optimal quantity and price determination in monopsonistic markets must account for the input elasticity of supply, but the inherent deadweight loss does not allow for resource allocation efficiency.

In navigating the intricate landscape of market structures within power systems, policymakers, regulators, and industry stakeholders must carefully consider the trade-offs, externalities, and specific conditions unique to the energy sector. Achieving the optimal balance between market competition, regulation, and societal welfare remains a central challenge in ensuring reliable, affordable, and sustainable power for all.

4.8 Proposed Exercises

Problem 4.1

The profit function of a company that offers in a perfectly competitive market is given by:

$$\pi(q) = pq - 2q^3 + 20q^2 - 80q - 10$$

1. Compute the short-run inverse supply function.
2. If the price is 50, what is the quantity the producer should sell? Compute the total revenue, average total cost, average variable cost, and profit in this situation.
3. Repeat for $p = 31$.
4. Compute the shutdown threshold and the profitability threshold.

Key Problem 4.1

1.

$$\pi(q) = R(q) - TC(q)$$

where $R(q) = pq$ is the total revenue and $TC(q)$ is the total cost.

We can therefore write:

$$TC\,(q) = 2q^3 - 20q^2 + 80q + 10 = C_V + C_F$$

The operating conditions for a perfectly competitive firm are:

$$\begin{cases} p = MC = \frac{dTC(q)}{dq} \\ p \geq AVC \end{cases}$$

where MC is the marginal cost and AVC is the average variable cost.

Developing these equations, we obtain:

$$\begin{cases} MC = 6q^2 - 40q + 80 = p \\ p \geq 2q^2 - 20q + 80 \end{cases}$$

$$\begin{cases} 6q^2 - 40q + 80 - p = 0 \\ 6q^2 - 40q + 80 \geq 2q^2 - 20q + 80 \end{cases}$$

$$\begin{cases} q = \frac{40+\sqrt{1600-24(80-p)}}{12} \\ 4q^2 - 20q \geq 0 \end{cases}$$

$$\begin{cases} q = \frac{40+\sqrt{24p-320}}{12} \\ q \geq 5 \end{cases}$$

For $q \geq 5$, we have $MC \geq 30$ and hence $p \geq 30$.

The inverse supply function is:

$$\begin{cases} q = \frac{40+\sqrt{24p-320}}{12} \text{ if } p \geq 30 \\ q = 0 \text{ if } p < 30 \end{cases}$$

2.

$$q\,(50) = 5.8054$$

$$TR = pq(50) = 290.27$$

$$TC = TC\,(5.8054) = 191.6935$$

$$ATC = \frac{TC(q)}{q} = 33.0199$$

$$AVC = \frac{VC}{q} = 31.2973$$

$$\pi = TR - TC = 98.5764$$

3.

$$q\,(31) = 5.0493$$

$$TR = pq(31) = 156.5274$$

$$TC = TC\,(5.0493) = 161.5027$$

$$ATC = \frac{TC}{q} = 31.9853$$

$$AVC = \frac{VC}{q} = 30.0049$$

$$\pi = TR - TC = -4.9752$$

There is no profit. However, the firm should not shut down in the short run because $p > AVC$, although $p < ATC$.

4.

Shutdown threshold: min AVC.

$$AVC = \frac{VC}{q} = 2q^2 - 20q + 80$$

$$\frac{d(AVC)}{dq} = 0 \overset{\text{yields}}{\rightarrow} 4q - 20 = 0 \overset{\text{yields}}{\rightarrow} q = 5 \overset{\text{yields}}{\rightarrow} \min\ AVC = 30$$

Profitability threshold: min ATC.

$$ATC = \frac{TC(q)}{q} = 2q^2 - 20q + 80 + \frac{10}{q}$$

$$\frac{d(ATC)}{dq} = 0 \overset{\text{yields}}{\rightarrow} 4q - 20 - \frac{10}{q^2} = 0 \overset{\text{yields}}{\rightarrow} q = 5.0963 \overset{\text{yields}}{\rightarrow} \min ATC = 31.9808$$

The solution of the cubic equation $4q^3 - 20q^2 - 10 = 0$ was obtained from an online cubic equation solver https://www.calculatorsoup.com/calculators/algebra/cubicequation.php

The plotting of the several quantities involved in this problem is illustrated in Fig. 4.18.

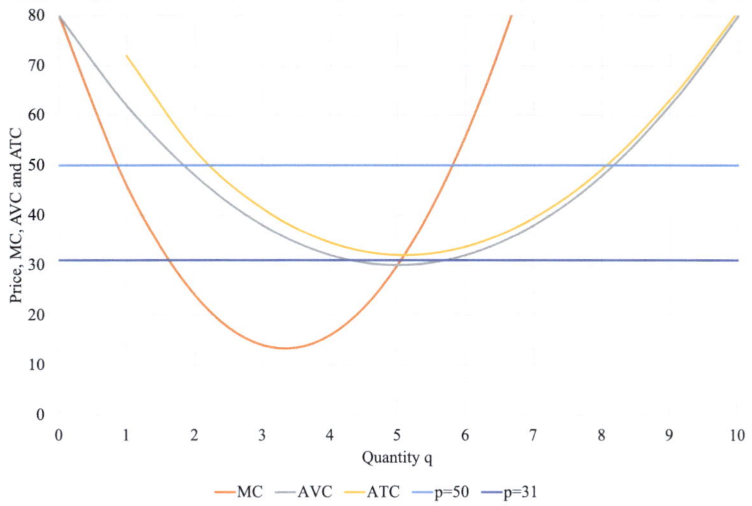

Fig. 4.18 Price, marginal cost, average variable cost, and average total cost. Problem 4.1

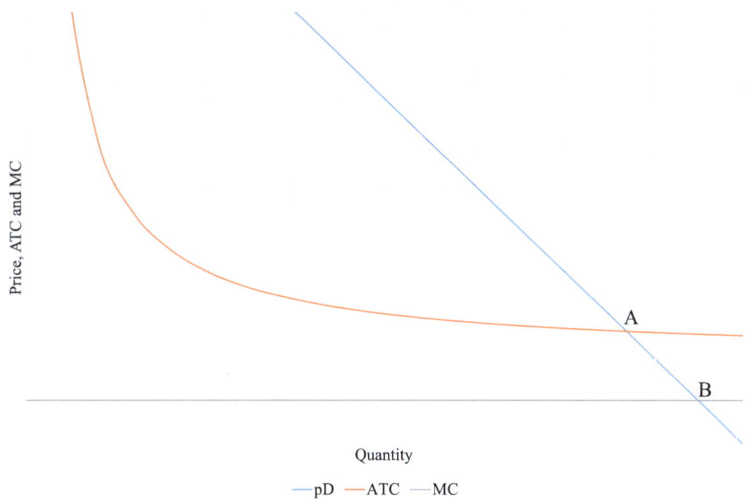

Fig. 4.19 Monopoly p_D, ATC and MC. Problem 4.2

Problem 4.2

Figure 4.19 shows a monopoly situation.

The total cost of the monopolist is given by: $TC(q) = 100 + 2q$

The market demand curve is given by: $p_D(q) = 140 - 0.5q$

1. In the regulated monopoly situation, the regulator tries to fix the price at point A or point B? Justify.
2. Determine the price and quantity characterising point A.
3. In a situation where the monopoly is unregulated, the monopolist maximises the profit. Determine the point (quantity and price) that maximises the monopolist's profit.

Key Problem 4.2

1.

The regulator fixes the price at point A so that to allow the monopolist to recover the average cost (in which a rate of return is implicit). At point B, the monopolist would simply recover the marginal cost, which would not be enough to recover all the costs.

2.

Point A is defined by:

$$ATC(q) = p_D(q)$$

$$\frac{100}{q} + 2 = 140 - 0.5q$$

$$0.5q^2 - 138q + 100 = 0 \overset{\text{yields}}{\rightarrow} q' = 275.273 \vee q_2 = 0.727$$

The solution to be retained is q' as q_2 corresponds to the other intersection point, not to point A. The solution to the problem is:

$$q' = 275.273$$

$$p' = p_D\left(q'\right) = 2.363$$

3.

The maximum profit is obtained at:

$$\max\left(R(q) - TC(q)\right)$$

$$\frac{dR(q)}{dq} = \frac{dTC(q)}{dq} \overset{\text{yields}}{\rightarrow} MR = MC$$

$$R\left(q\right) = p_D\left(q\right)q = \left(140 - 0.5q\right)q = 140q - 0.5q^2 \overset{\text{yields}}{\rightarrow} MR = 140 - q$$

$$TC\left(q\right) = 100 + 2q \overset{\text{yields}}{\rightarrow} MC = 2$$

$$MR = MC \overset{\text{yields}}{\to} q^* = 138$$

$$p^* = p_D\left(q^*\right) = 71$$

Note that the price is taken at the demand curve, not at the marginal cost curve. In fact, it is well above the marginal cost.

The situation is illustrated in Fig. 4.20.

We can now compute the maximum profit.

$$R = p^* q^* = 9798$$

$$TC = TC\left(q^*\right) = 376$$

$$\pi = R - TC = 9422$$

Figure 4.21 illustrates the situation.

It is interesting to verify the following.

The price elasticity of quantity is the ratio of relative changes in quantity and price.

$$\varepsilon_p(q) = \frac{\frac{dq}{q}}{\frac{dp_D(q)}{p_D(q)}} = \frac{dq}{dp_D(q)} \frac{p_D(q)}{q}$$

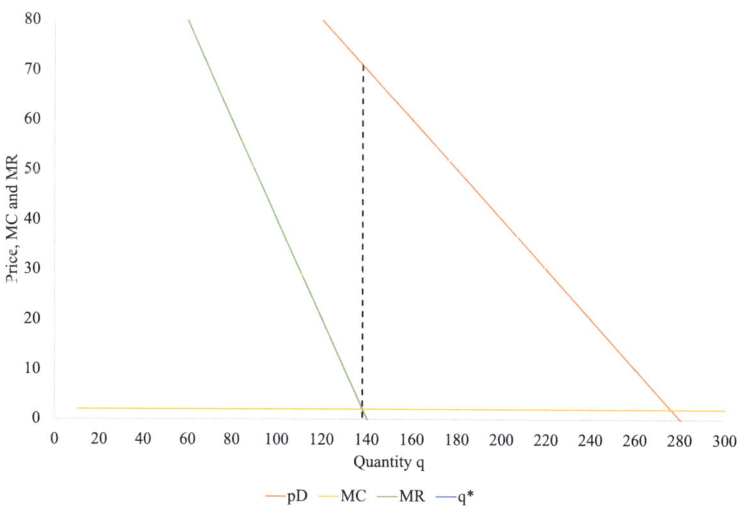

Fig. 4.20 Monopoly p_D, MC and MR. Monopolist operating point. Problem 4.2

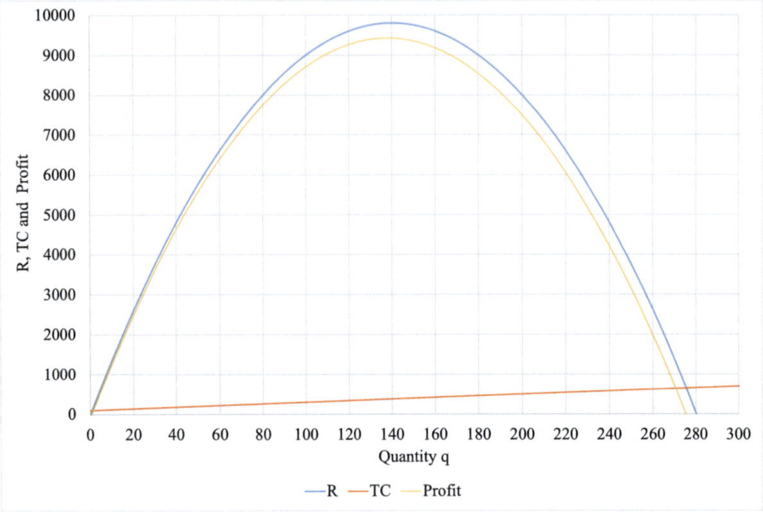

Fig. 4.21 Monopoly R, TC and profit. Problem 4.2

Applying to our problem, we have at the maximum profit point:

$$\varepsilon_p(q) = \frac{dq}{dp_D(q)} \frac{p_D(q)}{q} = -2\frac{p^*}{q^*} = -2\frac{71}{138} = -1.029$$

The monopolist price is higher than the marginal cost by a factor given by:

$$f = \frac{1}{1 + \frac{1}{\varepsilon_p}} = 35.5$$

In fact, in the maximum profit point, the monopolist price is 71 and the marginal cost is 2.

Costs in Power Systems

5

Abstract

The chapter "Costs in Power Systems" delves into the intricate web of expenses associated with the generation, transmission, and distribution of electrical energy. In today's rapidly evolving energy landscape, understanding these costs is paramount for policymakers, industry professionals, and consumers alike. We explore the multifaceted aspects of generation costs, from fuel procurement to maintenance and technological advancements. We highlight the importance of optimising energy sources to minimise financial burdens while meeting sustainability objectives. The concept of LCOE is examined in detail, shedding light on how it provides a standardised framework for comparing the long-term cost-effectiveness of different energy generation methods. It enables decision-makers to make informed choices regarding resource allocation. The chapter dissects the complexities of building and maintaining extensive transmission and distribution networks. We underscore the significance of efficiency enhancements and grid modernisation to reduce operational costs and enhance reliability. A critical discussion on the externalities of power generation, including greenhouse gas emissions and environmental degradation, emphasises the urgency of transitioning to cleaner and more sustainable energy sources. By delving into these facets of costs in power systems, this chapter equips readers with a holistic understanding of the economic and environmental considerations that underpin the energy sector's future. It provides a roadmap for fostering cost-effective, resilient, and eco-friendly power systems in an era defined by energy transition and sustainability imperatives.

5.1 Introduction

Power systems play a pivotal role in modern society, supplying the energy needed for various activities. However, the generation, transmission, and distribution of electricity come with associated costs. Understanding these costs is crucial for

R. Castro, *Engineering of Power Systems Economics*,
https://doi.org/10.1007/978-3-031-55251-9_5

policymakers, utility companies, and consumers alike, as they impact the efficiency, sustainability, and affordability of power systems.

Generation Costs: One of the primary costs in power systems is the generation of electricity. Power plants require significant investments in infrastructure and fuel sources, leading to capital and operational expenses. The costs vary depending on the type of generation technology employed. Conventional fossil fuel plants involve expenses for fuel procurement, operations, and maintenance, as well as environmental mitigation measures. Renewable energy sources like solar and wind have lower operational costs but higher upfront investments. The overall aim is to strike a balance between cost-effectiveness and sustainability by transitioning towards cleaner and more efficient generation methods.

Transmission and Distribution Costs: Once electricity is generated, it needs to be transmitted and distributed to end-users. This stage incurs costs related to infrastructure development, grid maintenance, and system losses. The transmission of electricity over long distances requires high-voltage transmission lines, substations, and control systems. These investments can be substantial, but they are necessary for ensuring efficient and reliable power delivery. Distribution costs include the construction and maintenance of local networks, transformers, and meters. To optimise costs, utilities employ advanced technologies such as smart grids and demand response systems to improve efficiency and reduce losses.

Environmental Costs: Power systems have environmental impacts, which come with additional costs. Conventional fossil fuel plants contribute to air pollution, greenhouse gas emissions, and climate change. These external costs also include health expenses, ecosystem damages, and climate adaptation costs. Governments and regulatory bodies are increasingly imposing environmental regulations and carbon pricing mechanisms to internalise these costs and incentives for cleaner energy alternatives. By promoting renewable energy sources and implementing emission reduction measures, the environmental costs can be mitigated over time.

The costs of power systems ultimately influence electricity pricing, which directly impacts consumers. Tariffs are designed to recover the costs of generation, transmission, and distribution, as well as cover the utility company's operational expenses and profit margins. Striking the right balance is essential to ensure affordability for consumers while allowing utilities to maintain a reliable power supply. Governments often regulate prices to prevent market abuses and protect vulnerable consumers. Moreover, promoting energy efficiency and conservation can help manage costs by reducing overall demand and the need for costly infrastructure expansions.

As power systems evolve, new challenges and opportunities arise. The integration of intermittent renewable energy sources presents challenges in maintaining grid stability and managing variability. Energy storage technologies, such as batteries, are being developed to address these issues but involve additional costs. The electrification of the transportation and heating sectors also requires substantial investments in infrastructure. To tackle these challenges, innovative solutions like decentralised microgrids, digitisation, and blockchain technology are being explored to optimise costs and enhance system flexibility.

In this chapter, we provide a holistic view of all the power system's costs previously outlined. We begin by addressing the *generation costs*, where we introduce the wide world used concept of *levelized cost of energy* (LCOE) and offer the groundings of the *economic assessment of projects*. Then, we analyse the generation costs in an *optimal power system* and compare them to the *cost recovery in the market* by the producers. *Transmission costs* are dealt with afterward, where the different components of the transmission system are addressed, and an estimation of their costs is provided. The *cost of the distribution system*, where the components are similar, but with very different costs, is explored. A note on the *penalty for excessive reactive power consumption* is offered. Finally, the chapter delves into the *environmental costs* problematic, where CO_2 emissions play a major role. As usual, the chapter ends with a set of proposed exercises and their solutions.

5.2 Generation Costs

Any power station can be summarily described by a production function, whose inputs are x and y and the output is z (Fig. 5.1). Variable x is the primary energy that is transformed into electricity, while y accounts for the size and operating conditions of the power station. The output variable z is the electrical energy produced.

In the design phase of the power station, variables x and y are selected. The primary energy associated with x can be hydro, coal, nuclear, natural gas, fuel, wind, biomass, and waste or photovoltaic. The variable associated with y may represent the power of the generator and operating conditions. Therefore, we can write the production function as:

$$z = f(x, y) \tag{5.1}$$

After the power station is built, the nominal power and operating conditions are established, thus the input variable y is determined and a simplification in the production function is allowed:

$$z = g(x);\ y = Y \tag{5.2}$$

The fixed cost, FC (€/MW/year), is associated with input $y = Y$, and accounts for the remuneration and depreciation of the investment and operation of the generating

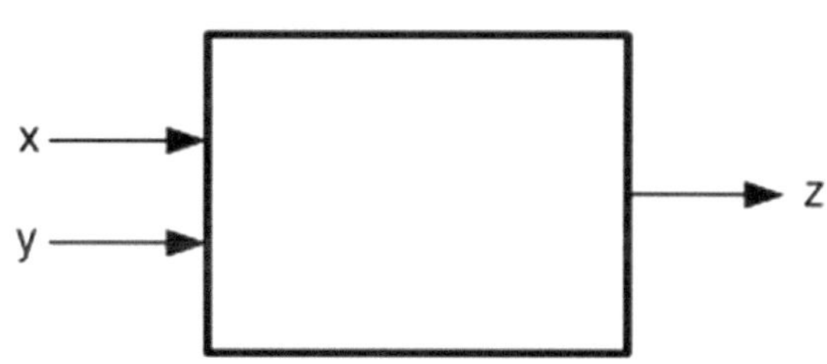

Fig. 5.1 Electrical energy production function

centre. The variable cost, VC (€/MWh), is associated with the input variable x (primary energy consumed).

In general, for a power station, the annual total production cost can be written as follows:

$$TC = FC + VC = CP + cE_a \tag{5.3}$$

In Eq. 5.3, it is:

- TC is the total annual cost of the production centre (€).
- FC is the annual fixed cost; it is a levelized cost (€), as we will see later.
- C is the annual fixed cost per MW; it is also a levelized cost (€/MW).
- P is the power plant-rated capacity (MW).
- VC is the annual variable cost (€).
- c is the variable cost per unit of energy produced (output) (€/MWh).
- E_a is the amount of electricity produced annually, i.e., output (MWh).

5.2.1 Levelized Cost of Energy and Marginal Cost

Now, let us take a look at what is meant by *levelized*.

In an investment project, cash inflows and outflows are spread over a period of time and time has a monetary value. The concept of *discount rate* is the way costs (or incomes) occurring in the future (or in the past) are reported to the present time. For instance, let us assume that F_{t1} (€) is a cost that occurs in time $t = t_1$. If the same cost is to be paid today ($t = 0$), the same cost is equivalent to F_0, given by:

$$F_0 = \frac{F_{t1}}{(1+a)^{t_1}} \tag{5.4}$$

in which a is the discount rate. F_0 is said to be the present value (or levelized value) of a payment (or a revenue) actually made in time $t = t_1$.

We note that the discount rate can be viewed as the minimum rate of return the investor demands to invest in a project.

It is usual to transform the investment cost (fixed cost) into a summation of constant annual payments over the lifetime of the generating centre. Each payment is usually referred to as the investment annuity, A_T. As the annuity is due each year of the power station's lifetime, each annuity has to be discounted to the present time. The sum of all the discounted annuities is the total investment, normally expressed per unit of capacity, I_0 (€/MW). We can therefore write:

$$I_0 = \frac{A_T}{(1+a)} + \frac{A_T}{(1+a)^2} + \cdots + \frac{A_T}{(1+a)^n} = A_T \sum_{j=1}^{n} \frac{1}{(1+a)^j} \tag{5.5}$$

where a is the discount rate, n is the power plant lifetime, and the salvage value has been neglected.

The sum of the series shown in Eq. 5.5 is known and given by a formula. We will now define:

$$k_a = \sum_{j=1}^{n} \frac{1}{(1+a)^j} = \frac{(1+a)^n - 1}{a(1+a)^n}$$
$$\alpha = \frac{1}{k_a} = \frac{a(1+a)^n}{(1+a)^n - 1} \tag{5.6}$$

where k_a is known as *discount factor*.

Therefore:

$$I_0 = A_T k_a$$
$$A_T = \alpha I_0 \tag{5.7}$$

The total investment per MW is also known as the *overnight cost*, that is, the cost of a construction project if no interest was incurred during construction as if the project was completed "overnight".

What has been said means that on the computation of the annuity—the investment cost due each year, there is an implicit minimum rate of return involved—the discount rate.

To obtain the annual fixed cost, the investment annuity must be added to the fixed operation and maintenance (O&M) costs, $c_{O\&M}$, which are assumed to be proportional to the investment per MW, the proportionality factor being β.

$$c_{O\&M} = \beta I_0 \tag{5.8}$$

Thus, the fixed annual generation cost is given by (we note that this is a levelized cost):

$$FC = CP = (\alpha + \beta) I_0 P \tag{5.9}$$

The unitary variable cost, c, includes the fuel cost and the cost of CO_2 emissions, when applicable.

Therefore, the annual total production cost is:

$$TC = FC + VC = (\alpha + \beta) I_0 P + c E_a \tag{5.10}$$

The annual total production cost per installed MW is:

$$TC_0 = (\alpha + \beta) I_0 + c h_a \tag{5.11}$$

where h_a is the *utilisation factor*. We recall that the utilisation factor is the number of hours per year a power plant should operate at rated capacity to produce the electricity it has actually produced. When the utilisation factor is computed as a percentage of the 8760 hours of the year, it is called *capacity factor*.

$$h_a = \frac{E_a}{P} \tag{5.12}$$

It is useful to know the *average total cost*, ATC, which is also known as (*Levelized Cost Of Energy*—LCOE), i.e., the cost per unit of electricity produced integrated along the lifetime of the power station, and the *marginal cost*, MC, i.e., the cost of the last unit produced.

$$LCOE = ATC = \frac{TC}{E_a} = \frac{(\alpha + \beta)I_0}{h_a} + c \quad (5.13)$$

$$MC = \frac{\partial TC}{\partial E_a} = c \quad (5.14)$$

The comparison of these unitary costs with the selling price value, p, obtained from the sale of electricity, allows the following statements (as we have seen in the previous chapters):

- To ensure the economic sustainability of the power station in the short term, the marginal cost must be recovered through the sale of electricity. This condition imposes:

$$p \geq MC \quad (5.15)$$

- To ensure the economic sustainability of the power station, in the long run, the average total cost must be recovered through the sale of electricity. This condition imposes:

$$p \geq ATC \quad (5.16)$$

Example 5.1

Consider the following average costs and usages of the different power generation technologies (Table 5.1). CCGT is the acronym for *Combined Cycle Gas Turbine*. Assume that the fuel costs for the different technologies are: coal: 30 €/MWh; natural gas (CCGT): 40 €/MWh; and nuclear: 8 €/MWh.
The rates of CO_2 emissions are 1.02 and 0.51 tons/MWh for, respectively, coal and natural gas. The costs for CO_2 emissions are estimated at 15 €/ton.
Consider the life span of the power plants to be 30 years and a discount rate of 7%.
Calculate the LCOE for the different technologies. ◀

Table 5.1 Average costs and capacity factors. Example 5.1

Technology	Capital costs (€/kW)	O&M costs (€/kW)	Capacity factor (%)
Solar PV	900	20	22
Nuclear	10,000	50	88
Coal	2800	25	85
Wind	1200	30	30
CCGT	900	10	70

Table 5.2 LCOE for Example 5.1

Technology	LCOE (€/MWh)
Solar PV	48.0
Nuclear	119.0
Coal	79.0
Wind	48.2
CCGT	61.1

Solution

The Levelized Cost Of Energy (LCOE) (€/MWh) is given by Eq. 5.13:

$$LCOE = \frac{(\alpha + \beta) I_0}{h_a} + c$$

where: α is the inverse of the discount factor, β is the O&M cost as a percentage of the unitary investment, I_0 (€/MW), αI_0 is the unitary annuity (€/MW), βI_0 is the O&M unitary cost (€/MW), h_a is the utilisation factor (h) (note that the capacity factor is $C_f = h_a/8760$), $c = c_f + c_{CO_2}$ is the variable cost (€/MWh), composed by the fuel cost, c_f, and the CO_2 emissions cost, c_{CO_2} (note that $c_{CO_2} = rate_{CO_2} c_{emiss}$, [tons/MWh × €/ton].
Application - see Table 5.2
Conclusion: Renewable technologies are today the cheapest way of producing electricity! ◄

5.2.2 Economic Assessment of Projects

Project finance is the long-term financing of infrastructure and industrial projects based on the projected cash flows of the project rather than on the balance sheets of its sponsors. Usually, a project financing structure involves a number of equity (own capital) investors, known as *sponsors*, as well as a *syndicate* of banks that provide loans to the operation.

Generally, a special purpose entity is created for each project, thereby shielding other assets owned by a project sponsor from the detrimental effects of a project failure. As a special-purpose entity, the project company has no assets other than the project.

In order to assess the profitability of a project, the *Net Present Value* (NPV) criterion is often used. NPV (€) is defined as the difference between the discounted incomes (Inc) and the discounted outcomes ($Outc$) (the so-called cash-flows) over the project lifetime:

$$NPV = \sum_{j=1}^{n} \frac{Inc_j}{(1+a)^j} - \sum_{j=0}^{n-1} \frac{Outc_j}{(1+a)^j} \tag{5.17}$$

In electricity generation projects, the incomes result from electricity sales; the outcomes include investment, fuel costs, O&M expenses, and so on.

In the event, the annual incomes and outcomes are constant and equal to R and E, respectively, year after year, during the project lifetime, and the total investment, I_T, is totally accomplished in $t = 0$, a simplified equation to compute the NPV can be written as:

$$NPV = (R - E)k_a - I_T \tag{5.18}$$

If $NPV > 0$, the project is economically profitable, meaning that all the costs are recovered, the minimum rate of return of the capital (discount rate) is achieved and a surplus is obtained. If $NPV = 0$, the project is feasible, i.e., all the costs are recovered and the minimum rate of return of the capital is achieved; in this situation, the profitability of the project is not straightforward because no surplus is obtained. If $NPV < 0$, the project is not economically profitable either because the minimum rate of return of the capital is not achieved, or because the costs are not even recovered.

Another useful economic assessment index is the *Internal Rate of Return* (IRR). IRR (%) is the discount rate that turns the NPV equal to zero. This index is handy because it portrays the real rate of return of the project. In general, the equation that defines the IRR is:

$$0 = \sum_{j=1}^{n} \frac{Inc_j}{(1 + IRR)^j} - \sum_{j=0}^{n-1} \frac{Outc_j}{(1 + IRR)^j} \tag{5.19}$$

In the general case, Eq. 5.19 is not easy to solve because the unknown is IRR. In the simplified conditions of Eq. 5.18, the IRR computation becomes simpler:

$$(R - E)\left[\frac{(1 + IRR)^n - 1}{IRR(1 + IRR)^n}\right] = I_T \tag{5.20}$$

This is a non-linear equation that needs to be solved iteratively, using, for instance, the Gauss method:

$$IRR^{(k+1)} = \frac{R - E}{I_T} \frac{(1 + IRR^{(k)})^n - 1}{(1 + IRR^{(k)})^n} \tag{5.21}$$

where (k) is the number of the iteration.

Gauss method implies we write the unknown in both sides of the equation and solve iteratively to successively find updated values for the unknown. An initial guess for the unknown (IRR) is needed. A typical value is $IRR^{(0)} = 0.1$. The iterative process stops when the difference between two consecutive computed values is less than a predefined tolerance. The tolerance depends on the desired accuracy. A good value for the tolerance is $TOL = 10^{-2}$.

This process is not handy, therefore a simplified method is available. NPV decreases as the discount rate increases. This is because a higher rate of return is being demanded. A typical variation of the NPV is shown in Fig. 5.2.

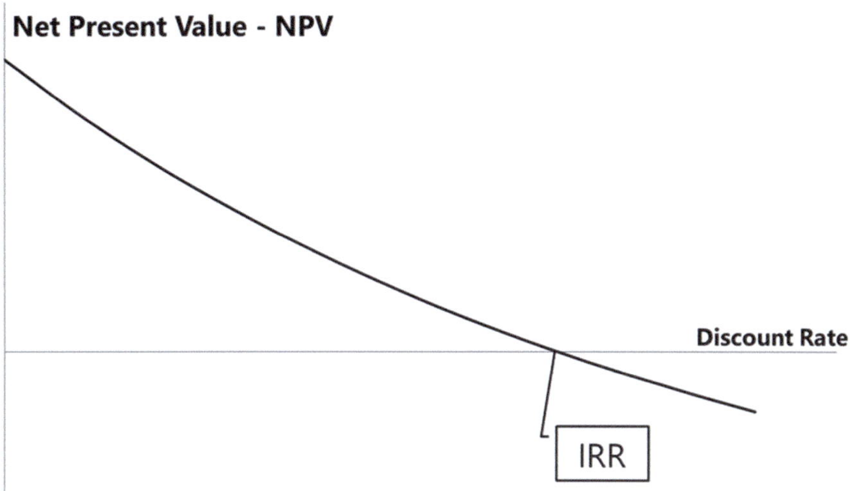

Fig. 5.2 Typical NPV versus discount rate curve

The goal is to determine the discount rate that turns the NPV equal to zero, i.e., the IRR. The method involves finding two points on the curve: one point, (a_1, NPV_1), corresponds to a discount rate, a_1, that leads to a positive NPV, NPV_1; the other point, (a_2, NPV_2), corresponds to a (higher) discount rate, a_2, that leads to a negative NPV, NPV_2. Then, we make a linear interpolation between the two points to find the x-intercept. The approximate IRR is given by:

$$IRR \approx a_1 - (a_2 - a_1)\frac{NPV_1}{NPV_2 - NPV_1} \tag{5.22}$$

If $IRR > a$, the project is economically viable because the real rate of return is higher than the minimum demanded by the investor. $IRR = a$ means $NPV = 0$. If $0 < IRR < a$, the minimum demanded rate of return is not achieved—the project is not economically attractive, but the costs are recovered. The costs will not be recuperated if $IRR < 0$.

Example 5.2

The capacity factor of a 10 MW wind park is 28.54%. The selling price of electricity is 50 €/MWh. The investment is 1.2 M€/MW, the expected lifetime is 20 years, and the annual O&M costs are 1.5%. Compute the NPV at a discount rate of 5% and the IRR. ◀

Solution

The NPV is given by Eq. 5.18:

$$NPV = (R - E)k_a - I_T$$

where $R = 50 \times 10 \times 0.2854 \times 8760$, $E = 0.015 \times 1.2 \times 10^6 \times 10$, $k_a = 12.4622$ and $I_T = 12 \times 10^6$. We conclude that $NPV = 1{,}335{,}213$ €.

As for the IRR, we apply Eq. 5.21:

$$IRR^{(k+1)} = \frac{1{,}070{,}052}{12 \times 10^6} \frac{(1 + IRR^{(k)})^{20} - 1}{(1 + IRR^{(k)})^{20}}$$

The initial guess may be $IRR^{(0)} = 0.1$. The obtained results are shown in Table 5.3.

The result is $IRR = 6.29\%$. Using the approximate method, the result is:

$$IRR \approx 0.05 - (0.07 - 0.05)\frac{1{,}335{,}213}{-663{,}854 - 1{,}335{,}213} = 0.0634$$

◀

Note that for $a_2 = 7\%$, $NPV_2 = -663{,}854$ €. The other required point is $a1 = 7\%$, $NPV_1 = 1{,}335{,}213$ €. The result is $IRR = 6.34\%$, which is a very good approximation of the real IRR.

The project NPV and IRR, which we have just learned how to compute, look at the project's profitability and not where the investment's money comes from. The investment may be partly financed by a bank through a loan, with a certain interest rate. In these circumstances, the loan annuity is to be calculated and is treated as a cost parcel. The equity (own capital) is normally made available at $t = 0$. The obtained NPV is called the *Equity NPV*.

Also, it is important to know the internal rate of return of the equity, the so-called *Equity IRR*. In this case, the IRR is calculated taking into account the cash flows net of financing, i.e., the cash flows considering the loan annuities and the equity

Table 5.3 IRR computation for Example 5.2

Iteration	IRR (%)
0	10.00
1	7.59
2	6.85
3	6.55
4	6.41
5	6.34
6	6.29
7	6.29

investment. If the project is fully funded by equity, the project IRR and Equity IRR are the same. If the project is fully funded by debt, equity IRR simply does not exist.

5.2.3 Costs in the Optimal Power System

In this Section, the optimal power system is going to be approached. We refer to the optimal power system as a system able to provide each producer with a remuneration level allowing it to recover all costs—fixed and variable—plus the minimum rate of return implicit in the computation of the fixed cost, and no more than that. This was, in theory, the procedure followed in the past, when the electricity sector was vertically integrated and a central dispatching agency determined said remunerations.

For this purpose, the model proposed by Paul Joskow[1] is generalised. It is worth mentioning that Joskow, a known believer in electricity market deregulation, highlights that well-behaved markets should reproduce the results of optimal planning.

Let us assume that the power system's monotonic duration demand curve is given by a straight line:

$$\begin{gathered} D(h) = b - mh \\ b > 0;\;\; m > 0; h \in [0; 8760] \end{gathered} \tag{5.23}$$

We recall that a duration demand curve shows the demand values ordered monotonically in descending order. Therefore, this straight line relates the demand, D, with the number of hours of the year in which the demand is exceeded, h.

The maximum and minimum demands are:

$$\begin{aligned} D_{\text{max}} &= D(0) \\ D_{\text{min}} &= D(8760) \end{aligned} \tag{5.24}$$

From the supply side, there are n power stations with different technologies that differ in fixed costs, C_i (\$/MW/year) ($C_1 > C_2 > \ldots > C_i > \ldots > C_n$) and variable costs, c_i (\$/MWh) ($c_1 < c_2 < \ldots < c_i < \ldots < c_n$). Both fixed costs and variable costs are supposed to be known.

It is intended to minimise the total cost, TC (\$), of the generating system to supply the demand $D(h)$, this problem being formulated as:

$$\begin{gathered} \min TC = \sum_{i=1}^{n} C_i P_i + c_i E_{ai} \\ s.t.\ C_{i+1} < C_i;\;\; c_{i+1} > c_i;\;\; D(h) = b - mh \end{gathered} \tag{5.25}$$

In Eq. 5.25, P_i (MW) is the installed capacity of power plant i and E_{ai} (MWh) is the annual electricity produced by power plant i.

---

[1] Joskow, Paul L., Capacity Payments in Imperfect Electricity Market: Need and Design, Utilities Policy, 2008, Vol. 16, Issue 3, pp 159–170. DOI: https://doi.org/10.1016/j.jup.2007.10.003.

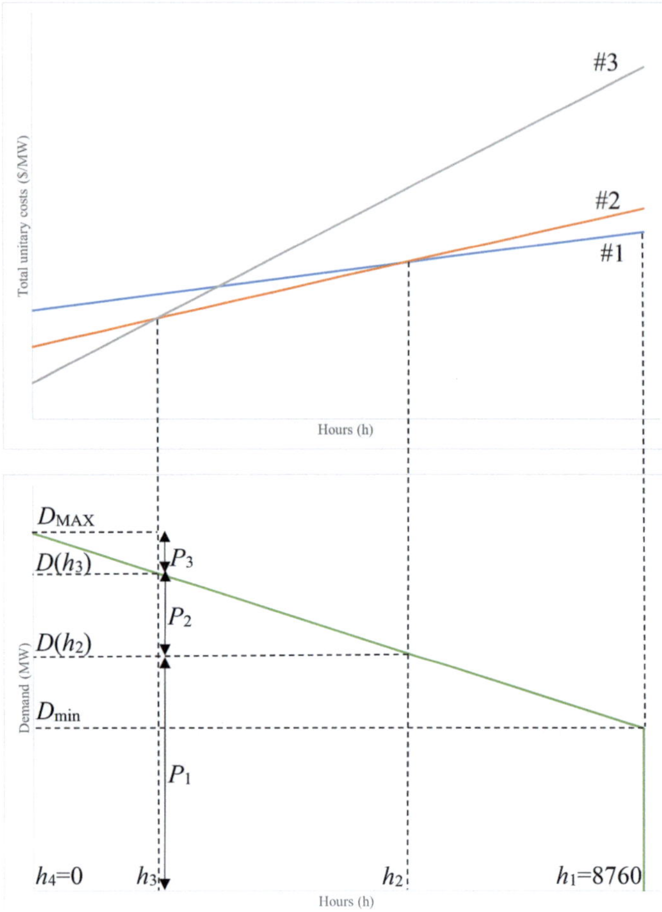

Fig. 5.3 Operating times and installed capacities of the generating technologies

This objective can be attained by using linear programming. For a small number of power stations, a graphic construction[2] can be used. Figure 5.3 portrays an example with three available power stations: Baseline (#1), Intermediate (#2) and Peak (#3).

In the upper part of Fig. 5.3, the total cost per MW of installed capacity for each technology is represented (Eq. 5.11).

$$TC_0 = (\alpha + \beta)I_0 + ch_a \approx C + ch \tag{5.26}$$

[2] Adapted from Joskow, Paul L., Competitive Electricity Markets and Investments in New Generating Capacity, The New Energy Paradigm (Dieter Helm, Editor), Oxford University Press, 2007.

In Eq. 5.26 an approximation is considered. It is assumed $h_a \approx h$, meaning we are considering that the utilisation factor, (h_a), is the operating hours, (h). We know that the utilisation factor, h_a, is the number of hours the power plant should operate at rated power to produce the same energy it actually produced. This is not the number of hours the power plant is operating, i.e., the operating hours, h. However, this method implies this approximation to make it possible to establish a link between both the total cost per MW and the demand curves, as depicted in Fig. 5.3.

The calculating process begins by determining the operating hours of each power plant, $h_1 > h_2 > \ldots > h_i > \ldots > h_n > h_{n+1}$:

$$\begin{aligned} h_1 &= 8760 \\ &\vdots \\ h_i &= \frac{C_{i-1} - C_i}{c_i - c_{i-1}} \\ &\vdots \\ h_n &= \frac{C_{n-1} - C_n}{c_n - c_{n-1}} \\ h_{n+1} &= 0 \end{aligned} \tag{5.27}$$

This is based on:

$$TC_{0i-1} = TC_{0i} \rightarrow C_{i-1} + c_{i-1}h_i = C_i + c_i h_i \tag{5.28}$$

For the example in Fig. 5.3, the operating hours are:

$$\begin{aligned} h_1 &= 8760 \\ h_2 &= \frac{C_1 - C_2}{c_2 - c_1} \\ h_3 &= \frac{C_2 - C_3}{c_3 - c_2} \\ h_4 &= 0 \end{aligned} \tag{5.29}$$

As illustrated in Fig. 5.3, at each time interval, the cheapest power stations should be operating. It is assumed that the cheapest power station ensures the base load, during the whole year ($h_1 = 8760\,h$).

Then, the installed capacities, P_i (MW), are to be determined by:

$$\begin{aligned} P_1 &= D(h_2) \\ &\vdots \\ P_i &= D(h_{i+1}) - D(h_i) \\ &\vdots \\ P_n &= D(h_{n+1}) - D(h_n) \end{aligned} \tag{5.30}$$

In the example that we are following, the installed capacities of each technology are:

$$\begin{aligned} P_1 &= D(h_2) \\ P_2 &= D(h_3) - D(h_2) \\ P_3 &= D(h_4) - D(h_3) = D_{\max} - D(h_3) \end{aligned} \tag{5.31}$$

Taking the example of Fig. 5.3, Baseline#1 power plant must be operating during the 8760 hours of the year. This power plant must have an installed capacity P_1 and will be alone sufficient to supply the load during $(h_1 - h_2)$ hours. Then, in addition to Baseline#1, the Intermediate#2 power station comes forward during $(h_2 - h_3)$ hours, because it is the cheapest one in this period. This Intermediate#2 technology shows an installed capacity P_2. All power stations are in operation at peak situation, which lasts h_3 hours; marginal technology at peak load has the installed capacity P_3.

In these circumstances, it can be stated that the system is balanced, that is, the system has the necessary and sufficient capacity, and this capacity is adequately distributed by each available power station so that the total system production cost is minimal.

To compute the minimal cost, the annual electricity, E_{ai} (MWh), produced by each power plant has to be determined through:

$$\begin{aligned} E_{a1} &= D_{\min}\frac{h_1 - h_2}{2} + P_1\frac{h_1 + h_2}{2} \\ &\vdots \\ E_{ai} &= P_i\frac{h_i + h_{i+1}}{2} \\ &\vdots \\ E_{an} &= P_n\frac{h_n + h_{n+1}}{2} \end{aligned} \tag{5.32}$$

The electrical energy produced by each technology (see Fig. 5.3) is:

$$\begin{aligned} E_{a1} = (h_1 - h_2)\frac{D_{\min} + P_1}{2} + P_1 h_2 &= D_{\min}\frac{h_1 - h_2}{2} + P_1\frac{h_1 + h_2}{2} \\ E_{a2} = P_2 h_3 + P_2\frac{h_2 - h_3}{2} &= P_2\frac{h_2 + h_3}{2} \\ E_{a3} &= P_3\frac{h_3}{2} \end{aligned} \tag{5.33}$$

Once the operating conditions—capacity and energy—are established, we can calculate the total cost corresponding to each power station as the sum of the fixed costs and the variable costs:

$$TC_1 = C_1 P_1 + c_1 E_{a1} = C_1 P_1 + c_1 \left[D_{\min} \frac{h_1 - h_2}{2} + P_1 \frac{h_1 + h_2}{2} \right]$$

$$\vdots$$

$$TC_i = C_i P_i + c_i E_{ai} = C_i P_i + c_i \left[P_i \frac{h_i + h_{i+1}}{2} \right] \tag{5.34}$$

$$\vdots$$

$$TC_n = C_n P_n + c_n E_{an} = C_n P_n + c_n \left[P_n \frac{h_n + h_{n+1}}{2} \right]$$

The total minimum system cost is:

$$TC = \sum_{i=1}^{n} TC_i = c_1 D_{min} \frac{h_1 - h_2}{2} + \sum_{i=1}^{n} \left[C_i P_i + c_i P_i \frac{h_i + h_{i+1}}{2} \right] \tag{5.35}$$

These are the costs that the central planning agent or single buyer must annually pay to the n power stations composing the electricity generating system in the situation of optimal minimum cost.

Regarding our example, the relevant costs are:

$$\begin{aligned} TC_1 &= C_1 P_1 + c_1 E_{a1} \\ TC_2 &= C_2 P_2 + c_2 E_{a2} \\ TC_3 &= C_3 P_3 + c_3 E_{a3} \\ TC &= \sum_{i=1}^{3} TC_i \end{aligned} \tag{5.36}$$

The situation described corresponds to an optimal situation, which is considered in traditional regulation. In a balanced system, technologies are dispatched according to the marginal costs ascending order (merit order effect) and the operating hours are obtained to achieve the minimum total system cost.

Nowadays, this is not the situation. Electricity is traded in a market, called a marginalist market. What is this? It is supposed that the suppliers bid at their marginal cost. The selling offers (supply) are upward ordered and the buying offers (demand) are downward ordered. Where the supply and demand curves meet the market price is defined. We note that the market price is established by the highest supply bid that is necessary to supply the demand. All the accepted producers receive the market price and all the accepted consumers pay the market price.

This is how it goes nowadays, optimal costs are no longer computed. Nevertheless, optimal costs are a benchmark that serves as a reference when performing a comparison with the results actually obtained from the regulation of the power system or, primarily, from the market.

5.2.4 Cost Recovery in the Market

Let us consider a situation in which the system is subject to the spot market clearing price which coincides with the marginal cost of the system. In this case, all the operating generators are remunerated by the variable cost of the marginal generator (the one with the higher marginal cost that is necessary to satisfy the load).

One must now compute the annual electricity supplied by each power plant in each marginal cost time interval, E'_{aij} (MWh). A matrix formulation may help in this regard.

$$\left[E'_{\text{aij}}\right] = \begin{bmatrix} (P_1 + D_{\min})\left(\frac{h_1-h_2}{2}\right) & \cdots & P_1\left(h_i - h_{i+1}\right) & \cdots & P_1\left(h_n - h_{n+1}\right) \\ \vdots & \vdots & \vdots & \vdots & \vdots \\ 0 & \cdots & P_i\left(\frac{h_i-h_{i+1}}{2}\right) & \cdots & P_i\left(h_n - h_{n+1}\right) \\ \vdots & \vdots & \vdots & \vdots & \vdots \\ 0 & \cdots & 0 & \cdots & P_n\left(\frac{h_n-h_{n+1}}{2}\right) \end{bmatrix} \tag{5.37}$$

The element (i, i) of this matrix is the electricity supplied by power plant i, in the time period in which power plant i is the marginal power plant. The element (i, j), $i \neq j$, is the electricity supplied by power plant i when the marginal power plant is power plant j.

In the market situation, the annual revenue, R_i ($), of power plant i is given by:

$$R_i = \sum_{j=1}^{n} c_j E'_{aij} \tag{5.38}$$

Or, putting it in another way:

$$R_1 = c_1\left[(P_1 + D_{\min})\left(\frac{h_1 - h_2}{2}\right)\right] + \sum_{j=2}^{n} c_j\left[P_1\left(h_j - h_{j+1}\right)\right]$$

$$\vdots$$

$$R_i = c_i\left[P_i\left(\frac{h_i - h_{i+1}}{2}\right)\right] + \sum_{j=i+1}^{n} c_j\left[P_i\left(h_j - h_{j+1}\right)\right] \tag{5.39}$$

$$\vdots$$

$$R_n = c_n\left[P_n\left(\frac{h_n - h_{n+1}}{2}\right)\right]$$

The total annual revenues, TR ($), provided for the n system power plants are:

$$TR = \sum_{i=1}^{n} R_i \tag{5.40}$$

Let us retake again our example with three generating technologies. With reference to Fig. 5.3, between $t = h_2$ and $t = h_1 = 8760h$, only generator#1 is operating. Therefore, the system marginal cost is c_1 and the energy supplied by the generator in this period is (see Eq. 5.37 and Fig. 5.3):

$$E_{a11} = (D_{\min} + P_1)\frac{h_1 - h_2}{2} \tag{5.41}$$

Between $t = h_3$ and $t = h_2$, generator#1 and generator#2 are operating. Both generators are remunerated by the marginal cost equal to c_2 because $c_2 > c_1$. During this time interval, the energy supplied by generator#1 is:

$$E_{a12} = P_1(h_2 - h_3) \tag{5.42}$$

and the energy supplied by generator#2 is:

$$E_{a22} = P_2\frac{(h_2 - h_3)}{2} \tag{5.43}$$

Finally, between $t = h_4 = 0$ and $t = h_3$, the three generators are running. In this period, all three generators are remunerated by the marginal cost of the most expensive generator, c_3, because $c_3 > c_2 > c_1$. The energies supplied by the three generators in this period are:

$$\begin{aligned} E_{a13} &= P_1(h_3 - h_4) \\ E_{a23} &= P_2(h_3 - h_4) \\ E_{a33} &= P_3\frac{(h_3 - h_4)}{2} \end{aligned} \tag{5.44}$$

The income of each one of the three generators is:

$$\begin{aligned} R_1 &= c_1 E_{a11} + c_2 E_{a12} + c_3 E_{a13} \\ R_2 &= c_2 E_{a22} + c_3 E_{a23} \\ R_3 &= c_3 E_{a33} \end{aligned} \tag{5.45}$$

At this point, one question arises. Do the producers in the market make a profit, i.e., $R_i > TC_i$? For each power plant i, the difference between the minimum production cost and the revenue provided by the market is given by:

$$\Delta CR_i = TC_i - R_i = C_i P_i + c_i P_i h_{i+1} - \sum_{j=i+1}^{n} c_j \left[P_i \left(h_j - h_{j+1}\right)\right];\; i = 1, \ldots, n \tag{5.46}$$

In our example, it is:

$$
\begin{aligned}
\Delta CR_1 = TC_1 - R_1 = C_1 P_1 + c_1 \left(P_1 h_2 + (D_{\min} + P_1)\frac{h_1 - h_2}{2} \right) - \\
c_1 \left((D_{\min} + P_1)\frac{h_1 - h_2}{2} \right) - c_2 P_1 (h_2 - h_3) - c_3 P_1 (h_3 - h_4) \\
= C_1 P_1 + c_1 P_1 h_2 - c_2 P_1 (h_2 - h_3) - c_3 P_1 (h_3 - h_4) \\
\Delta CR_2 = TC_2 - R_2 = C_2 P_2 + c_2 P_2 h_3 - c_3 P_2 h_3 \\
\Delta CR_3 = TC_3 - R_3 = C_3 P_3
\end{aligned}
\tag{5.47}
$$

Are these differences positive (loss) or negative (profit)? We cannot tell, it depends on market prices, but, for most producers, they should be negative, otherwise, they would not remain in the market. One thing we can tell for sure: generator#3 is at a loss because $\Delta CR_3 > 0$, this generator is unable to recover its fixed costs.

Profits are assured for generator#1 and generator#2 if:

$$
\begin{aligned}
C_1 < (c_2 - c_1) h_2 - (c_3 - c_2) h_3 \\
C_2 < (c_3 - c_2) h_3
\end{aligned}
\tag{5.48}
$$

Example 5.3

The load duration curve of a country is given by $D(h) = 22{,}000 - 1.4\,h$. D is the system load and h represents the number of hours that the system load is above a specific load level D (in MW). To satisfy the consumption of the country, we have two technologies, whose data is given in Table 5.4.

1. Determine the average cost of electricity in the country, in \$/MWh, for a balanced optimal system.
2. Assume that the consumption of the country, as well as the capacities of technologies #1 and #2 remain unchanged as computed in (1). Under these conditions, wind technology and solar PV technology are introduced in the generating system.

   The characteristics of wind and solar PV technologies are as follows in Table 5.5.

   Assume that wind and solar PV production have a uniform distribution throughout the year. Suppose that the costs incurred by the different technologies are recovered through a marginalist market, where it is assumed energy offers and market prices are based on marginal costs

   Determine the new average cost of electricity in the country, in \$/MWh.

◄

Table 5.4 Fixed cost and variable cost for Example 5.3

Technology	Fixed cost ($/MW/year)	Variable cost ($/MWh)
#1	240,000	20
#2	80,000	60

Table 5.5 Renewables data for Example 5.3

Technology	Capacity (MW)	Capacity factor
Wind	6,000	0.4
Solar PV	2,000	0.2

Solution

1.

We begin by determining the operating hours of each technology. Technology #1 has the lowest variable cost, so it is the baseline technology. Therefore:

$$h_1 = 8760 \text{ h}$$

The operating hours of technology #2 are:

$$h_2 = \frac{C_1 - C_2}{c_2 - c_1} = 4000 \text{ h}$$

Then, we compute the optimal installed capacity of each generator.

$$P_1 = D(h_2) = 16{,}400 \text{ MW}$$

$$P_2 = D_{\max} - P_1 = 5600 \text{ MW}$$

Now, it is time to compute the electricity supplied by each generator.

$$E_{a1} = D_{\min}\frac{h_1 - h_2}{2} + P_1\frac{h_1 + h_2}{2} = 127.8 \text{ TWh}$$

$$E_{a2} = P_2\frac{h_2}{2} = 11.2 \text{ TWh}$$

Now, we can calculate the total generation costs as:

$$TC_1 = C_1P_1 + c_1E_{a1} = 6.49 \times 10^9\$$$

$$TC_2 = C_2P_2 + c_2E_{a2} = 1.12 \times 10^9\$$$

The average total cost for the balanced power system is:

$$ATC = \frac{TC_1 + TC_2}{E_{a1} + E_{a2}} = \frac{TC}{E_a} = 54.76 \text{ \$/MWh}$$

2.

We begin by determining the electricity supplied by the renewables.

$$E_{aW} = P_W C_{fW} 8760 = 21.02 \text{ TWh}$$

$$E_{aPV} = P_{PV} C_{fPV} 8760 = 3.5 \text{ TWh}$$

It is supposed that renewables offer at zero marginal cost, therefore they are always selected to participate in the market. As renewable production is uniformly distributed throughout the year, this corresponds to a shift down in the demand duration curve, therefore creating a residual demand duration curve, i.e., the demand that must be supplied by the non-renewable technologies.
The residual demand is a straight line with the same slope as the demand curve and with a y-intercept given by:

$$b_r = D_{\max} - P_W C_{fW} - P_{PV} C_{fPV} = 19{,}200 \text{ MW}$$

We note that, for instance, in the case of wind technology:

$$P_{avgW} = P_W C_{fW}$$

where P_{avgW} is the wind average power supposed constant throughout the year.
The situation is illustrated in Fig. 5.4 where the original demand (blue) and the residual demand (orange) are depicted.
Now, we need to dispatch the non-renewable technologies according to the residual demand.

$$D_r = 19{,}200 - 1.4h$$

The first thing is to compute the new operating hours, as follows (looking at Fig. 5.4 will help).

$$h'_1 = 8760 \text{ h}$$

$$P_1 = D_r(h'_2) \rightarrow h'_2 = 2000 \text{ h}$$

The installed capacities remain unchanged, so the operating hours must change because the load that is to be supplied is lower.
In the time interval $h'_1 - h'_2 = 8760 - 2000$ h, three technologies are in the market: generator#1, wind and solar PV. All of them are paid at $c_1 = 20$ €/MWh, because this is the highest marginal price—wind and solar PV offer at zero. The energy supplied by each one is:

$$E_{a11} = (P_1 + D_{r\min}) \frac{h'_1 - h'_2}{2} = 78.88 \text{ TWh}$$

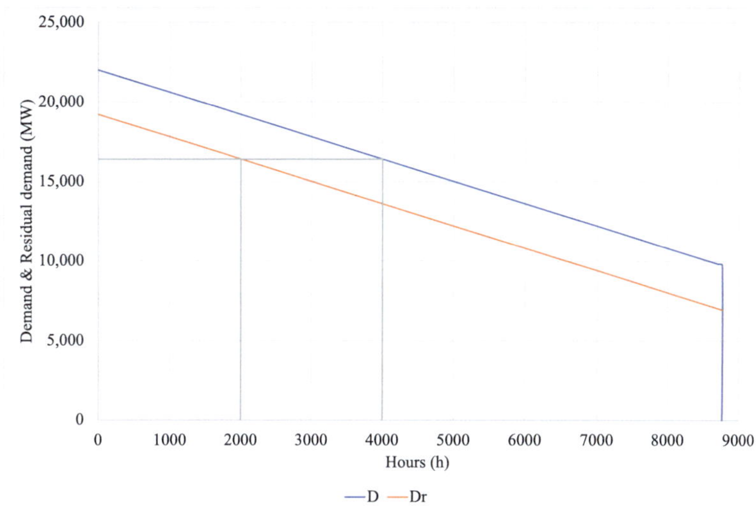

Fig. 5.4 Original demand and residual demand. Example 5.3

$$E_{aW1} = P_W C_{fW}(h'_1 - h'_2) = 16.22 \text{ TWh}$$

$$E_{aPV1} = P_{PV} C_{fPV}(h'_1 - h'_2) = 2.70 \text{ TWh}$$

In the time interval $h'_2 - h'_3 = 2000 - 0$ h, all technologies are in the market. All of them are paid at $c_2 = 60$ €/MWh, because this is the highest marginal price. The energy supplied by each one is:

$$E_{a12} = P_1 h'_2 = 32.80 \text{ TWh}$$

$$E_{a22} = (D_{r\max} - P_1)\frac{h'_2}{2} = 2.80 \text{ TWh}$$

$$E_{aW2} = P_W C_{fW} h'_2 = 4.80 \text{ TWh}$$

$$E_{aPV2} = P_{PV} C_{fPV} h'_2 = 0.80 \text{ TWh}$$

We point out that generator#2 is operating at:

$$P'_2 = D_{r\max} - P_1$$

The revenue of each producer is:

$$R_1 = c_1 E_{a11} + c_2 E_{a12} = 3.55 \times 10^9 \text{ \$}$$

$$R_2 = c_2 E_{a22} = 0.17 \times 10^9 \text{ \$}$$

$$R_W = c_1 E_{aW1} + c_2 E_{aW2} = 0.61 \times 10^9 \text{ \$}$$

$$R_{PV} = c_1 E_{aPV1} + c_2 E_{aPV2} = 0.10 \times 10^9 \ \$$$

The total revenue, which is equal to the total cost to be supported by the consumers, is:

$$R = TC' = \sum_{i=1}^{4} R_i = 4.43 \times 10^9 \ \$$$

and the ATC is:

$$ATC' = \frac{TC'}{E_a} = 31.86 \text{ €/MWh}$$

◄

5.2.5 A Note on Risk in Generation Costs

When generation costs are regulated (optimal power system), the producer's risk is minimised, as it is virtually reduced to the technical operation of the power station. The producer will receive always the fixed cost (including an adequate remuneration of the capital) and will receive the variable cost as well providing it produces. In this way, the risks arising from uncertainty in fuel prices, carbon emissions, oversizing of the generating system, the introduction of non-dispatchable sources of energy, and so on, do not affect the regulated producers: at the end of the day, all the risks are transferred to the consumers.

If production is non-regulated, i.e., if it is market-driven, the producers take risks: the technical operation risk and all the risks that, in the regulated system, were taken by the consumers.

The question of risk in generation costs is one of the most crucial aspects in determining the price of electricity. This is because the generation-related cost parcel is the most significant in the electricity overall cost and also because of the capital-intensive nature of generation investments, namely the baseline technology ones.

The issue of financing the baseline generation plants, namely, nuclear or even coal, puts all risk issues at stake: regulatory and policy, fuel costs, construction costs, operating performance, and organisation of the sector. The latter determines how risk is passed on to the various stakeholders: consumers, producers, traders, or equipment vendors.

5.3 Transmission Costs

The function of electrical networks is to ensure the flow of energy from the power stations to the final consumers. This task is subject to certain requirements, including the quality of service required by either the consumers or the secure operation of the power system itself.

The classification of electrical networks is typically based on the voltage levels that characterise them. The transmission network is designed to transport high amounts of energy over long distances. This energy transportation should be carried out with minimal losses. It is well-known that losses decrease with the square of the voltage. Therefore, the transmission network is operated in *Extremely High Voltage* (EHV) to assure reasonable losses. Typical EHV voltage levels include 400 kV, 220 kV, and 150 kV phase-to-phase voltages.

Most of the electrical power is transferred from the generators to the users via overhead lines suspended from large engineered pylons. These are ugly, but they are far cheaper than the alternative underground cables, which have to be provided with highly engineered insulation to withstand the extra high voltages and the arduous underground conditions. The suspended EHV overhead lines are generally based on steel cores surrounded by a layer of conducting aluminum. The overhead lines are generally supported from the arms of the pylons with engineered insulators.

Another important element of electrical networks is the transformer. The transformers are responsible for the voltage level change. For instance, inside the transmission network, a transformer is necessary to interconnect a 400 kV power line to a 220 kV one. Moreover, transformers are needed to interconnect the transmission network to the distribution network, which is operated at lower voltage levels.

Transformers are static electrical machines that "transform" one voltage level into another and that work both ways. A transformer is composed of two windings—a primary and a secondary—that are magnetically connected through an iron core. When power (roughly voltage times current) is supplied in the primary, it passes to the secondary through the magnetic link, apart from some minor transformer losses. Depending on the number of turns in the primary and secondary windings, the secondary voltage may be higher or lower than the primary's voltage. As a transformer is not a generator, the power in the primary must match the power in the secondary, apart from said minimal transformer losses. Therefore, if the voltage in the primary is lower than the voltage in the secondary, it turns out that the current in the primary is higher than the current in the secondary. The operating principle is the same whether the power is supplied in the primary or in the secondary.

One additional role of the transmission network is to establish electrical interconnections between countries. All the European transmission grid is interconnected, therefore improving the reliability of the whole European power system. Interconnections are of utmost importance for a bunch of reasons, the main ones being:

- Interconnections allow energy markets involving several countries to take place. For instance, Portugal and Spain are strongly interconnected, therefore allowing a single common energy market to be operated.
- Interconnections allow countries to be rescued in case of a serious fault in their power systems, for instance, a sudden loss of a power plant, therefore avoiding blackouts.
- Interconnections allow for a reduction in the operating reserves needed to assure the secure operation of the power system.

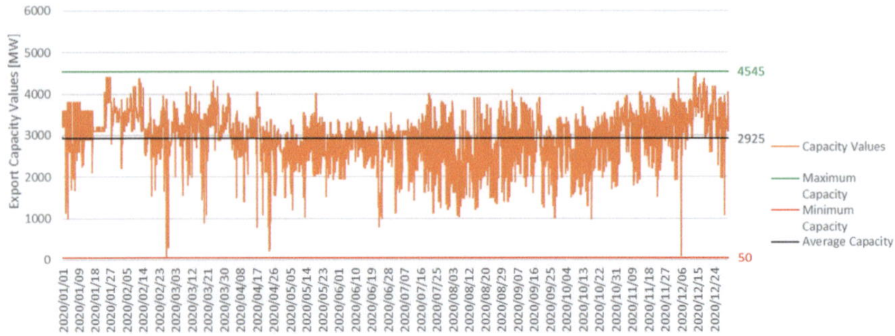

Fig. 5.5 Hourly ATC from Portugal to Spain in 2020. *Source* ERSE (Portuguese energy regulator)

The *Available Transfer Capacity* (ATC) at the interconnections is important for the operation of the markets. The ATC is made available by the grid operator considering the foreseen generation and load and the grid constraints imposed by electrical laws (the Kirchoff Laws). Moreover, the maximum capacity that can be carried by the lines depends on the ambient temperature where they are installed. Altogether, this means that the ATC changes a lot over the year. In the limit, it can be zero, due to technical restrictions.

Figure 5.5 shows the hourly ATC from Portugal to Spain in the year 2020.

Due to its central role in the power system, the transmission system is responsible for the overall integration and development of the power system, taking into account the global foreseen developments of both the generation and the demand.

The planning of the transmission network provides an overview of the evolution of the global power system and must be periodically updated. This is recognised in most European countries, which establish the obligation of preparation and publication of the transmission system development plan by the Transmission System Operator (TSO). In general, one can say that in Europe the transmission network development plans have the following aims:

- Network connection of new power stations.
- Power reinforcement to the distribution network or large consumers.
- Interconnection capacity reinforcement.
- Optimising the operation of the power system.
- Consideration of the energy policy objectives established by the Government.

The transmission network development plan implies investments to be made in the future, which have consequences for the economic regulation of the company that manages the transmission network.

As highlighted before, the transmission grid is the heart of the power system. Its failure has dramatic consequences leading to blackouts. Therefore, the transmission network is operated in closed mesh for security reasons and to improve reliability. The failure of any one of its elements (line, cable, or transformer) must not lead to a failure in consumers' supply. This feature of the transmission system is called the *N-1 criterion*.

The inputs and outputs of the transmission system, which can be inside the grid or in the interface with the generators or with the distribution grid, are generally concentrated in nodes, materialised by transformer substations. The high cost of each transmission system project requires the mobilisation of complex calculation models to ensure the proper functioning and security of equipment and people. Furthermore, the quality of service is quite rigorous as reflected by the N-1 criterion and even N-2, for more demanding situations. Examples of said complex assessment models are:

- Load flow, which is broadly the computation of the voltages (magnitude and angle) at every single node of the grid and the active and reactive power flowing in each line, cable, and transformer. This calculation is performed in steady-state operation of the grid and aims at verifying if the operational limits of voltage and power flow are respected.
- Transient behaviour, which is the determination of the time evolution of selected quantities of interest (voltage, current, frequency, and so on) following some disturbances in the grid, like short circuits and generation or load steps.
- Short-circuit currents, which is the computation of the maximum short-circuit currents at every single node of the grid, with the objective of sizing the protection equipment that eliminates the short-circuit currents.
- Studies under contingency, which is the assessment of the grid behaviour when one or more grid elements are unavailable, whether it is by maintenance or following some disturbance.

The transmission network is managed and operated by a company named the *Transmission System Operator* (TSO). The transmission system does not buy or sell energy. It just receives energy from the large power plants, transports it, and delivers it to the distribution grid. For this transmission service, the TSO receives an income, that takes the form of a regulated tariff, paid by the users, which are the consumers. The TSO can be a private company, but it is not a normal private company because its activity (and therefore, its profit) is regulated by an independent energy regulator.

The transmission network is one of the power system's sectors that is considered a natural monopoly, therefore, it is subject to economic regulation, as mentioned. We say that we have a natural monopoly when the service (in this case, the electric transmission service) is cheaper if provided by one company rather than by two or more. Due to the extremely high voltages involved in the transmission network, the related investments are extremely high. Therefore, it makes no sense to have two identical transmission systems, side-by-side, competing one with the other to provide the transmission service.

Tariffs to be paid by the consumers for the use of the transmission network are based on average costs and not marginal costs as was the case of generation, as seen before.

The investment in the transmission grid elements (mainly, overhead lines, underground cables, and transformers, but also the control and protection equipment) are hugely heavy. Just to serve as a mere example, a 400 kV overhead line cost can be as high as half a million euros per kilometre; this value rises to a couple of million if a 220 kV underground cable is considered; a top-of-the-line transformer with active power regulation capabilities (called a phase-shift transformer) can cost more than ten million; a regular 400/220 kV transformer costs half of this.

5.4 Distribution Costs

The distribution network is intended to deliver electricity to consumers. The distribution network establishes the link to final consumers: domestic, services, or industrial.

The distribution network is fed from the transmission grid through transformer substations that transform the EHV level to the *High Voltage* (HV) voltage level of 60 kV. This part of the distribution grid that is connected to the transmission grid is called the *High Voltage distribution grid*. The *Medium Voltage distribution grid* (MV) is regionally defined by voltage levels of 10, 15, and 30 kV, for example; it is fed from the HV distribution network through substations and provides power to the HV/MV transformer stations. The *Low Voltage distribution grid* (LV) is characterised in Europe by the voltage level 400/230 V (phase-to-phase voltage/line-to-neutral voltage), being fed through MV/LV transformer stations.

In what regards the final consumers of electricity (domestic, services, or industrial), they are separated according to the respective power they require. Generally, the higher the required power, the higher the voltage. In general, domestic consumers are connected in LV, services are connected in MV, and industrial consumers are connected in HV.

To manage and operate the grid, there is a *Distribution System Operator* (DSO). The DSO may be a private company, which is regulated by the energy regulator. In the same way that in the transmission grid case, distribution networks are a sector of the power system that is considered a natural monopoly. That is why it is subject to economic regulation. This means that their operating costs and investments, plus a regulated rate of return, are recovered through a tariff, fixed by the regulator, for the use of the distribution grid. This tariff is paid by the consumers.

Again, it is stressed that the DSO does not buy nor sell energy. It distributes the energy it receives from the transmission grid and from the renewable distributed generation to the end-users. For this distribution service, the DSO is paid through a regulated tariff.

This brief description of the distribution network operation suggests that energy flows from the level of higher voltages to the lower voltage levels. This is true for most of the energy carried over, however, as distributed generation, mainly renewable, is directly connected to the distribution grid, this may cause reverse power flow. This is an undesirable situation that needs to be avoided by the DSO.

In the past, undervoltages were a matter of great concern for the DSO, as the power absorbed by the consumers caused voltage drops. The voltage was regulated by the transformers and, in some cases, capacitors were required to inject reactive power and increase the voltage. Today, small dispersed production is spread all over the distribution grid which can cause the inverse phenomena—overvoltages. To fight against this new problem, transformers with regulating capacities are used as well as the voltage regulating capabilities of the power electronics devices that interface some renewables with the grid. This means that, despite the challenge, there are solutions for these new problems that arise from the connection of dispersed renewable generation to the distribution grid.

The cost of each distribution grid element is much cheaper than the equivalent in the transmission grid. This is because the voltage levels are lower. However, the distribution grid is much denser than the transmission grid, which implies a higher number of lines and transformers are required. As an example, an overhead line with a 25 mm^2 cross section capable of carrying 100 Amps may cost 7,000 €/km, while an underground cable, 35 mm^2, 130 Amps, costs 5 times more. A 100 kVA transformer may cost 10,000 €. The reader is invited to compare these values with the equivalent ones in the transmission grid. In general, the revenues of the DSO are higher than the revenues of the TSO, not because of the unitary price, but because of the quantity.

5.4.1 Kelvin Law

Distribution planning aims at choosing the project that meets the forecast demand at the lowest total discounted cost, considering investment, operation and maintenance, and network losses.

According to a rule known as *Kelvin law*, an electrical line is well designed when Joule losses discounted cost is equal to the installation cost, which depends on the conductors' cross section, as we will see right away.

The total cost per unit length, TC_i, of each line or branch, i, of the network, can be represented by:

$$TC_i = C_{0i} + C_{vi} V_i + C_{si} s_i + C_{omi} + C_{Li} \frac{I_i^2}{s_i} \tag{5.49}$$

The first three parcels correspond to the cost of cable installed and ready to use, and the last two parcels correspond, respectively, to the discounted cost of operation and maintenance and the cost of losses by Joule effect, discounted to the date of begin of the operation of the line.

The variables in Eq. 5.49 are:

- V is the operating voltage of the line (V).
- s is the section of the conductor, (mm^2).
- I is the current in the peak period, in the first year of operation (A).
- $C_0, C_v, C_s, C_{om}, C_L$ are coefficients.

Kelvin's law seeks to find the value of the section, s_i, which minimises TC_i, as depicted in:

$$\frac{\partial TC_i}{\partial s_i} = 0 \rightarrow C_{si} s_i = C_{Li} \frac{I_i^2}{s_i} \tag{5.50}$$

As we wanted to show, the most economical section is found when the installation cost equals the cost of losses.

5.4.2 The Penalty for Excessive Reactive Power Consumption

The reactive power flow increases the losses in the networks and decreases the capacity, expressed in kW, of generation, transmission, and distribution in the power system. Industrial clients are heavy reactive power consumers because they use induction motors that consume reactive power.

Let us recall that (ϕ is the angle between voltage and current, Q is the reactive power (kvar) and P is the active power (kW)):

$$\tan \phi = \frac{Q}{P} \tag{5.51}$$

In peak hours, customers do not pay for reactive power consumption, while the amount of reactive power (kvar) they absorb is below a threshold. In most countries, this threshold is equal to 30% of the active power (kW) consumed in the same period ($\tan \phi = 0.3$). For $\tan \phi > 0.3$, the tariff increases by steps: the higher $\tan \phi$, the higher the charges for reactive power consumption.

On the contrary, in off-peak hours, the consumption of reactive power is free, but the injection of reactive power is charged. This is due to the capacitive behaviour of power lines during off-peak hours, therefore any injection of further reactive power would increase the voltage profiles to unacceptable levels.

In the face of this, locally performing reactive power compensation (also called *power factor compensation*) by installing capacitor banks is easy to achieve and much cheaper than paying the penalties for excess reactive power absorption. This is why the clients' charging of reactive power by the distribution grid is nowadays a small percentage of the total electricity sales.

The procedure outlined above applies to services and industrial consumers. The LV domestic consumers are not expressly charged for the reactive power they absorb. But, as they are controlled in terms of the apparent power they consume, they are indirectly charged by the inherent reduction of active power. Let us recall that (S is the apparent power, in kVA):

$$S = \sqrt{P^2 + Q^2} \tag{5.52}$$

5.5 Environmental Costs

Power systems are essential for modern society, providing energy for homes, businesses, and industries. However, these systems also have significant environmental costs that must be addressed. These costs can manifest in a variety of ways, including greenhouse gas emissions, habitat destruction, and pollution.

Environmental costs in power systems are an intricate web of interconnected factors that arise from the generation, transmission, and consumption of energy. These costs encompass a wide range of environmental impacts, from the extraction of raw materials to the emissions released during power generation and the long-term effects on ecosystems and climate change.

One of the most significant environmental costs of power systems is their contribution to climate change. Traditional power sources, such as coal and natural gas, emit large quantities of carbon dioxide (CO_2) into the atmosphere during combustion. This carbon dioxide traps heat in the atmosphere, leading to global warming and its associated impacts, such as rising sea levels, extreme weather events, and changes in ecosystems.

In addition to carbon dioxide emissions, traditional power sources also release other pollutants into the air, such as sulfur dioxide (SO_2) and nitrogen oxides (NOx). These pollutants can contribute to acid rain, smog, and respiratory problems, among other health impacts.

Nuclear power, while relatively clean in terms of greenhouse gas emissions, has its own set of environmental costs. The mining and processing of uranium and other nuclear fuels can cause environmental damage, such as the contamination of water sources. The radioactive waste produced by nuclear power plants is also a significant concern, as it can remain dangerous for hundreds of thousands of years.

Renewable energy sources, such as solar and wind power, are often heralded as cleaner alternatives to traditional power sources. However, these sources also have environmental costs. Solar panels and wind turbines require large amounts of energy and resources to manufacture, and the mining and processing of rare earth minerals used in their production can cause environmental damage. Solar farms and wind turbines can also disrupt wildlife habitats and cause other ecological impacts.

Hydropower is another renewable energy source that has been used for decades, but it also has environmental costs. Large dams can have significant impacts on river ecosystems, altering the natural flow of water and impacting fish populations. Additionally, the creation of reservoirs can cause the displacement of communities and the loss of cultural heritage sites.

Transmission and distribution networks are essential components of power systems, but they also have environmental consequences. The construction of power lines and infrastructure can result in land clearance, habitat fragmentation, and the disturbance of natural landscapes. Additionally, transmission losses during electricity distribution contribute to inefficiencies in the system and the need for additional power generation, further exacerbating environmental impacts.

Carbon licenses and *emissions trading systems* (ETS) are powerful tools used to charge carbon emissions in power systems, encouraging a transition towards cleaner and more sustainable energy generation. These mechanisms are designed to internalise the environmental costs associated with greenhouse gas emissions, particularly CO_2, and incentivise power plants to reduce their carbon footprint.

Carbon licenses, also known as carbon permits or allowances, are a system where power plants are allocated a specific quantity of carbon emissions they are allowed to release. Each license represents a defined amount of CO_2 emissions. If a power plant exceeds its allocated license, it must purchase additional licenses from facilities that have reduced their emissions below their allocated limit. This creates a financial incentive for power plants to invest in emission reduction measures, improve efficiency, or adopt cleaner energy sources.

On the other hand, emissions trading systems (ETS), also known as cap-and-trade systems, set an overall cap on total emissions within a jurisdiction. The cap is divided into emission allowances, which can be bought, sold, or traded among power plants and other participants. ETS allows more flexibility in reducing emissions, as power plants that can achieve emissions reductions more easily can sell their excess allowances to those facing challenges in meeting their emission targets. This market-based approach encourages emissions' reductions at the lowest cost while maintaining an overall cap on emissions.

These mechanisms create a financial cost for emitting carbon dioxide, reflecting the environmental impact of greenhouse gas emissions. By placing a price on carbon, carbon licenses and ETS incentivise power plants to invest in cleaner technologies, improve energy efficiency, or shift towards low-carbon or carbon-free energy sources. Power plants facing higher costs due to emissions may choose to reduce their emissions or purchase additional licenses, making emission reduction a more economically viable choice.

Despite the significant environmental costs associated with power systems, there are steps that can be taken to mitigate these impacts. One key strategy is to transition to cleaner power sources, such as renewable energy and nuclear power with proper waste disposal protocols. Additionally, the implementation of energy-efficient technologies and practices can help reduce the overall demand for power, leading to less reliance on traditional power sources.

Carbon capture and storage (CCS) technologies can also help in this regard by capturing carbon dioxide emissions from traditional power sources and storing them in underground reservoirs. While still an emerging technology, CCS has the potential to significantly reduce greenhouse gas emissions from power systems.

The process of CCS begins with capturing CO_2 emissions from industrial processes, such as power generation, cement production, or steel manufacturing. Various capture technologies are employed, including post-combustion capture, pre-combustion capture, and oxyfuel combustion. Post-combustion capture removes CO_2 from flue gases after fuel combustion, while pre-combustion capture involves extracting CO_2 prior to fuel combustion. Oxyfuel combustion utilises oxygen instead of air to produce a flue gas predominantly composed of CO_2, simplifying the capture process. These technologies are continuously advancing to improve efficiency and reduce costs.

Once captured, the CO_2 is compressed and transported to a suitable storage site. Storage sites include deep saline aquifers, depleted oil and gas reservoirs, or unmineable coal seams. These geological formations provide secure and permanent storage options for CO_2, preventing its release into the atmosphere. The CO_2 is injected deep underground, and suitable seals and monitoring systems ensure its containment and prevent leakage. Over time, the CO_2 is naturally trapped and mineralised, further enhancing the long-term storage effectiveness.

However, challenges remain in the widespread adoption of CCS. One significant challenge is the high cost associated with the capture, transport, and storage of CO_2. Continued research and development, coupled with supportive policies and incentives, are crucial to reducing costs and improving the efficiency of CCS technologies.

Investments in research and development can also lead to advances in clean energy technologies, such as more efficient solar panels and energy storage systems, that have less of an impact on the environment. Similarly, smart grid technologies can help optimise the use of energy and reduce overall demand, leading to more sustainable power systems.

5.6 Conclusion

In the ever-evolving landscape of power systems, the intricacies of cost management are undeniably paramount. This chapter has journeyed through the multifaceted realm of costs in power systems, shedding light on various dimensions that collectively shape the present and future of our energy infrastructure.

Generation costs have shown us the importance of balancing affordability with sustainability, emphasising the need for smart resource allocation and continuous innovation. The concept of the levelized cost of energy (LCOE) has illuminated the path towards cost-effective and environmentally responsible energy generation, helping us make informed decisions about our energy mix. A graphical method to determine the generation costs in an optimally sized generation infrastructure was offered. The results are compared with what the producers receive in the market when they sell their electricity.

Transmission costs and the evolving landscape of Extremely High Voltage (EHV) lines and transformers remind us of the unseen arteries that channel power across vast distances. These infrastructural investments are essential for maintaining the reliability and efficiency of our grid, fostering a resilient energy future. An estimation of the transmission system cost, disaggregated by component, is offered as a rough figure to highlight the importance of proper transmission tariffs allowing transmission system operators to recover the high costs they incur.

The same procedure was followed regarding distribution system costs, where the costs of the components and O&M expenses justify the establishment of adequate tariffs to be paid by the consumers. Distribution costs, including the infrastructure of distribution lines and transformers, remind us that the last mile of the power journey is equally critical. Efficient distribution systems are the bedrock of reliable power supply for consumers.

Environmental costs and the looming shadow of CO_2 emissions have served as a stark reminder of the environmental toll of conventional energy sources. As we pivot towards greener alternatives, the imperative to reduce environmental costs and CO_2 emissions becomes paramount for a sustainable future.

In conclusion, the management of costs in power systems transcends mere financial considerations. It is about fostering resilience, sustainability, and reliability while safeguarding our environment.

5.7 Proposed Exercises

Problem 5.1

An entrepreneur wants to make an investment in wind power. Wind power is subject to a feed-in tariff (in €/MWh) and has an investment cost of 1100 €per installed kW. The annual O&M costs are estimated at 20 €/kW and the average capacity factor for this technology is 25%.

1. Assuming a required rate of return for the investment (discount rate) of 10% and a plant lifetime equal to the duration of the feed-in tariff (20 years) calculate the tariff level that will render the investment economically viable.
   Now, assume that the feed-in tariff is 75 €/MWh, and that 75% of the investment comes from a 20-year loan with an interest rate of 5% (the remaining 25% comes from equity).
2. Calculate the project NPV and IRR.
3. Calculate the equity NPV and IRR.
4. What if the interest rate of the loan is 10%?

Key Problem 5.1

1.

The LCOE is the discounted average production cost, i.e., it is the cost of producing one MWh, considering all the project's expenses, throughout the entire lifetime. For the project to be economically viable, the tariff level (selling price) should be, at minimum, equal to the LCOE.

The Levelized Cost Of Energy (LCOE) (€/MWh) is given by:

$$LCOE = \frac{(\alpha + \beta) I_0}{h_a} + c$$

where:

- $$\alpha = \frac{a(1+a)^n}{(1+a)^n - 1}$$

is the inverse of the discount factor, a is the discount rate and n is the project lifetime.

- β is the O&M cost as a percentage of the unitary investment, I_0.
- I_0 is the unitary investment (€/MW).
- αI_0 is the unitary annuity (€/MW) and βI_0 is the O&M unitary cost (€/MW).
- h_a is the utilisation factor (h). Note that the capacity factor is $a = h_a/8760$.
- $c = c_f + c_{CO_2}$ is the variable cost (€/MWh), composed by the fuel cost, c_f, and the CO_2 cost, c_{CO_2}.

Given the data supplied, we obtain:
$\alpha = 0.1175$, $\beta = 0.0182$, $h_a = 2190$ h, $c = 0$ and $LCOE = 68.13$ €/MWh.

As so, the minimum tariff level that renders the project economically viable is 68.13 €/MWh. If the selling price is higher than this value, the project is viable in the long run.

2.

The project NPV and IRR look at the profitability of the project itself and not where the investment's money comes from.

Considering that the investment is totally made in $t = 0$ and that the revenues and expenses are constant and equal each year of the project's lifetime, the NPV (Net Present Value) (€) and IRR (Internal Rate of Return) (%) are given by:

$$NPV = (R - E)k_a - I_T$$

$$IRR^{(k+1)} = \frac{R-E}{I_T}\frac{(1+IRR^{(k)})^n - 1}{(1+IRR^{(k)})^n}$$

where:

- R are the annual revenues (€)
- E are the annual expenses (namely, O&M expenses) (€)
- $R - E = CF$, CF being the annual cash-flow
- $k_a = 1/\alpha$ is the discount factor
- I_T is the total investment (€).

We note that, in this problem, we can use unitary values (€/kW) because all monetary data are given in (€/kW), the installed capacity being unknown. In this case, the NPV is given in € per kW and not in €, as usual.

With the problem's data, we have (FiT is the Feed-in Tariff):

$$CF = (FiTh_a - c_{O\&M}) = 144.25 \text{ €/kW}, I_T = I_0 = 1100 \text{ €/kW}, NPV = 128.08 \text{ €/kW}$$

As can be seen, the IRR's equation is a non-linear equation that must be solved by iterative methods. The given formula uses the Gauss method. We need an initial guess: the discount rate is an appropriate choice.

Starting with $IRR^{(0)} = 10\%$, after 4 iterations, the error is less than 0.03%, indicating that the iterative process has converged. The result is $IRR = IRR^{(4)} = 11.7\%$.

3.

The equity NPV and IRR consider the origin of the investment's money. In this case, we have a loan (75% of the investment), with the respective interest rate (5%) and depreciation period (20 years). To repay the loan, an annual payment, called a loan annuity, is due. The loan annuity, which is an additional annual expense to be added to the O&M expenses, is:

$$A_{loan} = \alpha_{loan} 75\% I_T = 66.20 \text{ €/kW}$$

We can now compute the equity NPV, using the following formula:

$$NPV_e = (FiTh_a - A_{loan} - c_{O\&M})k_a - 25\% I_T = 389.48 \text{ €/kW}$$

As for the IRR, the same reasoning is to be applied. The annual expenses increase by the loan annuity and the initial investment is reduced to the equity. After 4 iterations, the error is less than 0.01%. Therefore, $IRR_e = IRR_e^{(4)} = 28.2\%$.

The equity NPV and IRR are considerably higher than the project NPV and IRR. This is explained by a loan obtained in favourable conditions. The interest rate of the loan is 5%, whereas the discount rate (or the project's minimum rate of return) is 10%.

4.

Given the reasoning presented above, one could think that, if the loan's interest rate is 10%, the project and equity NPV and IRR would be equal. This holds true for the NPV, which is the summation of the discounted cash flows. At the end of the day, it is the same thing, except for a reordering of the terms.

The case of the IRR is different. The relationship is non-linear, and the IRR computation is sensible to the cash-flow sequence. The lower initial investment makes the equity IRR to be higher than the project IRR. In fact, it is $IRR_{e1} = IRR_{e1}^{(4)} = 16.4\% > IRR$.

Problem 5.2[3]

To supply a load, three technologies are available: baseline (#1), intermediate (#2) and peak (#3). These technologies feed a hypothetical electric system, characterised by the monotonous load duration curve given by:

$$D = 22{,}000 - 1.37h$$

D is the system load and h represents the number of hours that the system load is above a specific load level D (in MW). Table 5.6 contains the generation costs.

[3] Problems 5.2, 5.3 and 5.4 are adapted from P. Joskow 'Capacity Payments in Imperfect Electricity Market: Need and Design", Utilities Policy, 2008, https://doi.org/10.1016/j.jup.2007.10.003.

Table 5.6 Fixed and variable costs data. Problem 5.2

Generation technology	Fixed cost ($/MW/Year)	Variable cost ($/MWh)
Baseline (#1)	240,000	20
Intermediate (#2)	160,000	35
Peak (#3)	80,000	80

Calculate:

1. The installed capacity (MW) of the different technologies to minimise the total system production cost.
2. The total production cost ($) and the average production cost ($/MWh) for each technology.
3. Calculate the total production cost and the average production cost for the whole system.

Key Problem 5.2

1.

The graphic construction illustrated in Fig. 5.6 helps in finding a graphical solution for our problem.

The first step towards the solution to the problem is to determine the operating hours of each technology. We sometimes call this the utilisation of each technology but keep in mind that this has nothing to do with the utilisation factor that we have learned before. The utilisation is the number of operating hours, i.e., the number of hours the technology operates to supply power to the grid. We usually denote it by h.

At each time, the technology that should be operating is the cheapest one. This is the basic principle to determine the operating hours of each technology. Also, once a technology is launched, it will be left operating, even though a cheaper one is launched in the meantime. This is because we are talking about fossil fuel-fired power stations that have high start-up and shut-down costs.

Looking at the right-hand side of the upper graphic in the figure, we conclude that Baseline#1 technology is the cheapest one, until some point, when Intermediate#2 becomes cheaper. Therefore, Baseline#1 technology must be present during the 8760 hours of the year, i.e., $h_1 = 8760$ h. Now, we must compute the point where Intermediate#2 becomes cheaper. This is the interception point between the cost lines of Baseline#1 and Intermediate#2.

$$TC_{01} = C_1 + c_1 h_2$$

$$TC_{02} = C_2 + c_2 h_2$$

where: C_i is the fixed cost and c_i is the variable cost.

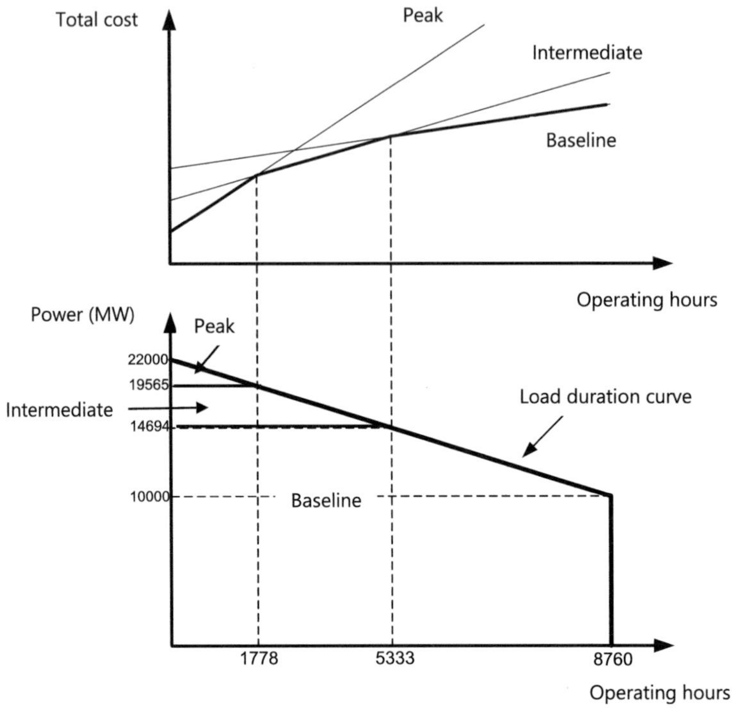

Fig. 5.6 Graphical construction. Problem 5.2

Applying this to our data, at the interception point $TC_{01} = TC_{02}$, we obtain $h_2 = 5333.33$ h. Using the same reasoning for the interception point between Intermediate#2 and Peak#3, we obtain $h_3 = 1777.78$ h.

The next step is to compute the installed capacities of each technology. To do that, we now look at the lower graphic of the figure and conclude that:

$$P_1 = D(h_2) = 14{,}693.33 \text{ MW}$$

$$P_2 = D(h_3) - P_1 = 4871.11 \text{ MW}$$

$$P_3 = D_{\max} - P_2 - P_1 = 2435.56 \text{ MW}$$

In these circumstances, it can be stated that the system is balanced, that is, the system has the necessary and sufficient capacity, and this capacity is adequately distributed by each available technology so that the total system production cost is minimal.

2.

Now, we proceed to the computation of the energy supplied by each technology. This is the area below the load diagram of each technology. These areas are marked in the figure by the name of the respective technology. The results are as follows:

$$E_{a1} = (P_1 + D_{\min})\frac{(h_1 - h_2)}{2} = 120{,}670{,}300 \text{ MWh}$$

$$E_{a2} = P_2 h_3 + P_2\frac{(h_2 - h_3)}{2} = 17{,}319{,}506 \text{ MWh}$$

$$E_{a3} = P_3\frac{h_3}{2} = 2{,}164{,}938 \text{ MWh}$$

Finally, the production cost of each technology is given by:

$$TC_1 = C_1 P_1 + c_1 E_{a1} = 5.940 \times 10^9 \$$$

$$TC_2 = C_2 P_2 + c_2 E_{a2} = 1.386 \times 10^9 \$$$

$$TC_3 = C_3 P_3 + c_3 E_{a3} = 0.368 \times 10^9 \$$$

If the producers receive the above-mentioned monetary quantities, they would recover all the production costs (fixed and variable) and would earn a rate of return implicit in the fixed costs given data (we note that the given fixed costs are an annuity). And no more than that, no extra-profits are allowed.

The average production cost for each technology is:

$$ATC_i = \frac{TC_i}{E_{ai}}$$

$ATC_1 = 49.22$ \$/MWh, $ATC_2 = 80.00$ \$/MWh, $ATC_3 = 170.00$ \$/MWh

These average costs show that baseline technologies are the cheapest ones (that is why they operate the whole year) and peak technologies are expensive (that is why they seldom work, they are called only to account for peak demands).

3.

The total system cost is:

$$TC = \sum_{i=1}^{3} TC_i = 7.693 \times 10^9 \$$$

and the average total cost is:

$$ATC = \frac{TC}{\sum_{i=1}^{3} E_{ai}} = 54.89 \text{ \$/MWh}$$

Problem 5.3

For the same balanced system as in Problem 5.2, suppose now that the costs incurred by the different technologies are recovered through the market, where it is assumed that energy offers and market prices are based on marginal costs.

1. Calculate the revenue obtained through the market by each technology.
2. Compare the values obtained in Problem 5.2 with the ones from the previous question, by computing the respective differences.
3. Divide the values computed in the previous question by the individual installed capacity. What is the meaning of the value that you have found?

Key Problem 5.3

1.

We are now supposing that the 3 producers are going to offer in the market. The market is a marginalist market. This means that the market price is, at every hour, the price of the marginal technology. In other words, at every hour, the market price is the highest of the prices each technology is offering at that particular hour. We recall that, as this is a marginalist market, each technology offers, in each hour, at marginal cost (equal to the variable cost, as can be seen below).

$$TC_i = C_i P_i + c_i E_{ai}$$

$$MC_i = \frac{\partial TC_i}{\partial E_{ai}} = c_i$$

To follow the solution of this problem, it is recommended to look at Fig. 5.6. During $(h_1 - h_2)$ = (8760–5333.33) hours, Baseline#1 is the only technology offering in the market. It is offering at $c_1 = 20$ \$/MWh. The energy supplied during this period is the area below the load duration curve, corresponding to that period of time.

$$E_{a11} = (P_1 + D_{\min})\frac{h_1 - h_2}{2} = 42{,}305{,}855 \text{ MWh}$$

The amount received is $R_{11} = c_1 E_{a11} = 0.846 \times 10^9\$$.

Now, let us go to the 2nd period. During $(h_2 - h_3)$ = (5333.33-1777.78) hours, Baseline#1 and Intermediate#2 are offering in the market, at $c_1 = 20$ \$/MWh and $c_2 = 35$ \$/MWh, respectively. The market price, which the two technologies are receiving, is $c_2 = 35$ \$/MWh. During this period, Baseline#1 supplies:

$$E_{a12} = P_1(h_2 - h_3) = 52{,}242{,}963 \text{ MWh}$$

Intermediate#2 supplies:

$$E_{a22} = P_2\frac{h_2 - h_3}{2} = 8{,}659{,}753 \text{ MWh}$$

Baseline#1 receives $R_{12} = c_2 E_{a12} = 1.829 \times 10^9\$$, and Intermediate#2 receives $R_{22} = c_2 E_{a22} = 0.303 \times 10^9\$$.

Table 5.7 Optimal costs and market revenue. Problem 5.3

Generation technology	Optimal cost	Market revenue	Difference
Baseline#1	5.940	4.764	1.175
Intermediate#2	1.386	0.996	0.390
Peak#3	0.368	0.173	0.195
TOTAL	7.693	5.934	1.759

Finally, let us analyse the 3rd and last period. During $h_3 = 1777.78$ hours, Baseline#1, Intermediate#2 and Peak#3 are offering in the market, at $c_1 = 20$ \$/MWh, $c_2 = 35$ \$/MWh and $c_3 = 80$ \$/MWh, respectively. The market price, which the three technologies are receiving, is $c_3 = 80$ \$/MWh. During this period, Baseline#1 supplies:

$$E_{a13} = P_1 h_3 = 26{,}121{,}481 \text{ MWh}$$

Intermediate#2 supplies:

$$E_{a23} = P_2 h_3 = 8{,}659{,}753 \text{ MWh}$$

and Peak#3 supplies:

$$E_{a33} = P_3 \frac{h_3}{2} = 2{,}164{,}938 \text{ MWh}$$

Baseline#1 receives $R_{13} = c_3 E_{a13} = 2.090 \times 10^9\$$, Intermediate#2 receives $R_{23} = c_3 E_{a23} = 0.693 \times 10^9\$$ and Peak#3 receives $R_{33} = c_3 E_{a33} = 0.173 \times 10^9\$$.

Now, it is time to sum up all the partial revenues and obtain: $R_1 = 4.764 \times 10^9\$$, $R_2 = 0.996 \times 10^9\$$ and $R_3 = 0.173 \times 10^9\$$

2.

Let us now compare the results of the previous problem (optimal cost) with the results of this problem (market revenue) in Table 5.7 (all costs in $10^9\$$).

It is apparent that, in the example we analysed, the technologies are unable to recover all costs in the market, i.e., the market may not remunerate all the costs involved in power generation. This is because marginalist markets are energy markets in the sense that they target to remunerate the variable costs (or energy costs). This may raise the question if there should exist capacity markets targeted to remunerate the fixed costs.

3.

If we divide the differences found in the table by the installed capacity of each technology, we would obtain a constant value of 80,000 \$/MW, which is the fixed cost of the Peak#3 technology (most expensive technology). This value is what is missing for the technologies to recover all costs.

Problem 5.4

Suppose that, with the system already planned in an optimal way (Problem 5.2), where the capacity of the different technologies was properly sized for a balanced system, a perturbation occurs, driving the system away from the equilibrium (optimality) condition. This perturbation was due to a change in the variable costs of both the Intermediate and the Peak technologies, their variable costs being now double what they were initially ($c'_2 = 70$ \$/MWh and $c'_3 = 160$ \$/MWh). The rest of the parameters remain unchanged.

1. Calculate the total production costs for each technology and for the whole system. Suppose now that the costs incurred by the different technologies are recovered through the market, where it is assumed that energy offers and market prices are based on marginal costs, like in Problem 5.3.
2. Calculate the revenue obtained through the market by each technology.

Key Problem 5.4

1.

Let us recall that the system is sized for a balanced system. Therefore, the capacity of each technology, as well as the respective utilisation, remain unchanged. As so, the fixed cost is equal to before.

The energy supplied by each technology is also the same. So, the total cost for Baseline#1 is the same, but the total variable cost of Intermediate#2 and Peak#3 changes, because the variable cost of these technologies has changed.

The total costs are now equal to:

$$TC_1 = C_1 P_1 + c_1 E_{a1} = 5.940 \times 10^9 \$$$

$$TC'_2 = C_2 P_2 + c'_2 E_{a2} = 1.992 \times 10^9 \$$$

$$TC'_3 = C_3 P_3 + c'_3 E_{a3} = 0.541 \times 10^9 \$$$

$$TC' = 8.473 \times 10^9 \$$$

The average production cost for each technology is:

$ATC_1 = 49.22$ \$/MWh, $ATC'_2 = 115.00$ \$/MWh, $ATC'_3 = 250.00$ \$/MWh

The average total cost is $ATC' = 60.45$ \$/MWh

Of course, the total cost of the unbalanced system is now higher (8.473 versus 7.693), due to the increase in the variable cost of Intermediate#2 and Peak#3 technologies.

2.

As far as the behaviour in the marginalist market is concerned, the computation process is exactly equal to Problem 5.3. We show hereafter the final results, the reader is invited to confirm them.

$R'_1 = 8.683 \times 10^9\$$, $R'_2 = 1.992 \times 10^9\$$, $R'_3 = 0.346 \times 10^9\$$

In Table 5.8, we compare the production costs with market revenues (all costs in $10^9\$$).

As we can see, the situation is different from the previous problem. Baseline#1 has a significant profit, Intermediate#2 manages to recover all costs and Peak#3 has a loss. We recall that this is only an example. But this example shows that in real life

Table 5.8 Production costs and market revenue. Problem 5.4

Generation technology	Production cost	Market revenue	Difference
Baseline#1	5.940	8.683	–2.743
Intermediate#2	1.992	1.992	0.000
Peak#3	0.541	0.346	0.195
TOTAL	8.473	11.021	–2.548

(the systems never operate in balanced situations) unfair profits and losses are prone to occur. Both producers and consumers may get harmed in a marginalist market that does not explicitly remunerate the fixed costs. This again justifies a second thought about capacity markets, as a way to minimise unfair situations.

This is important for the producers, who aim at recovering all the costs in the market, but also for the consumers who do not want to contribute to the unfair profits of some producers. We note that, in this case, the consumers paid more 2.548 G$ than the fair remuneration of the producers.

Regulation of Natural Monopolies

6

Abstract

The chapter on "Regulation of Natural Monopolies" delves into the intricate world of overseeing essential industries that, by their very nature, are monopolistic. It explores the compelling reasons for regulating these sectors and the mechanisms employed to ensure fairness, efficiency, and affordability. Natural monopolies, such as utilities and public infrastructure, possess characteristics that limit competition. This chapter elucidates the inherent risks of monopolistic power and the necessity of regulation to safeguard consumer interests, prevent abuse, and promote economic efficiency. A comprehensive overview of regulatory agencies and their roles as impartial watchdogs is presented. These agencies play a pivotal role in setting and enforcing rules that govern pricing, service quality, and market conduct. The chapter delves into the concept of cost-of-service regulation, which aims to ensure that prices charged by natural monopolies reflect the actual costs incurred in providing services. An exploration of price cap regulation is included, revealing how this innovative approach seeks to strike a balance between fostering efficiency and maintaining affordable prices. It focuses on controlling the rate at which prices can increase over time. The chapter also unravels the complex structure of electricity prices paid by the consumer. It demonstrates how regulation influences each aspect to ensure equitable pricing. A detailed examination of the billing process, which considers factors like usage, demand, and fixed charges, is offered. This transparency ensures consumers understand the basis for their bills. By demystifying the regulation of natural monopolies, this chapter equips readers with a deeper understanding of the vital role that regulatory frameworks play in balancing the interests of consumers and service providers. It underscores the importance of effective regulation in promoting fairness, efficiency, and accessibility in essential industries.

R. Castro, *Engineering of Power Systems Economics*,
https://doi.org/10.1007/978-3-031-55251-9_6

6.1 Introduction

The regulation, namely of natural monopolies, is a critical aspect of ensuring fair competition and protecting consumers in industries where monopolistic conditions exist due to inherent characteristics of the market. A natural monopoly occurs when a single firm can produce goods or services at a lower cost than multiple firms, thereby resulting in an inefficient allocation of resources if left unregulated. To prevent potential abuse of market power and to promote the public interest, governments implement regulatory frameworks to oversee and control natural monopolies.

Natural monopolies often arise in industries that require substantial infrastructure investments, such as water and sewage systems, natural gas pipelines, and electricity transmission and distribution networks, the latter being our focus here. These industries exhibit high fixed costs, economies of scale, and barriers to entry that make it impractical or uneconomical for multiple firms to compete effectively. As a result, a single firm typically dominates the market, leading to the need for regulation to ensure fair pricing and quality of service.

One common regulatory approach for natural monopolies is price regulation. Regulators set the prices that the monopolistic firm can charge to consumers, aiming to strike a balance between allowing the firm to recover its costs, earn a reasonable return on investment, and prevent excessive profits that would harm consumers. Price regulation can take various forms, including rate of return regulation, and price caps. In theory, these mechanisms provide incentives for efficiency improvements while ensuring that prices remain reasonable and affordable for consumers.

Another aspect of regulating natural monopolies is ensuring non-discriminatory access to essential facilities or services. In some cases, a monopolistic firm may control infrastructure that is necessary for competitors to enter the market or provide alternative services. To prevent unfair competition practices and promote market access, regulators often mandate that the monopolistic firm grants competitors access to its facilities under reasonable terms, ensuring fair competition and fostering innovation.

Quality of service regulation is also crucial in natural monopolies. Regulators set standards and performance targets to ensure that the monopolistic firm maintains a certain level of service quality and reliability. This can include measures such as response times, outage management, and customer satisfaction levels. By monitoring and enforcing these standards, regulators protect consumers from subpar services and encourage the monopolistic firm to continually improve its operations.

Regulatory bodies responsible for overseeing natural monopolies have various powers to enforce compliance. They have the authority to conduct audits, impose fines or penalties for non-compliance, and require the submission of regular reports and data. Additionally, regulators often have the ability to conduct investigations into the monopolistic firm's conduct and practices to ensure that they are acting in the best interests of consumers.

However, regulating natural monopolies is not without its challenges. Balancing the need to provide incentives for investment and efficiency improvements with the goal of protecting consumers and promoting competition requires a careful approach. Regulators must have a deep understanding of the industry, including its cost structures, technological advancements, and market dynamics, to make informed decisions.

Moreover, regulatory capture, where the regulated firm exerts undue influence over the regulatory process, can pose a significant challenge. Regulators must maintain independence and act in the best interests of the public, resisting any attempts to compromise their impartiality.

In this chapter, we explore the *regulation of natural monopolies*, defining what can be considered a natural monopoly and explaining *why natural monopolies must be regulated*. The pivotal role of the *regulatory agencies* is underscored to understand their vital importance in assuring the proper functioning of natural monopolies while protecting the consumers. The literature offers two main regulation models: *regulation for cost of service* and *price-cap regulation*. The outcomes of both models are analysed from the perspective of the producers and consumers. Then, the chapter addresses the *structure of selling prices of electricity*, explaining in detail how the *billing of a consumer* is built. To ensure transparency, nowadays, the billing of a consumer is obtained by adding the tariffs of each service (*tariffs additivity*) provided to the consumer—generation, transmission, distribution, and retail. The chapter offers, at the end, a collection of exercises alongside with the respective solution.

6.2 Regulation of a Natural Monopoly

The regulation of the sales price of a product or service and the restriction of any firm's entry into a sector are motivated by the existence of industries with natural monopoly characteristics. We say we have a *natural monopoly* when it is cheaper for one company to produce the product or service rather than two or more firms in competition.

6.2.1 Formal Definition of Natural Monopoly

If the total production cost is higher when two or more companies operate simultaneously, instead of a single one, the only company in the market operates a natural monopoly.

To put it formally, consider a set with n companies, each producing q_i units of output; therefore, the quantity produced by the all set (Q) is given by:

$$Q = \sum_{i=1}^{n} q_i \tag{6.1}$$

The definition of natural monopoly based on total production costs is given by the condition:

$$TC(Q) < TC(q_1) + TC(q_2) + \ldots + TC(q_n) \tag{6.2}$$

Thus, it is less costly to provide the amount Q with a single company than to split production between n companies. It is said that the company's production cost is *sub-additive* to the production level Q. If this property holds for all values of Q, it is said that the production function is globally sub-additive. In other words, there are *economies of scale*.

Consider that a company has the following cost function:

$$TC(q) = FC + VC(q) = FC + cq$$
$$ATC = \frac{FC}{q} + c \tag{6.3}$$

FC is the fixed cost of the company and c represents the unitary variable cost which is assumed constant.

We conclude that the average production cost decreases continuously with the increasing output quantity. The technology offers economies of scale, that is to say, revenues increase with increased quantities.

The technical definition of a natural monopoly can be generalised to take into account the case of one company with multiple output products. In the case of two products, we have:

$$TC(q_1, q_2) < TC(q_1, 0) + TC(0, q_2) \tag{6.4}$$

In this case, the production cost function exhibits *economies of scope*. Thus, it is more economical to produce the two products or services with only one company than to have two different companies produce each product or service.

In short, economies of scale focus on the cost advantage that arises when there is a higher level of production for one good, whereas economies of scope focus on the average total cost of production of a variety of goods. We, therefore, conclude that economies of scale and economies of scope are the attributes of the production cost function that define when it is more efficient to concentrate the production in one company than in a multitude of companies.

6.2.2 Why Regulate Natural Monopolies?

Regulation consists of requirements the governments impose on private firms to achieve certain fairness purposes by regulating and modifying economic behaviours. Regulation is put in place by setting rules that limit who can enter a business (entry controls) and what prices they may charge (price controls). Failure to meet regulations can result in fines, orders to cease doing certain things, or, in some cases, even criminal penalties.

The government's intervention through regulation poses some questions:

- What is the nature and extent of the problems that emerge in the absence of price regulation?
- What regulatory tools are available to stimulate the performance of the industry and what are the strong and weak points?
- What are the improved attributes of the industry that emerged as a result of the regulated environment?

In general, the regulation intends to:

- Address market failure.
- Ensure the most efficient allocation of resources.
- Ensure proper competition.
- Prevent the growth of a dominant group.
- Keep prices down and protect the consumer.

Price regulation of natural monopoly industries is justified if industries with natural monopoly characteristics exhibit poor (or even the worst) performance and it is possible, in theory, and practice, that the government implements price regulation with the objective of increasing the performance as compared to the unregulated situation.

In regards to the electricity-related natural monopolies, in particular, the regulation aims at:

- Constraining the exercise of monopoly power by incumbent suppliers, i.e., the suppliers of the service before regulation.
- Stopping subsidies to the electricity sector, which were often applied beforehand.
- Providing incentives for operating efficiency and quality of service.
- Optimising the structure of the sector.
- Promoting least-cost system expansion.
- Stimulating energy conservation and Research & Development.

If regulation is not active, a natural monopoly can lead to excessive prices, inefficient production, equipment or facilities duplication, and poor quality of service.

6.2.3 The Regulatory Agencies

The concept of a regulatory commission was first introduced in the United States. The regulatory commission is expected to be reasonably independent of the legislative and executive powers and is empowered to establish accounting and supervision rules for the regulated firms. The professional staff of the regulatory commission should be competent in the fields of engineering, economics, and law. One way of ensuring the regulator's independence is to allow the government to appoint the board of directors for a term, but not to dismiss them before the end of the term. This procedure is followed in several countries.

There are three basic principles of regulation that the regulatory agencies should comply with independence, transparency, and both investor and consumer protection. The regulatory body should act freely and independently from external sources (government, private sector, the public) influence. Transparency is crucial to legitimacy and is key to attracting and retaining investment as it creates confidence in the agents. To this end, robustness, quality, and predictability are expected from regulatory decision-making. Finally, the regulation should balance the conflicting interests of the government, service suppliers, and customers. Governments are subject to political pressures that may compromise objectivity, suppliers expect high returns and customers want reliable electricity at the lowest prices.

The main roles of the regulators are to protect consumers from abuse by firms with substantial market power, to support investment by protecting investors from arbitrary government action, and, all in all, to promote economic efficiency. At the end of the day, regulation should ensure functional market design, reasonable prices, and adequate quality of supply.

Some key functions of the regulator are listed below:

- Issue licenses related to regulatory functions.
- Set performance standards.
- Monitor the performance of regulated firms.
- Establish the level and structure of regulated tariffs.
- Establish a uniform accounting system.
- Arbitrate disputes among stakeholders.
- Perform management audits on regulated firms.

The guidelines of regulation include:

- Efficient products and services prices: Regulated prices (tariffs) shall provide customers with efficient price signals in order to guide them in their consumption decisions.
- Efficient production costs: The regulation should create incentives to induce the company to minimise the costs and increase the quality of service.
- Efficient investment levels: The regulated company must provide the number of services requested by the consumers and make the necessary investments, over time. Thus, the regulatory mechanisms must comply with the restriction that the company can only invest if the investment is guaranteed ex-ante with reasonable compensation.
- Efficient levels of service quality and product variety: Products shall be supplied with different levels of service quality and service guarantee.
- Adequate monopoly profit: Profits from the regulated firm should be adequate to induce efficiency.
- Social concerns: The regulation should be aware that the law may impose uniform tariffs and special rates and conditions for vulnerable consumers.

6.3 Regulation Models

6.3.1 Regulation for Cost of Service

The regulation for the cost of service works exactly as the name implies. The regulated company recovers the cost it incurred in providing the service plus a rate of return on the capital costs. This is why this type of regulation is also known as *Rate of Return Regulation*. The specifics of the method are developed below.

In this model, it is assumed that the regulatory commission knows all the information about the costs incurred by the regulated firm, which are a function of the demand. It is further considered that the regulated firm does not act strategically in order to increase the costs and distort the information held by the regulator.

This is the theoretical principle under which the cost of service regulation has been designed and is fully efficient. The real world is however a different story. In reality, the regulator does not hold all the information concerning the regulated firm. The company knows better than anyone its own cost structure and therefore can act strategically in order to obtain better-regulating conditions.

Let us take an example. The regulated company may tend to unnecessarily increase the capital costs—investment—because it knows the rate of return, in other words, the profit, will be higher. This is a known weak point of this method. Much of the evolution of regulatory commissions and regulatory procedures aimed at obtaining more and better information from regulated companies and overcoming this drawback.

Cost of service regulation, or rate of return regulation, has been the traditional regulation solution either in the USA or, more recently, in Europe. Thus, the results obtained with this methodology serve as a benchmark for other methodologies.

The traditional cost of service regulation has been often criticised as being inefficient for the aforementioned reasons. However, this criticism is somehow unfounded; the evolution of the real-world application of cost-of-service regulation tried to address the problem related to the fact that the regulator has imperfect and incomplete information.

In the practice of cost of service or rate of return regulation, the regulatory procedures set some rules to minimise the asymmetry of information. Regulators have adopted a uniform accounting system for each regulated sector, including detailed and exhaustive reports of capital costs, assets valuation, depreciation, management of taxes, operating costs categories, unbundling of regulated and non-regulated activities, and investments financing. Reports produced by the companies are audited and any falsehood is punished. As the same rules are used to deliver the reports, it is possible to perform comparative analyses between regulated companies and regulatory commissions. Thus, it is possible to establish a regulation by comparison (yardstick regulation).

Another way of preventing fraud is that, in the regulatory procedures, it is not mandatory that the regulator accept all the costs supplied by the company, even if the company is audited. The regulator can disregard some costs that, in its judgement, are not fair or reasonable. This regulatory risk is a pressure on the regulated company.

On the other hand, regulated prices may not be continually adjusted; sometimes there are long periods (the so-called regulatory lag) during which prices do not change.

The main mechanism of regulation is the setting of regulation parameters (rate case). This process is public and usually happens on a regular basis. A typical case has two phases: in the first step, the total company costs (or the total cost of service) are presented by the regulated company and determined (only used and useful costs are accepted) by the regulator; in the second step, the tariff structure is drawn.

Let us look at the first stage. The total cost of service can be divided into the following cost blocks:

- Operating and maintenance costs (OMC), e.g. fuel, labour, materials, etc.
- Capital costs, mainly equipment and machinery.
- Regulatory Asset Value (RAV).
- Annual assets depreciation (D).
- Allowed assets rate of return (r).
- Tax rate (TR).
- Other costs (O)

The total cost of service for year t is given by:

$$R_t = OMC_t + D_t + r_t(1 + TR_t)RAV_t + O_t \tag{6.5}$$

The basic legal principle governing price regulation is the following: regulated prices should be set so that the company is reasonably remunerated for the efficient investments it makes and which are strictly necessary for the satisfaction of service obligations, and no more than that.

A few notes on Eq. 6.5 are worth mentioning. OPEX (Operational Expenditure) is the cost related to maintenance and repairs, supplies of services, travel and vehicle expenses, leasing commissions, salary and wages, raw materials, etc.; asset remuneration relates to the net capital invested in the assets that are going to be paid at a certain rate of return; the depreciation is concerned with the fact that the payment of an asset is held over its useful life, for example, an annual payment equal to the asset value divided by the number of years; expenditure on assets like building, machinery, equipment or upgrading existing facilities so their value as an asset increases is called CAPEX (Capital Expenditure).

As it is apparent from Eq. 6.5, the utility cannot earn any profits on its operation and maintenance costs. The company just recovers the accepted O&M cost and no more than that. All utility profits come from investing in capital equipment, like power plants, transmission lines and substations (in the power system's case). This justifies the danger of over-investment that must be fought by the regulator by determining the size of the RAV. The regulator establishes the size of the RAV by selecting the items to be included on it using the criterion of *efficient investment*. This is very important because the RAV determines the utility's profits. Furthermore, the regulator sets the rate of return the RAV is going to be remunerated. This rate of return must be fair and reasonable, not too high, not too low, in order to induce efficiency and provide adequate signals to the company.

To determine the amount of capital equipment to be included in the RAV, there are two fundamental criteria that should be taken into account by the regulator. The first one is called the "Used and Useful" criterion. It states that capital equipment must be actively employed to provide the service to customers, and service would suffer if that equipment were not in place. The second is the "Prudence" criterion, i.e., the utility must have paid a fair price for any capital equipment and must have installed the equipment at a reasonable cost.

In theory, the "prudence" criterion keeps regulated companies from artificially inflating capital costs in order to increase profits. In practice, it may be a different story!

6.3.2 Price Cap Regulation

The conventional theory of the optimal price—regulation for the cost of service—considers that regulators are fully informed about the technology, costs, and consumer demand for the enterprises they regulate. In reality, this is not usually the case. Regulators may have incomplete information about the costs and behaviour of the regulated firm. The regulated firm can use their specific knowledge to act strategically in the regulatory process, in order to increase their profits to the detriment of the consumers.

The *price cap regulation* or *regulation by incentives* tries to overcome the problem of information asymmetry. The basics of the price cap regulation methodology involve fixing, ex-ante, the selling price of the product. Typically, this begins with the setting of a price formula. One assumes a given price, which is adjusted to take into account exogenous variations.

Under price cap regulation, the regulator sets an initial price (or a price vector), i.e., a baseline price cap, p_0. This price is then adjusted from one year to the next by taking into account the change in inflation (RPI—Retail Price Index) and a productivity objective designated by X. Thus, the price in year t is given by:

$$p_t = p_0(1 + RPI - X)^t \tag{6.6}$$

X reflects the productivity factor that the regulator expects to obtain. The productivity factor should pose a significant, but not insuperable, challenge to the regulated operator.

The fixed price mechanism (price cap) is in force over a predetermined period of time (e.g., 3 or 5 years), called the regulatory period. At the end of each regulatory period, a new starting price, p_0, and a new value of X are set for a new regulatory period. The regulator establishes p_0 and X, based on the computation of the effective cost of service.

The groundings of price cap regulation are to restrict the price rather than the profit, as opposed to the regulation by the cost of service. The mechanism encourages cost reduction and innovation in regulated companies because the firm is allowed to keep any profits obtained via cost reductions relative to the price cap fixed by the regulator. This differs from traditional regulation (cost of service), in which the regulator allows price increases based on cost increases.

Table 6.1 Comparison of regulation models

	Cost of service	Price cap
Pros	Prices stay in line with costs	Strong incentives for cost reduction and innovation
	Quality assured	
Cons	Likelihood of excessive investment	Quality may suffer
	Weak incentives for productivity improvement	Prices exceed costs on average

A numerical example may help. Suppose the regulator determined the cost of a product or service produced by a company to be equal to 100. The regulator expects to obtain from the company a productivity factor of 20%. Therefore, the regulator fixes the selling price at 80 (for simplification, let us not consider the RPI). Through innovation processes and cost reductions, the company manages to produce the product or service at 50. The profit of the company is 30.

The obvious advantage of price cap regulation is that, in theory, it increases efficiency by promoting innovation towards cost reduction. On the downside, one may point out that it can be expected that companies degrade the quality of service in an effort to cut costs.

We highlight that this is precisely the opposite of what happens in regulation by cost of service. In this case, the regulated company tends to over-invest because it knows the higher the investment, the higher the profit as the investment is remunerated at a certain rate of return. For sure, the quality of service will not suffer.

6.3.3 Comparison of Regulation Models

Table 6.1 outlines the main differences between the two regulation models.

6.3.4 Regulation Examples

Regulators try to benefit from the pros of each regulation method and oppose their drawbacks. To this end, mixed regulations are common in some European countries. For instance, CAPEX may be regulated through the cost of service and OPEX by price cap.

Moreover, it is usual that the RAV determination is based on reference costs set by the regulator after obtaining the minimum costs offered by the equipment suppliers. This means the regulator accepts these costs and not the ones provided by the company.

Also, the regulators often put in place schemes of rewards and penalties to be applied in addition to the regulation mechanisms. Examples of these are rewards and penalties for improvement of the quality of service, loss reduction, investment in smart grids, and extended equipment lifetime. The regulator sets standards for these objectives; if the company exceeds the standards it will be rewarded (the company is paid in addition to the revenue set by the regulation mechanism); on the opposite, if it fails a penalty will be applied (the company has to pay back the regulator that returns the revenue to the consumers).

6.4 Structure of Selling Prices of Electricity

Regarding the terminology question about prices versus tariffs, we can say that the concept of price implies a negotiated transaction, not necessarily in the market, while the concept of tariff or rate assumes values approved by the competent authorities.

The price or tariff for a given product is a signal that is transmitted to the client. It should reflect costs incurred by the company along the value chain. The price or tariff allows us to assess the economic efficiency of the supplier of the product, e.g. by comparison to the tariffs of other national or international companies.

The economic sustainability of enterprises depends on the obtained revenues. These are a consequence of prices or tariffs and of the quantities of products or goods purchased by customers. This shows that the methods of preparing the prices (or tariffs) and revenues are similar.

6.4.1 Terms and Concepts

In electricity billing, it is usual to use several terms, namely, power, measured in kW (active power) or kVA (apparent power), and active and reactive energy, measured in kWh and kvarh, respectively.

The power stations are called to operate in ascending order of variable production costs (regulated environment) or according to the price offered by the producers (market environment), thus, production costs are lower during off-peak and higher during peak hours. The time intervals with the same energy price, which are strongly correlated with the demand, are named tariff periods, or hourly periods. Sometimes, there is a seasonal classification of these periods, grouping months with the same prices, for instance, winter and summer.

The *contracted power* is the highest power value that the customer wishes to have available at any time. This value is of utmost importance for the design of the peripheral sections of the network shared by this client and a small number of neighbouring clients.

The *used power* should correspond to the amount actually used by the customer during the peak hours of the power system to which it belongs. The *billed power* corresponds to a weighted average between the contracted power and the used power.

There may exist also a *fixed parcel*, reflecting the costs of electricity reading, billing, collection, and client care.

As a result of the accumulation of costs, due to increased power flows through successive networks from generation (higher voltages) to delivery to customers (lower voltages), tariff levels grow strongly when the voltage decreases.

6.4.2 The Billing of a Customer

The equation that allows the establishment of the electricity selling price to end customers is said binomial because it explicitly separates the power and energy billing components. However, the number of parcels included in billing a client is usually greater than two. In the simplest case, the energy parcels may be broken down into two periods, peak hours and valley hours, according to the delivery time. In many countries, there are more periods, for instance, intermediate hours and super valley hours. The power parcels usually correspond to a weighted average between the contracted power and the used power. In general, the billing of an end customer can be given by the equation:

$$B = a + bP_c + cP_u + dE_p + eE_v \tag{6.7}$$

where:

- B: billing of the client.
- a: independent parcel parameter.
- b, c, d, e: parameters with unit price units.
- P_c: contracted power of the client.
- P_u: used power taken by the client.
- E_p: energy at peak hours of the client.
- E_v: energy at valley hours of the client.

To complete Eq. 6.7, one should add the reactive power-related parcels, depending on the voltage level, and according to the reactive power billing rules stated before, in Chap. 5. It is highlighted that the unit prices (b, c, d, e) depend on the voltage level and increase as the voltage level decreases.

With minor changes, the customer billing formula (Eq. 6.7) can also serve to represent the total sales revenue provided by a set of customers at the same rate:

$$R_s = aN + b\sum P_c + c\sum P_u + d\sum E_p + e\sum E_v \tag{6.8}$$

where:

- R_s: sales revenue provided by the set of customers.
- N: number of clients in the set.

The sum of the contracted powers of the clients of a network is always equal to or greater than the contracted power of this network, often much higher, due to the lack of synchronisation of the energy uses of each client.

6.4.3 Billing Variants

The general equation of billing a client (Eq. 6.7) can be adapted to different special cases. We will mention a few of these special cases:

No independent parcel
The independent parcel can be null, $a = 0$. Nevertheless, the cost of electricity reading, billing, and collection, though not explicit, must be present in the tariffs.

Merging the two power parcels
Instead of two power parcels, one proportional to the contracted power and another to the used power, one single parcel can be adopted, designated by billed power (P_b), defined as the weighted average of the mentioned powers:

$$P_b = \frac{b}{b+c} P_c + \frac{c}{b+c} P_u \tag{6.9}$$

Simple counting for domestic clients
Technological progress in counting equipment has provided successive cost reductions, with a particular focus on the roll-out of smart meters. Among other important functionalities, smart meters are able to separately count the energy consumed in different tariff periods (peak, valley, and other). However, smart meters are not yet available for all consumers, especially for small LV (Low Voltage) domestic consumers (actually, the great majority of the consumers). When it is not economically viable to separately bill the energy in different tariff hours, the energy parcels in Eq. 6.7 are merged as follows.

$$dE_p + eE_v = d_t E_t \tag{6.10}$$

where:

$$\begin{aligned} E_t &= E_p + E_v \\ d_t &= \frac{dE_p + eE_v}{E_t} \end{aligned} \tag{6.11}$$

Inclusion of the used power parcel in the energy parcel
Again, when smart metering is unavailable, it is not economical to measure the used power, especially for domestic LV consumers. Therefore, the used power parcel is transferred to the energy parcel, as follows:

$$cP_u + d_t E_t = \left(\frac{cP_u}{E_t} + d_t\right) E_t = \left(\frac{c}{h_t} + d_t\right) E_t \tag{6.12}$$

Where E_t/P_u was replaced by h_t, i.e., by using the utilisation factor of the used power taken by the domestic consumer. The contracted power parcel—bP_c—remains in the billing equation; its control is assigned to a circuit breaker, also necessary for safety reasons, that prevents the contracted power from being overcome.

Time of Use Tariffs (ToU)
Known broadly as *time of use* tariffs, these are specific pricing plans for energy use, aimed at encouraging consumers to use electricity at times when it is more

cheaply available. For instance, the consumer is offered to pay cheaper each kWh of electricity consumed at night, when electricity is cheaper and more available (it may even happen that wind power exceeds normal demand at night). This is reflected in Eq. 6.7 by the d and e parameters.

ToU tariffs can be divided into (for more information see https://www.irena.org/-/media/Files/IRENA/Agency/Publication/2019/Feb/IRENA_Innovation_ToU_tariffs_2019.pdf):

- *Static ToU pricing*—Different prices for different hours (time blocks) are defined in advance. The simpler instance is to define a higher energy price for peak hours and a lower energy price for valley hours. A higher granularity can be established, as well as seasonality.
- *Real time pricing*—Wholesale electricity markets are the basis for determining the prices applicable close to real-time. This is a dynamic pricing tariff.
- *Variable peak pricing*—This is a hybrid tariff where the energy price for the valley hours is fixed in advance, but the energy price for on-peak hours is dynamically fixed based on the signs provided by the wholesale electricity market.
- *Critical peak pricing*—Extremely high peak energy prices are fixed in advance for a few days, based on historical information from the wholesale electricity markets.

6.4.4 Tariffs Additivity

For transparency reasons, each activity or business should reflect the costs that were held to allow electricity to be delivered to the consumer. The consumer will pay for a set of parcels corresponding to the different activities. The sum must be consistent, i.e., the additivity of prices of the different activities shall be conducted with no cross-subsidies between activities or between groups of consumers.

With the additivity of prices or tariffs, one can consistently make responsible either the consumer or the business activity for the consumption of electricity or for the production of the good or service, respectively.

To supply electricity to the end-consumers, the following costs, or activities, may be identified:

1. Production costs; electricity is usually traded in a market (see Chap. 7 for details), so, this activity is non-regulated.
2. Transmission costs; regulated activity, held by a private or public company, designated the Transmission System Operator (TSO).
3. Overall management of the system; regulated activity is usually also held by the TSO.
4. Distribution costs, regulated activity carried out by a private or public company, named the Distribution System Operator (DSO).
5. Retail costs; retail is the selling of energy to the end-consumers, which happens in an open market, therefore, it is a non-regulated activity.

The consumers have to pay all these costs (plus a rate of return, in the case of regulated activities; for non-regulated activities, it is expected that the market works). Therefore, a tariff for each activity is set:

1. Energy tariff, set by the electricity (wholesale) market.
2. Use of the transmission grid tariff, set by the regulator.
3. Overall management of the system tariff, set by the regulator.
4. Use of the distribution grid tariff, set by the regulator; it is normally divided into three components, according to the voltage level: use of the distribution grid in HV, use of the distribution grid in MV, and use of the distribution grid in LV.
5. Retail tariff, set by the retail market.

Tariffs 2., 3. and 4. are regulated tariffs set by the regulator. They are called *access to networks tariffs*. Network access tariffs may be paid through four terms: contracted power, active energy, reactive energy, and average power during peak hours. Tariffs 1. and 5. are market-driven.

The consumer pays the sum of all five tariffs.

As mentioned before, the unit price (€/MWh) an LV consumer pays is higher than an HV consumer's. This is because an LV consumer has to pay for the use of the distribution grid in HV, MV, and LV, whereas an HV consumer only pays for the use of the distribution grid in HV. Of course, all the other tariffs are paid by both consumers.

6.5 Conclusion

In the complex and critical realm of natural monopolies, where competition is limited and essential services are at stake, regulation stands as the fulcrum upon which the interests of consumers, service providers, and the broader economy are delicately balanced. This chapter has navigated the multifaceted landscape of regulating natural monopolies, unveiling the myriad reasons behind such regulation and the mechanisms employed to ensure fairness, efficiency, and affordability in these sectors.

The journey began with a compelling exploration of why we regulate natural monopolies in the first place. These industries, characterised by high fixed costs and the absence of effective competition, wield immense power that can be easily abused. Without regulation, consumers are left vulnerable to arbitrary pricing, subpar service quality, and inadequate infrastructure development. The imperative to regulate is rooted in the fundamental principle of safeguarding the public interest, ensuring that essential services are accessible, reliable, and reasonably priced.

Central to effective regulation are the regulatory agencies, impartial guardians tasked with overseeing the conduct of natural monopolies. These agencies play a pivotal role in establishing rules and enforcing them to create a level playing field. By monitoring pricing, service quality, and market behaviour, regulatory agencies act as the custodians of fairness, protecting consumers from monopolistic exploitation while fostering an environment conducive to investment and innovation.

The chapter illuminated the concept of cost-of-service regulation, a bedrock principle in overseeing natural monopolies. This approach ensures that the prices charged by these entities reflect the actual costs incurred in providing services. By scrutinising expenses related to operations, maintenance, and capital investments, regulators ensure transparency in pricing and prevent monopolies from reaping windfall profits at the expense of consumers. Cost-of-service regulation is the linchpin that aligns the interests of the service provider and the consumer, trying to compel fair pricing while ensuring the financial viability of the utility.

Our exploration extended to price cap regulation, a dynamic approach that endeavours to balance efficiency and affordability. Price caps cap the rate at which prices can increase over time, compelling natural monopolies to enhance operational efficiency and minimise cost escalation. While fostering cost-consciousness among service providers, price cap regulation also shields consumers from arbitrary price hikes, promoting stability and predictability in essential service costs.

The intricate web of selling prices for electricity was unveiled to reveal the complex interplay between the different parcels—generation, transmission, distribution costs, various ancillary charges, and retail—that must be added to the consumers' bill. Regulation, in this context, serves to ensure equity in cost allocation and prevent any single segment from monopolistic excess. By scrutinising each component of the electricity price, regulators seek to maintain a delicate balance, encouraging investment in infrastructure, and promoting competition in markets where it is possible.

Lastly, the chapter delved into the customer experience, offering insights into how the consumers are billed. Billing is not just about receiving a monthly statement but understanding the rationale behind the charges. Effective regulation ensures transparency in billing practices, allowing consumers to comprehend the basis for their bills and ensuring they are not subject to unfair or hidden charges. To this aim, tariffs additivity is employed.

In conclusion, the regulation of natural monopolies is a delicate dance, one that requires balancing the need for a reliable and accessible service with the necessity of fostering efficiency, innovation, and reasonable pricing. The multifaceted mechanisms and principles explored in this chapter underscore the critical role that regulation plays in shaping the landscape of natural monopolies.

Effective regulation, carried out by diligent regulatory agencies, safeguards the interests of consumers and the broader economy. It prevents monopolies from exploiting their privileged positions while incentivising them to invest in infrastructure, enhance service quality, and innovate. By striking this regulatory balance, we can ensure that essential services remain accessible to all, reliable in delivery, and reasonably priced—ultimately contributing to the well-being and progress of society.

6.6 Proposed Exercises

Problem 6.1

A residential consumer has an annual consumption of around 3000 kWh and a contracted power of 6.9 kVA. The tariffs are as are shown in Table 6.2.

The on-peak time periods are 14 h in Winter and 14 h in Summer.

1. Assuming that this consumer is using the flat tariff, compute the cost for his electricity consumption. What is the average cost of energy for this consumer?
2. This consumer is considering the adoption of a time-of-use (ToU), two-rate tariff. What should be the conditions, in terms of the energy consumed in on-peak and valley periods, to make the adoption of this type of tariff more attractive from an economic point of view?
3. In practice, how is the contracted power of 6.9 kVA controlled?

Key Problem 6.1

1.

The annual electricity bill (p_c are the power charges and e_c are the energy charges) is:

$$B_a = 12p_c + e_c E_a = 600.12 \text{ €}$$

The average cost of electricity is:

$$ATC = \frac{B_a}{E_a} = 200 \text{ €/MWh}$$

2.

The power charges are independent of the tariff type: flat or ToU. The annual energy charges in the flat tariff are:

$$b'_a = e_c E_a = 490.2 \text{ €}$$

Let us consider that x is the consumption during valley hours. So, we can write (e_v are the energy charges at valley hours and e_{o-p} are the energy charges at on-peak hours):

$$e_v x + e_{o-p}(E_a - x) < b'_a$$

Table 6.2 Tariffs' data for Problem 6.1

Power charges	kVA	€/month
Flat and ToU	6.9	9.16
Energy charges	Time period	€/kWh
Flat		0.1634
ToU	On-peak	0.1909
	Valley	0.1002

$$x > 909.59 \text{ kWh} = 30\% E_a$$

For the ToU tariff to become economically attractive, the consumer must consume, at least, 30% of the energy during valley hours.

3.

Domestic LV consumers, contrary to what happens to industrial consumers, cannot exceed the contracted power. Actually, it is controlled by a circuit breaker that interrupts the supply when the consumer intends to switch on, at the same time, appliances that exceed the contracted power. Industrial consumers may exceed the contracted power. However, they are penalised for that because the contracted power is automatically updated to the value of the used power.

Problem 6.2

An MV industrial consumer has a constant 200 kW, $\tan(\phi) = 0.6$ lagging, annual load diagram. The consumed active and reactive energy tariffs are (disregard the fixed term and power charges) shown in Table 6.3.

For $0 < \tan(\phi) \leq 0.3$, the grid-supplied reactive energy is always free. For $\tan(\phi) > 0.3$, the reactive energy consumption is charged in steps (for example, for $\tan(\phi) > 0.5$, the reactive energy consumed is charged in 3 steps). The multiplicative factors are applied to the reactive energy charges in each step as shown in Table 6.4.

Consider that the number of annual hours in each tariff period is given in Table 6.5.

The firm is considering making two investments:

(a) To install a capacitor bank (reactive power compensation system) to avoid reactive power charges during peak and intermediate hours, with an investment equal to 100 €/kvar and negligible O&M costs.

Table 6.3 Active and reactive energy charges for Problem 6.2

	Peak	Intermediate	Valley	Super Valley
Active (€/kWh)	0.1382	0.1101	0.0777	0.0666
Reactive (€/kvarh)	0.0248	0.0248	0	0

Table 6.4 Multiplicative factors for the reactive energy tariff in steps. Problem 6.2

STEP1: $0.3 < \tan(\phi) \leq 0.4$	STEP2: $0.4 < \tan(\phi) \leq 0.5$	STEP3: $\tan(\phi) > 0.5$
0.3333	1	3

Table 6.5 Number of annual hours in each tariff period. Problem 6.2

Peak	Intermediate	Valley	Super Valley
1305	3496	2499	1460

(b) To install a PV plus batteries system to supply the active energy consumed during peak and intermediate hours, with the following characteristics: investment is 2500 €/kW (60% correspond to the PV panels and 40% to the batteries); utilisation factor is 2000 h; annual O&M costs are 1% of the investment; lifetime is 20 years and the discount rate is 5%; batteries should be replaced at year 10 and begin operation at year 11.

We recall the following relationship $Q = P\ \tan(\phi)$.
Compute:

1. Total annual reactive power charges, before the reactive power compensation.
2. Required kvar of the capacitors bank so that the consumer is not charged for reactive power consumption.
3. Simple payback time of the reactive power compensation system project.
4. Active energy charges during peak and intermediate hours, before the PV system installation.
5. Rated capacity of the PV system.
6. NPV of the PV system project.

Key Problem 6.2
1.
The reactive power is billed in steps as instructed in the text.

$$C_{Q1} = 3c_Q P(0.6 - 0.5)(1305 + 3496)$$

$$C_{Q2} = 1c_Q P(0.5 - 0.4)(1305 + 3496)$$

$$C_{Q3} = 0.3333c_Q P(0.4 - 0.3)(1305 + 3496)$$

$$C_Q = \sum C_{Qi} = 10{,}319\ €$$

2.
For reactive power consumptions lower than the ones corresponding to $\tan(\phi) = 0.3$, reactive power is freely supplied by the grid, i.e., $Q_{free} = 0.3P$. So, there is no necessity to compensate for $\tan(\phi) < 0.3$. The minimum capacity of the capacitors bank is therefore:

$$Q_C = P\ \tan(\phi) - Q_{free} = P(0.6 - 0.3) = 60\ \text{kvar}$$

With these capacitors, no reactive power charges are to be paid.
3.
The investment in the reactive power compensation system is 6000 €. The annual saving is 10,319 €.
The simple payback is:

$$P_b = \frac{I_T}{C_Q} = 0.58\ \text{year}$$

After half a year the investment is paid. Installing capacitors for compensating the power factor of industrial installations is far more profitable than paying the reactive power charges to the grid. The reactive power compensation system is a very profitable investment.

4.

The active charges during peak and intermediate hours are:

$$C_{P_{pi}} = c_P P h_p + c_i P h_i = 36{,}070.02 + 76{,}981.92 = 113{,}052.12 \text{ €}$$

5.

The energy supplied by the PV system is:

$$E_{aPV} = P(h_p + h_i) = 960.200 \text{ MWh}$$

As the utilisation factor is $h_{aPV} = 2000$ h, the rated capacity (peak power) of the PV system is:

$$P_{PV} = \frac{E_{aPV}}{h_{aPV}} = 480 \text{ kWp}$$

6.

The annual revenue is the saved energy charges due to the installation of the PV system. Hence,

$$R = 113{,}052 \text{ €}$$

The annual expenses are O&M costs equal to O&M=1% of the total investment. The total investment is:

$$I_T = I_0 P_{PV} = 1{,}200{,}250 \text{ €}.$$

The battery investment is:

$$I_{bat} = 40\% I_T = 480{,}100 \text{ €}$$

and the battery is to be replaced at year 10.

The discount factor is ($a = 5\%$):

$$k_a = 12.46.$$

Therefore, the NPV is:

$$NPV = (R - O\&M I_T)k_a - \frac{I_{bat}}{(1+a)^{10}} - I_T = -235{,}688 \text{ €}$$

We can see that, in this case, the PV system is not a profitable investment and should be rejected as the NPV is negative.

Problem 6.3

A big consumer in High Voltage (60 kV) has a constant 20 MW annual load diagram. The applicable tariffs are in Table 6.6 and the daily time periods are in Table 6.7.

Table 6.6 Tariffs' data for Problem 6.3

Fixed term	€/month
	79.93
Power charges	€/kW.month
Peak hours	6.378
Contracted power	0.870
Energy charges	€/kWh
Peak	0.1206
Intermediate	0.0972
Valley	0.0740
Super valley	0.0622

Table 6.7 Time periods for Problem 6.3

	Peak	Intermediate	Valley	Super valley
Weekdays	5	12	3	4
Saturdays		7	13	4
Sundays			20	4

1. Compute the cost associated with this energy consumption pattern. What is the average cost of energy?
2. In order to reduce costs, this consumer is thinking about making an investment (for a lifetime of 20 years with a discount rate of 6%) that will allow the installation to run only in the valley and super-valley hours. What should be the maximum investment level?

Key Problem 6.3

1.

This is just a matter of replacing values in the formula to obtain the annual bill:

$$B_u = 12(FT + e_{P_p} P_p + e_{P_c} P_c) + P(h_p e_p + h_i e_i + h_v e_v + h_{sv} e_{sv})$$

As the load diagram is flat, the peak power, P_p, is equal to the contracted power, P_c, and to the installed power, $P = 20$ MW. The number of hours in each tariff period, h_j, $j = p, i, v, sv$, is:

- $h_p = 1305$ h
- $h_i = 3496$ h
- $h_v = 2499$ h
- $h_{sv} = 1460$ h

The annual energy consumed is:

$$E_a = P8760 = 175{,}200 \text{ MWh}$$

After replacing the appropriate values, we obtain:

$$B_a = 17{,}199{,}123 \text{ €}$$

$$ATC = 98.17 \text{ €/MWh}$$

The average total cost is much lower in HV than in LV (compare with Problem 6.1). This is because the tariff for the use of the grid is lower in HV than in LV. LV consumers must pay for the use of the distribution grid in HV, MV and LV and HV consumers do not pay all this cost, only the use of the distribution grid in HV.

2.

The installation operates only during the valley and super-valley hours, but, of course, the annual energy consumed is the same. Therefore, the new installed capacity is:

$$P_{new} = P_{(c-new)} = \frac{E_a}{h_v + h_{sv}} = 44.2536 \text{ MW}$$

The new formula to compute the annual bill is:

$$B_{(a-new)} = 12(FT + P_{(c-new)}e_{P_c}) + P_{new}(h_v e_v + h_{sv} e_{sv})$$

The new annual bill and average total cost are:

$$B_{(a-new)} = 12{,}665{,}365.73 \text{ €}$$

$$ATC_{new} = 72.29 \text{ €/MWh}$$

A notable reduction in both values is observed because no energy is consumed during peak and intermediate hours.

The maximum investment level of the new equipment that allows the installation to operate solely in the valley and super-valley hours is defined by NPV = 0.

$$NPV = (R - E)k_a - I_T = 0$$

The revenue is the savings allowed by the new equipment:

$$R = B_a - B_{(a-new)} = 4{,}533{,}757.43 \text{ €}$$

We consider that the new equipment's O&M cost, E_{new}, is equal to the old one, E_{old}, so no saving in this item.

$$E = E_{old} - E_{new} = 0$$

Also, $k_a = 11.47$, is easily computed from $a = 6\%$, $n = 20$ y.
The maximum investment is, therefore:

$$I_T = Rk_a = 52{,}001{,}841 \text{ €}$$

Energy Markets

7

Abstract

The energy landscape is ever-evolving, with electricity emerging as a special commodity that drives economies and sustains modern life. This chapter delves into the complex realm of energy markets, with a particular focus on the electricity sector. We explore how electricity has transformed from a simple utility to a multifaceted and indispensable resource. The evolution of the electricity sector is a central theme, tracing its roots from centralised generation and distribution to the diversification of energy sources, the rise of renewable, and the integration of smart grids. This historical context sets the stage for understanding the intricate dynamics of today's energy markets. The chapter provides a comprehensive overview of the general structure of wholesale electricity markets, dissecting the key components that govern its functioning. This includes forward markets, where long-term contracts are traded to ensure future supply and demand; day-ahead markets, which facilitate the scheduling of electricity for the next day; intraday markets, allowing for real-time adjustments; balancing markets, which maintain system stability; imbalance settlement, which reconciles deviations from scheduled energy; and ancillary services, critical for grid reliability. Moreover, financial markets play a pivotal role, allowing participants to hedge against price fluctuations and manage risk. By examining these facets of energy markets, readers will gain a profound understanding of the complex interplay between supply and demand, pricing mechanisms, and the critical role of electricity in our modern world. This chapter serves as a valuable resource for anyone seeking insights into the intricate web of energy markets and the vital role of electricity within them.

R. Castro, *Engineering of Power Systems Economics*,
https://doi.org/10.1007/978-3-031-55251-9_7

7.1 Introduction

Energy markets play a vital role in our modern society, serving as the backbone of economic growth and development. They encompass the complex interplay of supply, demand, and pricing dynamics for various energy sources. From fossil fuels to renewables, the energy market landscape has undergone significant transformations in recent years, mainly driven by:

Supply and Demand Dynamics: Energy markets are shaped by the delicate balance between supply and demand for different energy sources. Traditionally, fossil fuels, such as oil, natural gas, and coal, have dominated the market. However, the growing concerns over climate change and the push for cleaner, more sustainable alternatives have led to an increased focus on renewable energy sources like solar, wind, and hydropower. This shift has significantly impacted the energy market, triggering a transition towards greener solutions.

Price Fluctuations and Global Factors: The prices of energy commodities are subject to a myriad of global factors. Geopolitical events, such as conflicts in oil-producing regions or sanctions on energy-exporting countries, can disrupt the supply chain and cause price volatility. Additionally, economic growth, technological advancements, and weather patterns also influence energy prices. For instance, a severe winter or summer can lead to spikes in demand for heating or cooling, respectively, impacting the market's pricing dynamics.

Government Policies and Regulations: In many countries, governments have implemented renewable energy targets and incentives to promote the growth of renewable energy sources. Some governments have also implemented carbon pricing mechanisms, which put a price on carbon emissions and encourage the transition to cleaner forms of energy.

Renewable Energy and Decentralisation: Renewable energy sources have witnessed remarkable growth in recent years, driven by falling costs, government incentives, and public awareness of environmental concerns. The integration of renewable energy into the grid has facilitated a shift towards decentralised energy systems, where consumers can generate their own power through rooftop solar panels or wind turbines. This trend has not only reduced reliance on traditional energy sources but has also empowered individuals and communities to take charge of their energy needs.

Digitisation and Smart Grids: The digital revolution has penetrated the energy sector, giving rise to smart grids and advanced energy management systems. These technological advancements enable real-time monitoring, optimisation, and control of energy distribution, leading to greater efficiency, reliability, and resilience. Smart grids also facilitate the integration of intermittent renewable sources into the existing infrastructure, ensuring a smoother transition towards a more sustainable energy future.

Energy Storage and Electrification: The integration of energy storage technologies, such as batteries, is crucial for enhancing the reliability and flexibility of renewable energy sources. Energy storage allows surplus power to be stored and used during times of high demand, reducing the reliance on conventional backup systems.

Moreover, the electrification of various sectors, including transportation and heating, is gaining momentum. The rise of electric vehicles and the adoption of electric heat pumps are reshaping energy consumption patterns and opening new opportunities for market players.

When we talk about energy markets, we can address two different categories of markets: the wholesale market and the retail market. The wholesale market refers to the buying and selling of energy at the production level, while the retail market involves the sale of energy to end-users such as households and businesses.

In the retail market, retailers compete for end customers just like in a regular market. The retail energy market is where end-users such as households and businesses purchase energy from utilities or retail energy providers. The prices of energy in the retail market are determined by a range of factors, including the cost of production, transportation, and distribution, as well as government taxes and regulations.

This chapter is about the wholesale market, which shows several specific procedures that are worth to be studied in more detail. In the wholesale market, energy is typically bought and sold through an auction-based system known as the power exchange. The power exchange is a marketplace where producers and buyers of energy come together to trade electricity and other forms of energy. The prices of energy in the wholesale market are determined by supply and demand factors, which can fluctuate depending on weather conditions, fuel prices, and other market variables.

Looking ahead, energy markets will continue to evolve in response to climate goals, technological advancements, and changing consumer preferences. The transition towards a low-carbon economy will accelerate, with renewable energy sources occupying a more prominent share. Governments and regulatory bodies are expected to play a crucial role in shaping the market through supportive policies, incentives, and carbon pricing mechanisms. The development of innovative energy storage solutions, the expansion of smart grids, and the advancements in digital technologies will further revolutionise the sector. In this context, energy markets need to adapt to this ongoing evolution, and current market procedures must be rethought.

One of the key challenges facing the energy market is the integration of intermittent sources of energy such as wind and solar power. These sources of energy can be unpredictable and may not always be available when needed, which can create instability in the energy grid. To address this issue, new technologies such as battery storage and demand response systems are being developed to help balance the grid and ensure a reliable supply of energy.

In the present chapter on "Energy Markets", we intend to deliver an overview of how the different energy markets, namely electricity markets, are run. We begin by showing why *electricity is a special commodity* and proceed to explore the *evolution of the electricity sector* from the vertical integration to the unbundling of the different activities and the digitalisation towards the smart grids. The players that act in the energy markets, grid operators, sellers and buyers, balancers, regulators and operators, are introduced in the *general structure of the wholesale electricity market. A note on price mechanisms*, *pay-as-bid* and *pay-as-clear*, is offered. We finish the chapter by presenting the different markets that compose what is commonly known as "The electricity market", namely, *forwards*, *day-ahead*, *intraday*, *balancing*, *imbalance settlement*, *ancillary services* and *financial*. The chapter ends with some application exercises on the topic.

7.2 Electricity: A Special Commodity

Electricity is a versatile form of energy which shows a lot of advantages:

- *Convenience*: Electricity is readily available and can be easily transmitted to homes and businesses. This means that people can use electricity to power their homes and devices with relative ease and convenience.
- *Efficiency*: Compared to other forms of energy, such as fossil fuels, electricity is much more efficient. It can be produced and distributed with minimal waste, and modern electrical devices are designed to use energy efficiently.
- *Clean energy*: Electricity can be generated from clean sources, such as wind, solar, and hydropower. This means that it can be a renewable and sustainable source of energy that does not contribute to pollution or climate change.
- *Versatility*: Electricity can be used to power a wide range of devices and appliances, from light bulbs to electric vehicles. It is also an important component in many industrial processes and is essential for the functioning of modern society.
- *Safety*: Compared to other forms of energy, such as fire or gas, electricity is relatively safe. It can be easily controlled and contained, and modern electrical devices are designed with safety features to prevent accidents and injuries.

Electricity can be consumed in a wide variety of ways, for example:

- Lighting (light bulbs, halogen lamps, televisions)
- Mechanical work (hoovers, washing machines, electric vehicles)
- Heating (cooking, resistive room heating, heat pumps)
- Cooling (refrigerators, air conditioning)
- Electronics (computation, data storage, control systems)
- Industry (electrochemical processes)

However, electricity has its particularities or specificities that must be taken into account when dealing with it. Here are some of them:

- The operation of the power system, composed of the generating facilities, the transformers, the power lines, and the consumers, implies a strict balance between generation and consumption. This means that, at every single time instant, the electricity being consumed must be produced at that very same time instant. A mismatch between demand and supply leads to dramatic blackouts. Therefore, a

minute-by-minute real-time assessment of the equilibrium between demand and supply is required.
- To transport the electricity from the places where it is produced and distribute it to the consumers, a physical infrastructure, the power grid, is needed. The power grid comprises overhead lines and underground cables, that carry the electricity, and the transformers, which adapt the voltage levels to minimise the transmission losses.
- The need for a constant equilibrium between generation and consumption is due to the difficulties in electricity storage. Electricity storage has been known for ages for the widespread use of batteries to store small amounts of electricity. However, even if the technology for storing huge quantities of electricity already exists, it is uneconomical, as of today. The best way to store large volumes of power is hydro-reservoirs, which, however, require proper geographical conditions.
- A large part of the electric demand is of a must-serve nature (residential, hospitals, etc.), a significant restriction to the power system operation.
- The final consumers cannot really differentiate the origin of the product (as well as its quality and nature). The power grid acts as a large buffer, where all the generators deploy their electricity and then it is distributed to the consumers.

Electricity transmission satisfies specific laws, the well-known Kirchhoff's laws. This means that, in a meshed network, power will go from one point to another according to specific rules. So far, the power flow paths are hardly controllable. However, recent advances in power electronics, namely on a family of new devices called FACTS (Flexible AC Transmission Systems), promise to be able to control active and reactive power according to the needs.

The European electrical grid is interconnected, which significantly improves power system operation, as it allows emergency help in case of failures and provides cross-border trading opportunities, up to the available transfer capacities. In Europe, the available Net Transfer Capacity (NTC) is managed and published by the Association of TSO—ENTSO-E (European Network Transmission System Operator for Electricity) –, publicly available in the transparency platform (https://transparency.entsoe.eu/).

The *law of conservation of energy*, also known as the *first law of thermodynamics* states that energy cannot be created or destroyed in an isolated system, only transferred from one form to another. This means that the total amount of energy in a system remains constant, even if it is converted from one form to another. Despite being widely used, the expression "generation" or "production" of electricity is not accurate, because electricity is not produced, it is actually transformed from other forms of energy.

The primary sources of energy which are converted into electrical energy are:

- Chemical energy, accessed by combustion (coal, gas, oil, biomass).
- Nuclear energy, accessed by fission reactions.
- Hydroelectric energy, allowing water to flow downhill.
- Wind energy, converting wind speed.

- Solar energy, accessed with photovoltaic (PV) panels or concentrating solar power (CSP).
- Geothermal energy.

The electrical power supplied by the different power stations does not hold the same features.

Power can be firm or variable. It is said the power supplied by a power point is firm if it is supplied whenever necessary, meaning that the primary energy is controllable in order to supply electrical power at the time it is effectively required. For instance, a Combined Cycle Gas Turbine (CCGT), burning natural gas, delivers firm power because if power is to be supplied at a particular time, it is just a matter of burning gas at the required moment. On the contrary, supplied power is said to be variable if the primary energy is not controllable, meaning that the supply of power is delivered independently of the needs. An example is wind power which delivers power as a function of the input wind speed (an uncontrollable resource) and not exactly when needed. Synonyms for *firm* are *dispatchable* or *controllable*, and for *variable* are *non-dispatchable* or *non-controllable*.

On the other side, electrical power can be delivered with different levels of flexibility. Flexibility means the ability to regulate the amount of power delivered. For instance, a hydropower plant with a dam (reservoir) delivers highly flexible power because the output power is easily regulated by opening or closing the gates that control the water flow input into the turbine. PV power shows a very low degree of flexibility because the output power depends on the input solar power (irradiance) which is not controllable. It should be mentioned that it is possible to deliberately operate a PV power plant or a wind power plant below the output power corresponding to the given meteorological conditions so as to have a reserve margin to increase the output power if necessary. This operation mode is made possible by appropriate control of the power electronics converters. This possibility is to be used as an emergency resource as it implies wasting renewable resources. Of course, the opposite situation —to operate above the output power determined by the meteorological conditions and having a margin to decrease—is not possible at all.

Table 7.1 shows said features of the main types of power plants.

It is noted that biomass is considered a low-carbon technology as it is assumed that the regrowth of biomass compensates for the CO_2 emissions associated with the combustion of the biomass.

7.3 Evolution of the Electricity Sector

In the past, and still in some countries today, the generation and consumption of electricity were done centrally by "vertically-integrated" monopoly utilities that owned generating assets, networks, and retailing. It was the time when, in each country, a single big state-owned utility managed all the power system. These utilities owned all the infrastructure, so it was hard for third-party generators to compete.

Table 7.1 Generation technologies features

Type	Fuel	Firmness	Flexibility
Coal	Fossil	Firm	Medium
Natural gas	Fossil	Firm	High
Biomass	Renewable	Firm	Medium
Nuclear	Nuclear	Firm	Low
Hydro w/ dam	Renewable	Firm	Very high
Hydro run of river	Renewable	Variable	Medium
Solar PV	Renewable	Variable	Very low
Wind	Renewable	Variable	Very low
Geothermal	Renewable	Firm	High

From the 1980s onwards, countries began to liberalise their electricity sectors with the goal of introducing competition and reducing costs. The activities of generation, transmission, distribution and retail were unbundled, and regulated competition was allowed. Transmission and distribution were considered natural monopolies and became regulated activities. Market-driven features were recognised in generation and retail that become competitive activities.

At the same time, it was recognised that electricity should be traded in the framework of an organised market—an arrangement to transfer electric energy from producers to consumers –, governed by supply and demand laws. This is called the wholesale market or power exchange. On the other end of the value chain, a retail market, where retailers compete to supply the customers, also exists. Regulatory agencies with the task of regulating the electricity sector were created. We will see later on that power can be negotiated in another way, as bilateral contracts which are traded Over The Counter (OTC).

Power exchange in Europe refers to the trading of electricity between different countries in the region through various market mechanisms. The main power exchanges in Europe include:

- EPEX SPOT–This is a European electric power exchange operating in Austria, Belgium, Denmark, Finland, France, Germany, Great Britain, Luxembourg, the Netherlands, Norway, Poland, Sweden, and Switzerland. In 2015, EPEX SPOT integrated its business with the former APX Group, which operated the power spot markets in Belgium (through Belpex), the Netherlands, and the United Kingdom.
- Nord Pool–This is the largest power exchange in Europe and operates trading platforms in the Nordic countries, Germany, the UK, and many other geographies.
- OMIE–This is the Iberian power exchange and operates a trading platform for electricity in Spain and Portugal.

In many countries in Europe, there are several power exchanges in operation in the same country. The European Union Agency for the Cooperation of Energy Regulators (ACER) publishes information about the situation in each European country.

This vast transformation of the electricity sector was carried out in the framework of a shift towards renewable energy sources, such as wind and solar power. Advances in technology have made renewable energy more cost-competitive with traditional sources, and government policies have encouraged their adoption.

In recent years, the electricity sector has been undergoing a digital transformation, with the development of smart grid technologies that allow for greater integration of renewable energy sources and more efficient use of electricity. This has the potential to reduce costs and improve reliability but also raises concerns about cybersecurity and data privacy.

Overall, the electricity sector has undergone significant changes over the years, driven by technological innovation, environmental concerns, and changing market conditions. As we move forward, the sector will likely continue to evolve in response to these and other factors.

7.4 General Structure of Wholesale Electricity Markets

It is not possible to run an electricity market for every single second, for practical reasons, the most important of which is that the network must be checked for stability, i.e., to check if the grid is stable given the power flows that result from the market.

Time is needed to perform this check, so electricity is traded in blocks of time, e.g. hourly, or quarter-hourly, well before the time when it is consumed, normally in the day before. This is called the *day-ahead market*, where the pre-trading is based on forecasts, both for the generation and demand.

Forecasts have always an associated uncertainty because it is impossible to predict, with 100% certainty, what the generation and demand will be the next day, additional intraday markets allow market agents to adjust the day-ahead market's resulting schedule according to expected needs in real-time. Near real-time, further market trades in backup balancing power. This real-time market is called the *balancing market*.

There are several players in the market, each with a particular role.

Grid operators

The Transmission System Operator (TSO) operates the transmission assets and is responsible for the power balance on the transmission system, ensuring the quality and security of the electricity supply, and coordinating the operation of the transmission system with other electrical systems. It should be independent of other market players as producers and consumers.

Normally, each country has a single TSO, either a private company operating under a public concession, or a state-owned company. When there is more than one

TSO in one country, each one manages a different (regional) part of the transmission grid.

In addition to managing the extremely high voltage transmission grid, the TSO is required to maintain a continuous balance between electricity supply from power stations and demand from consumers, and also ensure the provision of reserves to solve for sudden failures in the power system. This function makes the TSO to be the overall manager of the power system. Its service is normally specified in grid codes established as part of the electricity market.

In Europe, the TSO are brought together in the European Network of Transmission System Operators for Electricity (ENTSO-E) association.

The other grid operator is the Distribution System Operator (DSO). A DSO is an entity responsible for operating, maintaining, and developing the electricity distribution system (involving high, medium and low voltages) that delivers electricity from the transmission system to end-users, such as households, businesses, and industrial consumers.

The DSO also plays a role in facilitating the integration of renewable energy sources (RES) into the distribution system. This includes coordinating the connection of RES, storage systems, and electric vehicles (EV) on the distribution system, and ensuring that it can accommodate the increasing use of these distributed resources.

Depending on the country, there may exist only one DSO for the whole country or regional DSO which operates regional parts of the distribution grid. As in the TSO case, the DSO can be a private company operating under a public concession contract or a state-owned company and should be independent of the other market participants. It can happen, but it is not frequent in Europe, that the DSO can additionally act as a retailer, buying electricity in the market and selling it to its clients.

Both the TSO and the DSO are typically regulated by regulatory independent agencies to ensure that they operate in the public interest and meet the requirements of the electricity market.

Sellers and buyers

The generating companies, named *Genco* in the USA, own the production assets and offer to sell in the market a specific amount of energy (MWh) for a specific price (€/MWh), in different time frames.

The generating companies can own a single big controllable power plant (coal, nuclear, gas, large hydro) or, more often, own a portfolio of production facilities. The portfolio is especially important to efficiently manage RES. Let us explain why. In the day-ahead market, the producer commits to sell a certain amount of electricity for a certain price at a certain hour of the next day. If the producer is unable to fulfil his promise, he has to pay penalties for non-compliance with the promise. RES production is difficult to predict, therefore a producer owning only RES will hardly survive in the market because RES are non-controllable. If a producer owns a portfolio composed of different technologies, controllable and non-controllable, he can compensate for the uncontrollability of the RES with controllable resources, namely, gas, or storage facilities, in the future. This enforces the role of storage in the future power system.

The retailer is an entity that buys electricity from the wholesale electricity market, to then sells it to the end consumers.

The retailers offer to buy a specific amount of energy (MWh) for a specific price (€/MWh) at a certain time frame, for instance, one hour. In general, retailers are private companies that act on behalf of their consumers. Retailers compete with each other in another market—the retailing market—to provide electricity supply service to the customers.

In many countries, there is a special retailer, the *Last Resort Trader* (LRT), also known as *Supplier of Last Resort* (SoLR), which represents the consumers that, for some reason, for instance, vulnerable consumers, are unable to participate in the retailing market. The LRT is a regulated private company operating under a public concession agreement or a state-owned company. In the wholesale market, the LRT is a retailer like any other.

Finally, we have the consumers (large and small), which are those who eventually use electricity for any purpose, from watching TV to heating to industrial production processes. In general, they are not allowed to directly participate in the market. They participate indirectly in the market, through the retailers, which buy electricity on their behalf.

Balancers

As stated before, in the day-ahead market, both producers and consumers commit to sell and buy, respectively, a certain quantity of energy for a certain unit price. The compromise has to be honoured, otherwise, penalties apply. This is what ensures the balance between generation and consumption, or, in other words, between supply and demand.

Producers and retailers are required to either be balance responsible themselves or arrange for some other player to be balance responsible on their behalf. The entities responsible for ensuring that what has been promised is actually delivered are called *Balance Responsible Parties* (BRP), who are in charge of all financial responsibilities.

Regulators and market operators

The regulator is responsible for the market design and for issuing its specific rules. It also monitors the market to spot and prevent misbehaviour (collusion, abuse of market power, etc.). The regulator works to promote competition in the market by ensuring that all market participants have equal access to the market and that no participant is given an unfair advantage. The regulator does not interfere in the formation of the market price, which is strictly market-driven and non-regulated.

Overall, the role of the regulator in the wholesale electricity market is to ensure that the market operates in a fair, efficient, and transparent manner. Regulators oversee the operation of the market and ensure that it operates in a way that promotes competition, encourages investment in new generation capacity, and provides consumers with reliable and affordable electricity.

The role of the market operator in the market is to manage and operate the market platform where electricity is bought and sold. The market operator is responsible for ensuring that the market operates efficiently and effectively and that all participants have equal access to the market. This may include the definition of bid products and bid forms, set up and maintenance of the trading platform, daily matching of supply and demand offers, developing the rules and protocols for trading, establishing the technical infrastructure for the market, and managing the flow of information between market participants.

When referring to the wholesale market, we are basically considering two types of markets:

- Energy markets, where energy is exchanged between producers and retailers and is the central place for the optimal (from the market point-of-view) scheduling and settlement of energy exchanges.
- Ancillary services market, where any type of service that supports power system operations is traded directly by the system operator.

Another type of market, run outside the wholesale market, is the *capacity market*, where the system operator ensures that enough generation capacity is present for reliable system operation in future years and at competitive prices.

In a capacity market, generators bid for long-term contracts to provide a certain amount of firm power for a specified period of time, typically several years in advance. These contracts guarantee payment to the generator regardless of whether or not the capacity is actually used to produce electricity. The goal of a capacity market is to ensure that there is enough electricity generation capacity available to meet peak demand, thereby reducing the risk of blackouts or other grid failures.

However, the use of capacity markets is a subject of ongoing debate, as some critics argue that they can lead to inefficiencies and higher costs for consumers.

The general structure consisting of two overall types of markets inside the wholesale market, mentioned before, can be further detailed as depicted in Fig. 7.1, where the electricity markets are sequenced by maturity.

Two kinds of transactions can happen in electricity markets:

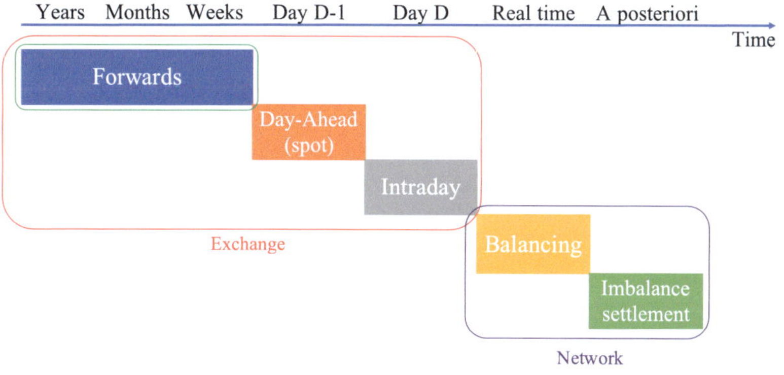

Fig. 7.1 Electricity markets sequenced by maturity

- Trades on energy exchanges (trading markets), where standard products with a standardised bid/offer procedure and clearing mechanism (price formation following market rules) are used.
- Bilateral contracts traded Over The Counter (OTC) where either standard or specially designed products are used.

Electricity may be traded well in advance in relation to Day D when it is really generated and consumed. This exchange takes place in the *forwards markets*. In this market, energy trading can be realised either using standard or tailor-made products.

On the day before, producers and retailers place their offers for the next day in the *day-ahead market*. Normally, markets are run on an hourly basis, meaning that the offers are placed for each hour of the next day. Usually, most of the electricity is traded in this market, but it depends on the specific market. Both generation and demand offers are based on the best forecasts available at the moment.

Between day D-1 and day D, producers and retailers are allowed to make adjustments to their initial bids (made in the day ahead) based on more recent and enhanced forecasts. These adjustments can be made in the *intraday market*, on several moments (called sessions), or continuously.

All the above-mentioned markets are managed by the market operator.

And then the real time comes and the TSO steps forward. The Balance Service Providers (BSP), such as generators or demand facilities, balance out unforeseen fluctuations in the electricity grid by increasing or reducing their power output as requested by the TSO. Each balancing energy trade from a BSP is assigned to a Balance Responsible Party (BRP).

The TSO manages the *balancing market* (other names are adjustments market or real-time market). *A posteriori*, the entities that caused the imbalances must pay the bill plus a penalty.

7.5 A Note on Price Mechanisms

Pay-as-bid (PAB) and pay-as-clear (PAC) are two pricing mechanisms commonly used in energy markets, particularly in the context of electricity markets. These mechanisms determine how generators are compensated for the electricity they produce and how consumers are charged for the electricity they consume. Each mechanism has its own advantages and disadvantages, and the choice between them can significantly impact the functioning of the energy market. Below, we will elaborate on the key differences between pay-as-bid and pay-as-clear:

Pay-as-Bid (PAB)

Fixed Pricing: Under the PAB mechanism, generators are paid the prices they bid into the market. Each generator submits a bid specifying the quantity of electricity they are willing to supply at a certain price. These individual bids are then ranked from lowest to highest price.

Generators Get Their Bid Price: Generators are compensated based on their own bid price, regardless of the market-clearing price. If a generator's bid price is higher than the market-clearing price, they are paid at the higher bid price.

Price Certainty for Generators: PAB provides price certainty for generators, as they are guaranteed to receive the price they bid, even if the market-clearing price is higher.

Risk for Consumers: PAB can lead to higher electricity prices for consumers if many generators submit high bids. Consumers may end up paying more than they would under a PAC mechanism.

Inefficient Outcomes: PAB can lead to inefficient outcomes if generators strategically submit high bids to maximise their profits, resulting in higher prices for consumers.

Pay-as-Clear (PAC)

Market-Clearing Price: In the PAC mechanism, generators are paid the market-clearing price, which is the price at which supply and demand in the electricity market are balanced. The market-clearing price is typically determined through an auction or a clearing process.

Efficient Pricing: PAC is designed to encourage efficient pricing by ensuring that generators are paid the market price, which reflects the true cost of supplying electricity at a given point in time.

Price Volatility: PAC can lead to price volatility, as the market-clearing price can change rapidly in response to fluctuations in supply and demand. This volatility can benefit consumers during periods of low demand and low prices but can also result in high prices during periods of scarcity.

Revenue Uncertainty for Generators: Generators face revenue uncertainty under PAC, as they are not guaranteed a specific price for their electricity. Their earnings depend on market conditions.

Risk Management: PAC may incentivise generators to engage in risk management strategies to mitigate revenue uncertainty, such as entering into financial contracts or using advanced forecasting techniques.

In summary, the main difference between pay-as-bid and pay-as-clear in energy markets is how generators are compensated. PAB provides price certainty for generators but can lead to inefficient outcomes and higher prices for consumers. PAC, on the other hand, pays generators the market-clearing price, which can result in efficient pricing but introduces revenue uncertainty and price volatility.

Usually, the day-ahead and the intraday markets are operated using the pay-as-clear mechanism, and the balancing market using the pay-as-bid mechanism.

7.6 Electricity Markets

7.6.1 Forwards Markets

Forward electricity markets are markets where electricity is bought and sold for delivery at a future date, typically ranging from a few days to several years in advance. These markets allow electricity suppliers and consumers to manage their exposure to price fluctuations and hedge their risks.

Forward electricity markets can be used by generators to lock in a price for their future electricity production, which can provide stability and predictability for their business operations. Similarly, large consumers such as industrial companies or utilities can use forward markets to secure a price for their future electricity needs and manage their budget.

The forward markets are usually divided into two categories:

- Futures—long-term standardised instruments traded in organised venues.
- Forwards—products that are unstandardised and traded OTC (i.e., on a bilateral basis).

7.6.2 Day-Ahead Market

The day-ahead market is usually the market where most of the energy is traded. In the future, the situation may change with new market design proposals. As mentioned before, we assume a pay-as-clear price mechanism.

At a predefined hour, the producers and retailers place their bids and offers for the 24 hours of the next day. For each hour, each bid/offer is composed of a pair of values: the amount of energy to be generated, or consumed (MWh), depending if it is a producer or a retailer, and the corresponding unit price they are willing to receive (producers) or pay (retailers) (€/MWh). It is noted that sometimes the energy is expressed in MW. This is because, for each hour, MW and MWh are represented by the same number (1 MWh = 1MW × 1 h).

All generation bids and consumption offers are placed at the same time and each participant does not know about the other participants' bids and offers.

The market operator ranks the generation bids in ascending (upward) order, from the bid with the lowest price to the bid with the highest price. The consumption offers are ranked in the opposite way, i.e., downward, from the offer with the highest price to the offer with the lowest price. This is called the *merit order*.

The point where the two curves, the supply curve (generation bids) and the demand curve (consumption offers), meet defines the *market price*, or *clearing price*, or *spot price*.

All producers that have placed bids and all retailers that have placed offers to the left of the interception point are accepted to participate in the market. Each one will receive or pay if it is a producer or a retailer, respectively, the market price and must supply (consume) the amount of energy it has committed to.

Eventually, the TSO is informed about the trades that occurred, because the network must be checked for stability and security taking into account the power flow that resulted from the market. The TSO must check if the voltages at the buses are in between the operating limits if the power flow in each line and transformer does not exceed the technical maximums and if the power flow in the interconnections (that connect two adjacent bidding zones) is lower than the maximum defined for that hour. If it is higher than the maximum, we say the *interconnection is congested.*

If a security issue is identified by the TSO (interconnection congestion), in the operational planning phase, the TSO can impose restrictions on the operation of the market (before or after market clearance), in order to solve said problems. If non-compliance persists and is impossible to settle with such restrictions, the market has to be split into two bidding zones (see below on what is market splitting).

All producers' bids and all retailers' offers located to the right of the interception point are excluded from the market. What happens to the bids and offers that were excluded? They are not accepted in the day-ahead market, but respective producers and retailers are free to apply to the intraday markets (see next section) or to the balancing market (see below).

The process that has just been described is called the *clearing mechanism* and, as said before, is carried out by the market operator. The market price is established by the highest bid that is necessary to supply the demand. In a more complex way, it is said that all participants pay or are paid the highest price to clear the incremental MW of demand. Bids higher than the market price are not needed to satisfy the demand, therefore, they are excluded from the day-ahead market.

As seen in Chap. 3, the area below the demand curve and above the supply curve, till the equilibrium point, measures the *social surplus*, also known as *social welfare*. Also as seen the equilibrium point (market price) is where the social welfare is maximised, because any buyer is to pay at most what he was ready to pay and any seller will get at minimum the price he was ready to sell for.

The day-ahead market was designed considering the generation bids are made at marginal cost, which is why this type of market is called *marginalist market*. We recall that the marginal cost is the variable cost, i.e., the focus is solely on the costs that vary based on the amount of energy produced, comprising the fuel cost, CO_2 emissions cost, and variable O&M cost. This is supposed to happen, but, of course, producers may bid at any price they want (except if the regulator finds the bids to be abusive and blocks them) and take risks for the options they make (one of which is ending outside the market).

Renewable producers (wind and PV) usually bid at 0, or even negative, if the specific market rules so allow it, because they have 0 marginal costs. Of course, they know they will not receive 0 but will receive the spot price instead. Wind and PV producers are usually named vRES—Variable Renewable Energy Sources, precisely due to their inability to regulate the output power. Hydropower is a Renewable Energy Source (RES) that usually does not bid at 0 because they have 0 variable costs but they profit from an opportunity cost given its storage capabilities.

Table 7.2 Supply bids. Example 7.1

No.	Company	Amount (MWh)	Price (€/MWh)
1	P#1	120	0
2	P#2	50	0
3	P#3	200	15
4	P#1	400	30
5	P#2	300	50
6	P#1	70	60
7	P#2	50	70
8	P#4	160	110

Table 7.3 Demand offers. Example 7.1

No.	Company	Amount (MWh)	Price (€/MWh)
1	R#1	250	140
2	R#2	400	90
3	R#3	250	70
4	R#3	100	40
5	R#1	100	20

Example 7.1

Consider the generating bids and consumption offers as depicted in Tables 7.2 and 7.3, respectively.

1. Compute the market price (€/MWh) and the total energy traded (MWh).
2. Repeat considering that, ceteris paribus, generation bids at 0 double the amount of energy.

◀

Solution

1.

A graphical representation of the clearing mechanism would help in understanding what is happening (see Fig. 7.2).

Producers are willing to sell 1350 MWh and retailers are willing to buy 1100 MWh. The supply and demand curves meet at a point (900 MWh; 50 €/MWh), which defines the total energy traded and the respective spot price.

Demand offers No. 1, 2, and 3, will pay 50 € per each MWh each one takes to deliver to their end-consumers. Demand offers No. 4 and 5 are offered at a price lower than the spot price, so they take nothing (and pay nothing, of course).

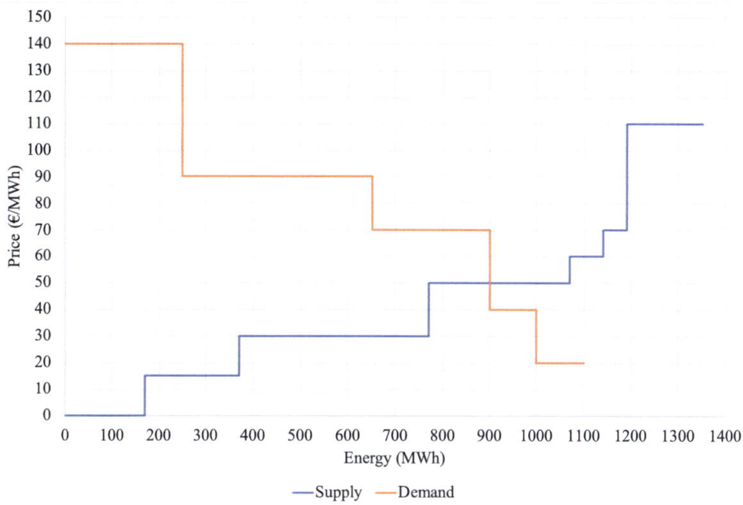

Fig. 7.2 Clearing mechanism. Example 7.1

Generation bids No. 1, 2, 3, and 4 will receive 50 € per each MWh each bid. Bid No. 5 was willing to sell 300 MWh for 50 €/MWh; however, they are going to sell only 130 MWh (900–770) and will receive 50 € per each. Bids No. 6, 7, and 8 bid at a price higher than the spot price, therefore they are excluded from the day-ahead market.

2.

Now, the bids at 0 €/MWh total $2 \times 170 = 340$ MWh. This represents an increase in bids from RES producers. All the rest remains unchanged. In Fig. 7.3, we have added to Fig. 7.2 the new generation bids ranked upward from the lowest to the highest.

It is noted that the supply curve was shifted to the right. This implies the new spot price is reduced from 50 €/MWh to 40 €/MWh as a consequence of more RES being offered in the market. Furthermore, 940 MWh of energy is now traded instead of 900 MWh. Bid No. 5 is now excluded from the market as it bids higher than the new market price. Demand offer No. 4 is now able to buy 40 MWh out of 100 for 40 €/MWh. ◄

Market coupling and market splitting

An electricity market involves typically more than one country, at least two, as is the case of the Iberian Electricity Market (MIBEL), which is composed of Portugal and Spain. Therefore, power flows in the interconnection lines: from country A to country B, or the opposite, depending on the market outcome. The TSO of both countries are responsible for fixing, and making it available to the market beforehand, the *Available Transfer Capacity* (ATC), i.e., the maximum power that can flow in the

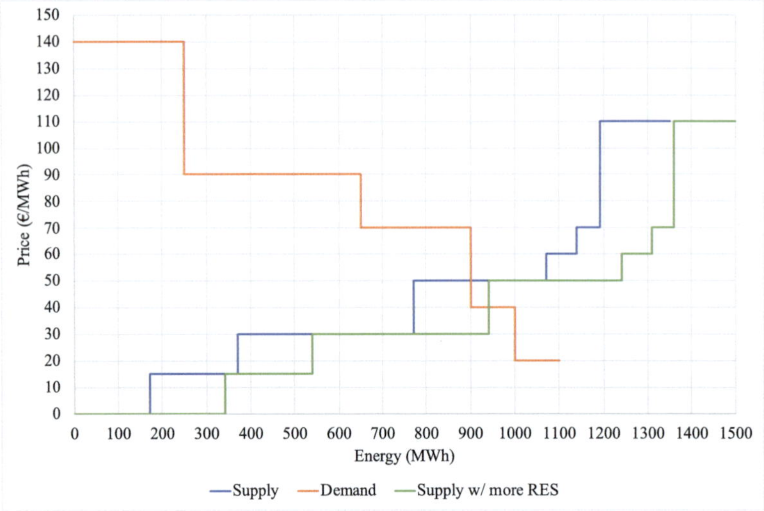

Fig. 7.3 Clearing mechanism with more RES. Example 7.1

interconnection lines. There is an ATC for importing and, normally, a different one for exporting, reflecting the different conditions in both countries.

If, following the market clearing process, no transfer constraints violations (no *congestions*) occur, i.e., the projected power flow in the interconnection is less than the ATC, there will be a single price for the clearing area and the market is coupled. On the contrary, if at least one transfer capacity constraint is violated (the interconnection is *congested*), two prices emerge, one for each bidding zone, and the market is said to be split. In this case, the power flow at the interconnections is fixed and equal to the ATC and each country (bidding zone) determines its own market price, separately but taking into account that the power at the interconnection is the ATC. Depending on the generation and demand conditions, this can happen in one hour and not in the next hour.

The direction of the power at the interconnection is always from the country with the lower price to the country with the higher price. This means that the country with the higher price is taking an unfair advantage because it is buying power at the interconnection at a lower price and reselling it to its consumers at a higher price. This unfair advantage is withdrawn because the country importing power is due to pay a rent, called the *congestion rent*, to a common fund that will be used to expand the interconnection capacity between the two countries and other applications. The congestion rent is given by the product of the price difference by the ATC.

Table 7.4 Generation bids. Example 7.2

Technology	Portugal		Spain	
	MW	€/MWh	MW	€/MWh
Renewables	5,000	0	15,000	0
Hydro	5,000	0	10,000	0
Nuclear			5,000	10
CCGT	5,000	50	20,000	50

Example 7.2

At a certain hour of a certain day, the producers deliver to MIBEL (the electricity market of Portugal and Spain) market operator the following bids (Table 7.4).
At the same hour, the consumption in Portugal and Spain was 5,000 and 32,000 MWh, respectively, and the interconnection capacity—ATC—was:

- PT → SP: 4,000 MW
- SP → PT: 2,000 MW

1. Compute the prices and quantities set in MIBEL at this hour, knowing that Spain is exporting 1,000 MW to France and there is no power flow to Morocco.
2. If, in this case, a congestion rent exists, compute its value.

◀

Solution

1.
First, we should simulate the clearing mechanism without considering the interconnection constraints given by the ATC. The market situation is represented in Fig. 7.4. The total aggregated demand is 38,000 MWh, from Portugal, Spain, and exports.
We conclude that, following the admission of all renewables and hydro (from both countries) in the market, a power flow of 5,000 MW in the interconnection, from Portugal to Spain, is forecast. As this exceeds the ATC (PT → SP), the market has to be split. The power flow at the interconnection (fixed at 4,000 MW from Portugal to Spain) is leaned against this limit and a separate market is cleared for each country.
Figures 7.5 and 7.6 show the market clearing in Portugal and Spain, respectively. Portugal is unable to export to Spain all the hydro production because only 4,000 MW are allowed at the border. This hydro resource will be lost, assuming that it is a run-of-river (no storage) resource. The price in Portugal is 0 €/MWh.

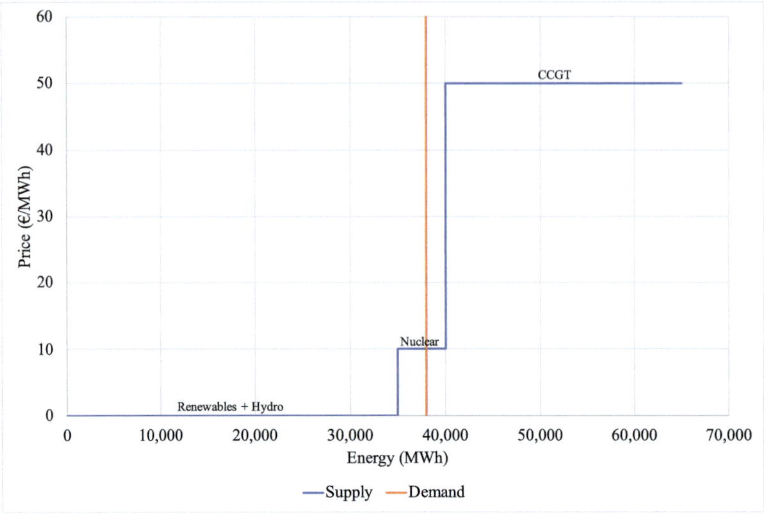

Fig. 7.4 Clearing mechanism with no consideration of ATC. Example 7.2

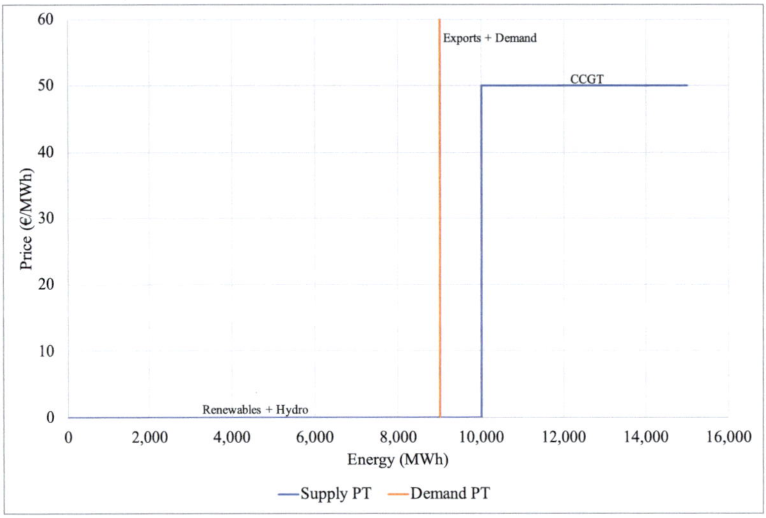

Fig. 7.5 Clearing mechanism in Portugal. Example 7.2

Spain imports 4,000 MW from Portugal, dispatches all the renewable and hydro production, and needs another 4,000 MW of nuclear energy to satisfy the total demand, including exports, of 33,000 MWh. The price in Spain is 10 €/MWh.

2.

The market has been split, so there is a congestion rent to be paid, in this case, by the Spanish TSO.

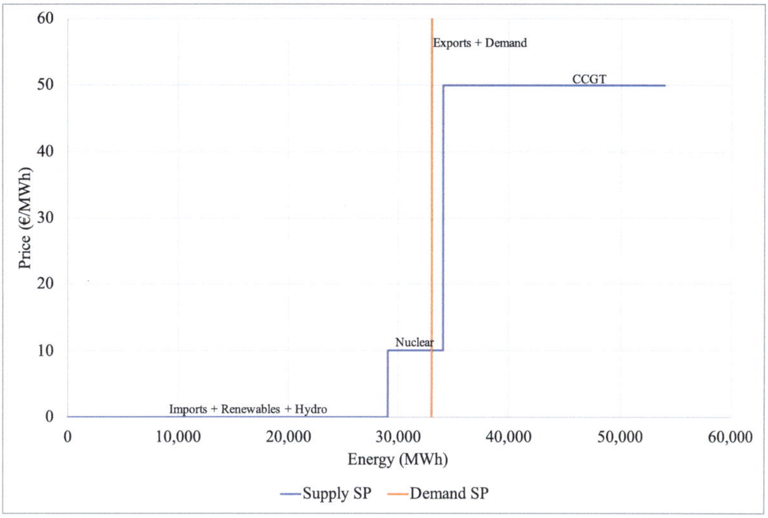

Fig. 7.6 Clearing mechanism in Spain. Example 7.2

The congestion rent is:

$$CR = \Delta p \times ATC = (10 - 0) \times 4{,}000 = 40{,}000 \text{ €}$$

◀

7.6.3 Intraday Market

The intraday market is the marketplace where participants can buy and sell electricity on short notice (for the next hours), usually for delivery on the same day. In relation to the day-ahead market, the intraday market provides an additional layer of flexibility, allowing participants to adjust their bids and offers throughout the day to account for changes in demand and supply that in the meantime were bought to their knowledge due to updated forecasts.

The intraday market is particularly important for renewable energy sources, which can be unpredictable and fluctuate rapidly. By participating in the intraday market, generators can adjust their positions to account for recently known changes in renewable energy output.

There are several sessions of the intraday market, in addition to the *continuous intraday market*, where both generators and retailers can adjust their bids and offers for the next hours. The clearing mechanism applied to the intraday sessions is similar to the day-ahead market. In the continuous intraday market, incoming orders are executed one by one based on a "first come first serve" (FCFS) principle. Any partial execution of an incoming order concludes one commercial transaction resulting in one trade with a specific trade price.

7.6.4 Balancing Market

As we know, the most important restriction of the power system is that, at every instant, the production must equal the consumption plus the losses, i.e., the system must be balanced.

Until real-time, several markets have already occurred—day-ahead and intraday (several sessions and continuous)—allowing for a closer match between generation and demand, based on increasingly improved forecasts. In real-time, previous predictions are still not certain, and, in addition, last-minute network congestions or loss of some power generation units might occur. These events trigger the need for both production and consumption adjustments. These last-minute adjustments are performed in the *balancing market*, which, moreover, sets a transparent price for the cost of the imbalances.

The balancing market happens close to real-time and is operated by the TSO. It is run as follows in two stages. BSP submit bids to increase or decrease their production (or consumption). This is a declaration of availability. For instance, in a particular hour, BSP#1 submits a 90 MW regulation band offer, divided into 60 MW upward and 30 MW downward, at a price of 100 €/MW. This means that this BSP is available to increase production by 60 MW, or to decrease it by 30 MW, if necessary. The TSO selects the cheaper bids that are necessary to ensure the safe operation of the system and pays-as-bid. This is a payment for availability, not for actually supplying electricity.

In the second stage, if, for instance, more electricity is really needed, the TSO mobilises the available reserves and instructs either a power station to generate more, or a flexible customer to consume less. If less electricity is needed, the TSO accepts a bid, and either a power station will turn down or a flexible customer will consume more.

The TSO receives all the bids and ranks them by economic precedence, from the cheaper to the more expensive ones. The TSO then selects all the bids necessary for the real-time production-consumption equilibrium and additionally pays the actually used electricity to the BSP as a bid (pay-as-bid). Later the participant that caused the imbalance will receive the bill.

7.6.5 Post-Market Imbalance Settlement

After the trading period, measurements of actual generation and consumption are collected. For each market participant, the system operator calculates a balance, for example as balance = generation + purchase – consumption – sales. The imbalance price (purchase and sales) is based on prices in real-time trading, i.e., on up- and down-regulation prices.

The post-market imbalance settlement ensures the costs of any shortage or surplus of electricity, bought or sold as balancing energy by each TSO, are allocated to the market participants that caused the imbalance. A penalty is added to emphasise the strong pressure on avoiding imbalances.

7.6.6 Ancillary Services Market

The power system is to be operated in absolutely secure conditions. Relevant quantities, such as voltage and frequency, must be controlled strictly. To control the voltage, reactive power trading is necessary; to control the frequency, operating reserves should be managed. All these services are provided in the ancillary services market, managed by the TSO. The ancillary services market can be jointly operated with the balancing market or be a separate market.

An operating reserve is a generation that can be mobilised with a short-term notification, helping to recover system balance by compensating for an expected loss of a large generator or a grid outage. The aim is frequency control. There are several types of reserves:

- FCR–Frequency Containment Reserve. FCR intervenes automatically within seconds to restore the balance between supply and demand. The FCR, also known as primary reserve, is the first response to frequency disturbances. If the deviation persists, the aFRR subsequently replaces the primary control reserve.
- aFRR–Frequency Restoration Reserve with automatic activation, also known as secondary reserve, provides the reserve within 5 min. As the secondary reserve, aFRR replaces the FCR gradually after 30 s.
- mFRR–Frequency Restoration Reserve with manual activation. The mFRR, also known as tertiary reserve, helps to restore the 50 Hz frequency when there are longer-lasting deviations that cannot be resolved solely by the other upstream balancing services (FCR or aFRR). mFRR must be fully deployable after 12.5 min.
- RR–Replacement Reserve is the last resource to keep frequency under control. It includes operating reserves with activation time from 15 min up to hours.

Voltage is controlled through reactive power trading. In general, to increase the voltage, it is necessary to inject reactive power; to decrease the voltage, reactive power consumption is needed. In traditional generators, called alternators, reactive power control is achieved by the rotor field current (excitation system). Modern power electronics interfaced generators can easily control reactive power flow with the grid. Also, there are power electronics devices, belonging to a family called FACTS–Flexible AC Transmission Systems, for instance, a STATCOM (STATic synchronous COMpensator), that allow for reactive power control and, therefore, are voltage supporters.

7.6.7 Financial Markets

The several types of markets that we have introduced before are called *physical markets* because they involve the physical trading of the electricity product. There are other types of electricity markets that do not involve physical trading of the product but instead are related to financial trading—financial markets.

Examples of financial markets are:

- Options—Suppose company B buys power from the spot market. If the price exceeds X €/MWh, then Company A pays Company B the difference between the spot price and X. Company B pays company A a fee for this agreement.
- Futures—Again, suppose company B buys power from the spot market. If the price exceeds X €/MWh, then Company A pays Company B the difference between the spot price and X. If the price is less than X €/MWh, then Company A receives the difference between the spot price and X from Company B.

7.7 Conclusion

Electricity's uniqueness as a commodity cannot be overstated. Unlike other goods, it is uneconomical (as of today) to be stored in vast quantities, and its production and consumption must be balanced instantaneously. It powers our homes, fuels our industries, and drives technological innovation. This special status compels us to delve deeper into the mechanisms governing its trade.

Historically, centralised generation, transmission, distribution, and retail were the norm. However, as we embarked on a journey toward sustainability and efficiency, the sector evolved dramatically. The integration of renewable energy sources, the deployment of smart grids, and the proliferation of distributed energy resources are reshaping the sector. The sector's transformation not only reflects technological advancements but also society's growing commitment to sustainability. As we continue to prioritise renewable energy sources and transition to low-carbon alternatives, the electricity sector stands at the forefront of global efforts to combat climate change.

Within the realm of electricity markets, a complex ecosystem emerges, comprised of various actors and mechanisms. Grid operators oversee the physical infrastructure and ensure the reliable transmission of electricity from generators to consumers. Generators sell electricity to the market, while utilities and end-users are the buyers. The interactions between these entities establish supply and demand dynamics. Balancing market operators play a crucial role in maintaining grid stability by matching real-time electricity supply with demand, and making minute-by-minute adjustments to ensure equilibrium. Regulatory bodies set the rules and oversee market operations, while market operators facilitate trading, enforce regulations, and ensure fair competition.

In what concerns price mechanisms, pay-as-bid (PAB) and pay-as-clear (PAC) are two distinct pricing mechanisms prevalent in energy markets. Under PAB, electricity generators receive compensation based on the prices they individually submit in their bids, regardless of the market-clearing price. In contrast, PAC compensates generators at the market-clearing price, which fluctuates with supply and demand dynamics.

Understanding the broader structure of electricity markets leads us to dissect the various market segments that make up the whole. These segments are interconnected and serve distinct purposes. Forward Markets allow participants to enter into long-term contracts for the future delivery of electricity, offering price stability and predictability. In the day-ahead markets, market participants schedule electricity generation and consumption for the following day, ensuring a balance between supply and demand. Real-time adjustments are made in intraday markets, accommodating unforeseen changes in supply or demand. Balancing markets are instrumental in maintaining grid stability, and addressing fluctuations in electricity supply and demand as they happen. Imbalance Settlement reconciles deviations from scheduled energy, ensuring that market participants adhere to their commitments and maintaining market integrity. Ancillary services, such as frequency regulation and voltage support, are vital for grid reliability, ensuring that the system operates smoothly and safely. Finally, participants in financial markets can hedge against price volatility, manage risk, and engage in speculative trading, adding depth and liquidity to the market.

As we conclude our journey through the labyrinth of energy markets, it becomes evident that the sector's evolution will persist. The energy transition, with its emphasis on renewable energy sources and grid modernisation, presents both challenges and opportunities. Balancing the growing demand for electricity with environmental sustainability remains a paramount challenge. Market design, regulatory frameworks, and technological innovation will play a crucial role in addressing these complex issues. Additionally, the integration of electric vehicles, energy storage solutions, and demand response programs will reshape the landscape further.

In conclusion, energy markets are a dynamic and multifaceted domain where the pursuit of reliable, sustainable, and affordable electricity intertwines with the intricacies of supply and demand. As we move forward, a collaborative effort among stakeholders, driven by innovation and a commitment to a greener future, will shape the trajectory of energy markets, ensuring that electricity remains the special commodity that powers our world, today and tomorrow.

7.8 Proposed Exercises

Problem 7.1

At a certain hour, producers in the Iberian Electricity Market (MIBEL) finalised the following bids (see Table 7.5). We note that, currently, Portugal has closed the coal-fired power plants, so this market situation occurred before said closing.

At that time, the demand was 6,800 MWh and 33,000 MWh in the Portuguese and Spanish territories, respectively. The available transfer capability (ATC), previously established by the TSO, was:

- Spain → Portugal = 1,200 MW
- Portugal → Spain = 2,000 MW

Table 7.5 Generation bids. Problem 7.1

Technology	Portugal		Spain	
	MW	€/MWh	MW	€/MWh
Renewables	5,000	0	13,500	0
Coal	1,500	30	5,000	30
Hydro	1,800	40	4,700	40
Nuclear			6,000	10
CCGT			4,200	50

Note: Spain also has interconnections with France and Morocco.

1. In this situation, is the interconnection between Portugal and Spain enough to allow the necessary cross-border flow?
2. Calculate the electricity price in Portugal and Spain.
3. What would be the prices in Portugal and in Spain if the interconnection capability was 1,200 MW in both directions?
4. If it exists, calculate the congestion rent.

Key Problem 7.1

1.

The demand in Portugal and Spain is 6,800 and 33,000 MW, respectively. The bids in Portugal and Spain are 8,300 and 33,400 MW, respectively. Therefore, the cross-border power flow from Portugal to Spain is 1,500 MW, making the demand in Portugal equal to 8,300 MW. As the ATC from Portugal to Spain is 2,000 MW, it follows that there is no congestion at the border. Spain has a consumption of 33,000 MW and is receiving 1,500 MW from Portugal. As the production in Spain is 33,400 MW, it is concluded that Spain is exporting to France and/or Morocco 1,900 MW.

2.

We must build up the aggregated offer in the Iberia (Portugal + Spain). For the same price, the quantities are summed up. The aggregated bids are then ordered in an ascending way, from the lowest to the highest price. The aggregated demand is (6,800 + 1,500 + 33,000 – 1,500 + 1,900) = 41,700 MW. The graph in Fig. 7.7 is obtained.

We note that the Iberian aggregated load demand is 41,700 MW, the power flow in the France/Morocco interconnection being a load for the Iberian system.

The market price is the interception between the supply curve (blue) and the demand curve (orange). The value is 50 €/MWh.

3.

If the interconnection capacity from Portugal to Spain is limited to 1,200 MW, a redispatch must be performed. In this case, the capacity at the interconnection is fixed at the limit (1,200 MW). Consequently, the most expensive 300 MW in Portugal

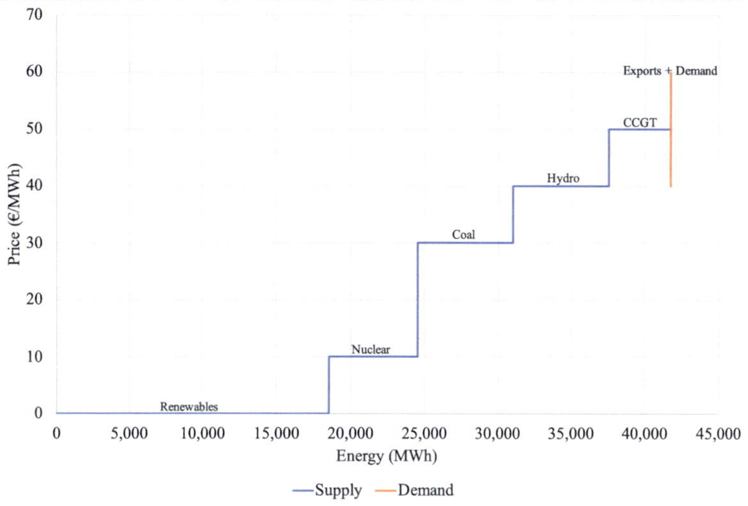

Fig. 7.7 Clearing mechanism in the Iberia (Portugal + Spain) in market coupling. Problem 7.1

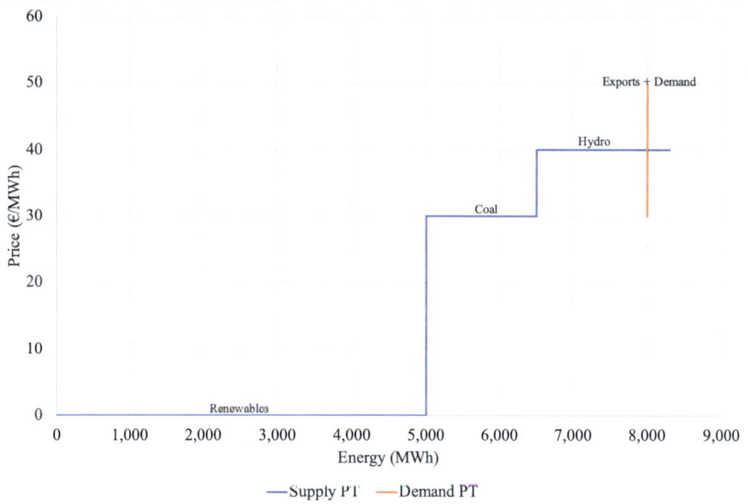

Fig. 7.8 Clearing mechanism in Portugal in market splitting. Problem 7.1

will be out of the market: the hydro production in Portugal is limited to 1,500 MW, instead of the 1,800 MW bid. We say that the interconnection is congested, and the market is to be split. A separate supply/demand curve is built for each country and two different market prices will arise. The curves for Portugal in market-splitting are displayed in Fig. 7.8.

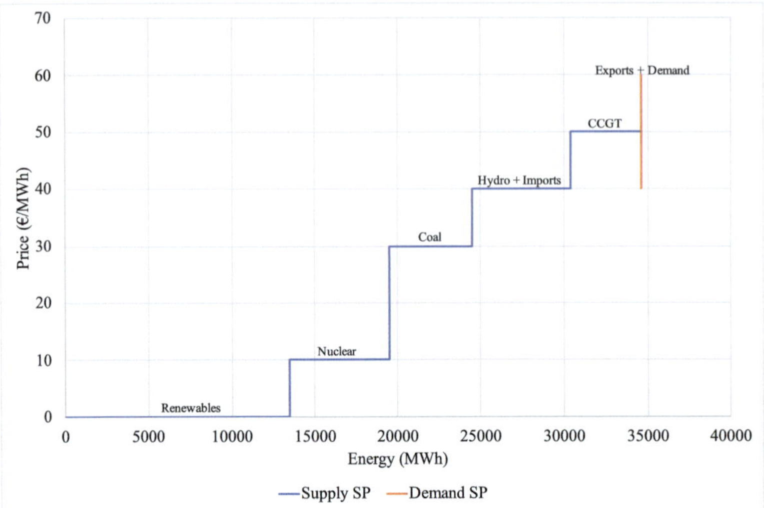

Fig. 7.9 Clearing mechanism in Spain in market splitting. Problem 7.1

Note that the bids total 8,300 MW, but the demand (including exports, which is a load) is only 8,000 MW because the available transfer interconnection capacity limits it. There are 300 MW in Portugal that are not selected to produce. The spot price in Portugal is 40 €/MWh.

The supply and demand curves for Spain in market-splitting are shown in Fig. 7.9.

The demand in Spain is 34,600 MW. The internal demand is 33,000 MW and Spain is exporting 1,600 MW to France/Morocco. The spot price in Spain is 50 €/MWh. We highlight that power is always flowing from the country with the lower price to the country with the higher price.

4.

Spain is buying energy at 40 €/MWh at the border and selling it at 50 €/MWh to their consumers. This is an unfair rent, caused by the congestion of the interconnection. In this case, the Spanish TSO must pay 1200 MW × (50-40) €/MWh = 12,000 € to a common fund aimed at improving the interconnection infrastructures. In case the situation was reversed, it would be the Portuguese TSO to pay the congestion rent.

In the past, market-splitting situations occurred very often in MIBEL, because the Portuguese/Spanish interconnection was very weak. Nowadays, market splitting occurs very seldom, due to the current robustness of the interconnection.

Problem 7.2

Consider an interconnected market composed of two zones, Zone 1 and Zone 2, with no other interconnections. At a particular hour, the market agents submitted the following bids (Table 7.6) and offers (Table 7.7).

Table 7.6 Generation bids. Problem 7.2

Seller	Zone	Energy (MWh)	Price (€/MWh)
G1	2	15	75
G2	1	100	15
G3	2	32	0
G4	2	25	42
G5	1	70	10

Table 7.7 Demand offers. Problem 7.2

Buyer	Zone	Energy (MWh)	Price (€/MWh)
D1	2	35	65
D2	2	23	78
D3	2	12	10
D4	2	38	46
D5	1	43	63
D6	2	16	32
D7	1	57	50

1. Considering no ATC restrictions, compute the system market price and the energy traded in the market.
2. Determine the market price, the energy traded in Zones 1 and 2, and the congestion rent, considering the exchange power capacities are Zone 1 → Zone 2 = 30 MW and Zone 2 → Zone 1 = 35 MW.

Key Problem 7.2

1.

If no interconnection capacity constraints are considered, we have a coupled market, all the generation bids must be ordered upwards and all the demand offers must be ranked downwards. The market clearing is depicted in Fig. 7.10.

The energy traded and the market price is defined where the supply curve meets the demand curve. It is possible to see in Fig. 7.10, that 202 MWh of energy are traded at a price of 32 €/MWh. Generators G4 and G1 are excluded from the market because they bid at a price higher than the spot price. Demand D3 is out of the market as it is offered below market price. In what concerns demand D6, it offered to buy 16 MWh, but it was allowed to buy just 6 MWh.

2.

If interconnection capacity constraints are active, we must look at what happens in the cross-border. The demand in Zone 1 is 100 MWh and the generation is 170 MWh. So, Zone 1 is willing to sell 70 MWh to Zone 2. However, only 30 MWh are allowed to flow in the interconnection from Zone 1 to Zone 2. This

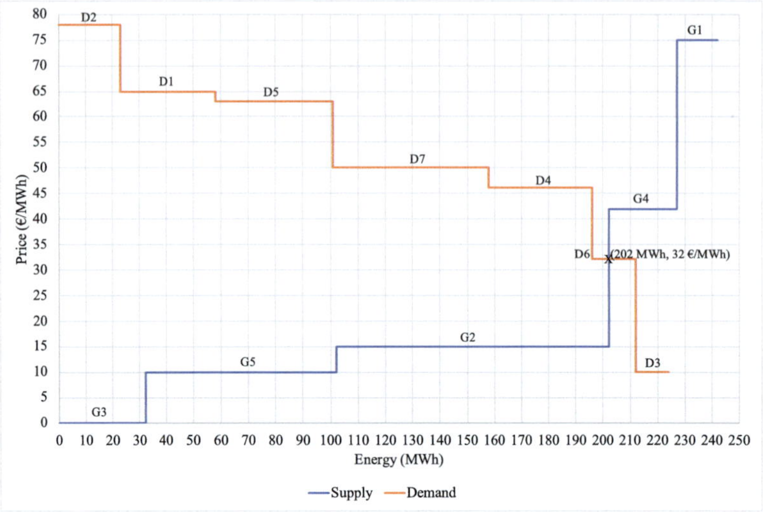

Fig. 7.10 Clearing mechanism in market coupling. Problem 7.2

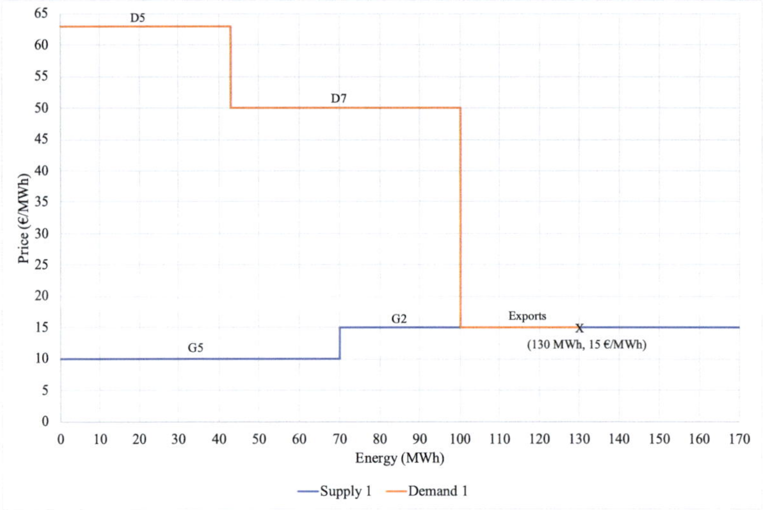

Fig. 7.11 Clearing mechanism in Zone 1 in market-splitting. Problem 7.2

results in a market-splitting situation, where the power flow at the interconnection is fixed at 30 MW and the two zones need to be separated.

Figures 7.11 and 7.12 show the market clearing in Zone 1 and Zone 2, respectively.

In Fig. 7.11, we can see that, in Zone 1, 130 MWh are traded at a price of 15 €/MWh. Please note the export of 30 MWh from Zone 1 to Zone 2, at a price of

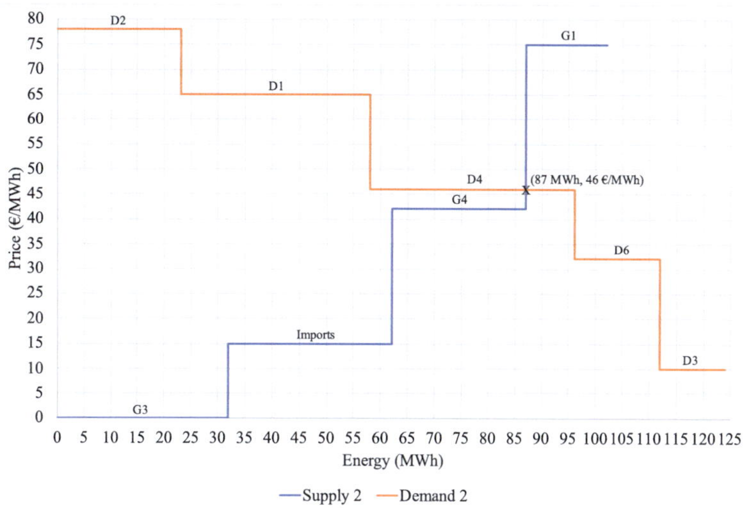

Fig. 7.12 Clearing mechanism in Zone 2 in market-splitting. Problem 7.2

15 €/MWh. The export is an equivalent load from the Zone 1 perspective. Generator G2 was willing to sell 100 MWh but they just sold 60 MWh, due to interconnection congestion.

In Fig. 7.12, the price is 46 €/MWh and 87 MWh are traded in Zone 2. We highlight that the 30 MWh imports are an equivalent generator from the Zone 2 perspective. Generator G1, demand D6, and D3 are out of the market. Demand D4 was willing to buy 38 MWh, but only 29 MWh were feasible.

Emerging Energy Markets

8

Abstract

In the realm of energy, transformation is the new constant. This chapter offers a concise exploration of "Emerging Energy Markets", where innovation and adaptability take centre stage, in the context of the proliferation of renewables. Possible evolutions of energy markets include market segmentation, where the dynamic interplay between renewables and dispatchable energy sources is explored. This strategic division lays the foundation for a resilient and balanced energy ecosystem. Further, we unveil co-optimisation strategies that incentivise flexibility providers to ensure grid stability in the face of renewable intermittency. We illuminate the rise of innovative business models like on-site power-to-x, placing power production and distribution in the hands of businesses and communities, and physical and Virtual Power Purchase Agreements (PPAs), enabling direct renewable energy procurement. A disruptive force emerges in Peer-to-Peer (P2P) energy markets, reshaping energy governance and consumer empowerment. Flexibility markets, bolstered by diverse service models, address grid congestion, voltage control, and grid services' provision in general to keep the stability of the grid, while powered by renewables. This chapter serves as a gateway to the exciting, ever-evolving landscape of "Emerging Energy Markets", inviting readers to explore the transformative potential of these dynamic energy systems.

8.1 Introduction

In the dynamic landscape of the global energy sector, a profound transformation is unfolding, ushering in a new era of "Emerging Energy Markets". This chapter delves into the promising horizons of this evolving landscape, exploring visionary approaches poised to shape the future of energy markets dominated by a high share of renewables.

R. Castro, *Engineering of Power Systems Economics*,
https://doi.org/10.1007/978-3-031-55251-9_8

As the world grapples with the imperatives of climate change and the quest for sustainable energy sources, the perspectives for future development in energy markets are nothing short of revolutionary. The traditional paradigms of centralised energy production and distribution are being challenged by innovative concepts that promise to redefine the way we generate, consume, and manage energy.

Current marginalist energy markets are not appropriate for extremely high shares of renewables. One of the pivotal strategies on the horizon is the segmentation of energy markets. Instead of treating all energy sources as one uniform entity, there is a growing recognition that a split between renewables and dispatchable resources could be the key to unlocking a cleaner and more resilient energy future. By creating distinct markets for intermittent renewables like wind and solar and dispatchable sources such as natural gas and energy storage, we can foster more efficient allocation of resources and incentivise the growth of sustainable energy.

Compensating for flexibility is another critical aspect of emerging energy markets dominated by renewables. With the increasing integration of variable renewables into the grid, maintaining system reliability becomes paramount. One innovative solution on the horizon is the concept of Equivalent Firm Power Capacity Auctions. This mechanism compensates providers not just for the raw capacity they offer, but for their ability to deliver power when it is most needed. This would require renewable producers to join firm power providers like energy storage systems. This approach encourages investments in flexible resources and ensures a reliable energy supply even as the grid becomes greener.

Furthermore, co-optimisation is emerging as a linchpin in the transformation of energy markets. Rather than treating various energy services as separate silos, co-optimisation envisions a holistic approach where different facets of the energy system are optimised together. Traditionally, electricity and ancillary services have been treated separately, but the future beckons a more integrated approach. This integrated strategy not only enhances overall system efficiency but also fosters synergy between renewables, dispatchable resources, and grid management.

The pursuit of sustainable, efficient, and resilient energy systems has spurred the emergence of groundbreaking business models that challenge conventional paradigms. As the world's commitment to combat climate change intensifies, these models offer a glimpse into the future of energy markets.

One of the most compelling developments is the rise of on-site Power-to-X (PtX) solutions. Traditionally, energy production and consumption have been geographically separated, but on-site PtX technologies are changing the game. These systems empower businesses and communities to generate renewable energy locally, converting excess electricity into valuable resources like hydrogen or synthetic fuels. This not only enhances energy self-sufficiency but also opens up new revenue streams while contributing to a greener energy landscape.

Physical Power Purchase Agreements (PPAs) are another pillar of innovation in the emerging energy markets. These agreements, whether local or sleeved, allow consumers and businesses to directly source renewable energy from nearby generators. This localised approach promotes sustainability and strengthens local economies while reducing transmission losses and enhancing grid resilience.

Virtual Power Purchase Agreements, particularly those employing Contract for Difference (CfD) mechanisms, are also at the forefront of change. They enable organisations to support renewable projects indirectly, entering into long-term agreements that guarantee a fixed price for electricity generated by these projects. CfDs provide financial stability for renewable developers, foster market competition, and give consumers the satisfaction of knowing their energy consumption contributes to a cleaner future.

Traditionally, energy flowed in one direction, from centralised generators to passive consumers. However, the rise of Peer-to-Peer (P2P) energy markets represents a seismic transformation. P2P markets introduce a democratic and decentralised approach, where individuals and communities become active participants in both energy production and consumption.

Full P2P markets epitomise this decentralisation. Here, energy users can directly buy and sell excess electricity to one another, bypassing traditional intermediaries. This not only fosters a sense of energy independence but also encourages the proliferation of renewable energy sources as users share surplus green power with their neighbours.

Community-based P2P markets take this concept further, creating localised microgrids where residents and businesses collaborate to generate, share, and trade energy within their communities, overseen by a community manager.

As the energy transition accelerates, the importance of flexibility in the energy system has never been clearer. Flexibility, the ability to quickly adjust supply and demand, is at the heart of making renewable energy sources like wind and solar more reliable and efficient. Flexibility markets represent a visionary approach to ensuring the adaptability of the energy grid in a rapidly changing world.

Congestion management and voltage control are two critical facets of flexibility markets, which are expected to be better addressed in the framework of flexibility markets. Managing congestion ensures that energy flows smoothly through the grid, preventing bottlenecks and optimising transmission. Voltage control, on the other hand, regulates voltage levels to enhance grid stability and protect sensitive equipment. Together, these aspects are pivotal in maintaining the reliability and resilience of the energy system.

The main players in flexibility markets span a wide spectrum, from traditional utilities and grid operators to emerging aggregators and technology providers. The diversity of participants fosters innovation and competition, driving the development of more efficient and responsive markets.

Clearing methods in these markets come in various forms, including day-ahead and real-time auctions, bilateral contracts, and algorithm-based platforms. These methods facilitate the matching of supply and demand for flexibility services and ensure the reliable operation of the grid.

Furthermore, emerging market platforms are serving as the bedrock for local flexibility markets. These platforms provide the digital infrastructure for market participants to interact, trade, and optimise their flexibility resources efficiently. They enable seamless integration of distributed energy resources into the grid while ensuring secure and transparent transactions. NODES, ENERA, PICLO, and IREMEL are pioneering flexibility market platforms, fostering a dynamic energy ecosystem. They connect diverse stakeholders, optimise distributed energy resources, and enhance grid reliability, contributing to a sustainable and resilient energy future.

In this chapter, we embark on a journey into these emerging energy markets, delving deep into the potential benefits, challenges, and regulatory implications. As our energy landscape continues to evolve, understanding these dynamic markets is imperative for policymakers, industry stakeholders, and all who are passionate about shaping a more resilient and sustainable energy future.

Firstly, we analyse several *evolution options for energy markets* in the future. Key alternatives include *split the market*, *compensate for flexibility* and *co-optimisation*. Then, we delve into the emerging *renewable business models for variable renewable energy sources*, exemplifying with *on-site power-to-X*, *physical power purchase agreements*, and *virtual power purchase agreements*. *Peer-to-Peer (P2P) markets*, where producers and consumers become prosumers, are addressed, focusing on the different P2P models: *full P2P*, *community-based P2P* and *hybrid P2P*. The emerging energy markets' path goes on with *flexibility markets*. Flexibility is the buzzword of future energy markets, as the grid is to be kept stable when powered essentially by renewables. Grid services such as *congestion management and voltage control* are addressed as key services provided by proper *flexibility players*. The different flexibility *models* and *clearing methods* are discussed. The chapter ends with some details given on the different *market platforms* available to trade flexibility.

This chapter is deeply inspired by the master's thesis by Sanha de Almeida, "Redesign the European energy markets to better allocate the increment of renewable energy suppliers", Técnico Lisboa, 2023, and by Tiago Felizardo, "Local flexibility market for congestion management at the distribution-level", Técnico Lisboa, 2023.

8.2 Evolution Options for Energy Markets

Back at the beginning of the century, governments encouraged the development of variable Renewable Energy Sources (vRES) by paying economically attractive feed-in tariffs (FiT) to the producers in order to induce investment in the renewable energy sector. At the time, outside the market paid FiT were significantly higher than the average market prices. Moreover, vRES benefited from dispatch priority, meaning the grid was obliged to purchase all the renewable production. All these beneficial conditions allowed the producers to recover capital costs and make significant profits.

As these contracts are coming to an end, renewable producers are being forced to enter the wholesale market as regular producers. To hedge the risk of forecasting errors, renewable producers are often integrated into larger power plants' portfolios including controllable energy resources.

On the other hand, FiT have ended in most countries, because there is no economic justification for them, as renewables are currently the cheapest way of producing electricity. This means that newcomers will have to compete in the wholesale market as any other producer. It is expected a high share of renewable production will take part of the electricity markets in the near future, also due to the ambitious environmental targets the governments are committed to.

Renewable production, namely vRES, is usually offered at 0 €/MWh in electricity markets because the renewables' marginal cost is 0. Despite bidding at 0, vRES receive the system's marginal price set by the most expensive technology needed to satisfy the demand. This allows for the recovery of fixed costs. As the number of 0 bids increases, the market price tends to decrease, as previously demonstrated. In the limit, if all producers bid at 0, the market price is 0, which is a highly undesirable situation for the producers who are unable to recover the fixed costs. Therefore, if the market gives unsatisfactory signals, there will be less investment and, eventually, no investment at all.

It is becoming more and more apparent that a marginalist market, based on marginal prices, is not appropriate to deal with high shares of renewable electricity in the market.

Solutions have been proposed in the literature to redesign electricity markets and address the issues raised. The following sub-sections present some of these proposals. Some are wholesale market-focused solutions, and others are retail market-focused solutions. They aim to increase the profitability of vRES generators and prevent these producers from exiting the electricity market.

8.2.1 Split the Market

The market division, or segmentation, is one of the solutions proposed. The wholesale electricity market should be divided into two types of supply—"on-demand" and "as available". In other words, there should be a market only for dispatchable plants and another for renewables. It is proposed that plants would have the freedom to choose in which market they would like to operate. The criterion for entering the "on-demand" market could be to provide power on demand, e.g., wind generators could participate if they recur to storage.

The system operator dispatches "on-demand" generators, and they receive their remuneration from the market. These generators can be remunerated for capacity, flexibility, energy, and other products/services. The purpose of this market is to be used when it is necessary—to guarantee a balance between supply and demand. The price would be determined through short-run marginal cost, as it happens nowadays. Price distortion caused by the presence of plants with zero marginal cost will no longer exist, and it is expected that the market will give necessary and proper price signals.

Regarding the "as available" market, the production is automatically accepted by the system operator, and they receive a relatively low and fixed remuneration, compared to the "on-demand" generators, in the short-run. This payment could be based on the current and future Levelized Cost Of Energy (LCOE) of each technology. In the long run, it is intended that the price is influenced by demand in the retail market.

The benefits of this proposal for renewable energy producers would be to provide investment signals to future producers and to allow renewable energy investments to be remunerated solely from the market. However, some equity questions could arise, as this solution implicitly suggests that only the consumers who can pay more could have a secure and reliable electricity supply.

More on this proposal in:

- https://doi.org/10.1016/j.jup.2021.101312
- https://doi.org/10.1016/C2013-0-16038-2
- https://www.oxfordenergy.org/publications/decarbonised-electricity-system-future-two-market-approach/

8.2.2 Compensate for Flexibility

The explicit flexibility approach can be a solution to address some market failures affecting renewable producers. In addition, it may encourage investment in electricity storage services which adds a revenue stream.

This strategy could be implemented through "Equivalent Firm Power Capacity Auctions" that would substitute the capacity markets and renewable subsidies. In this proposal, variable renewable generators may join conventional or storage operators to negotiate equivalent firm power supply contracts.

The idea is that the future of energy markets is based on a large share of decentralised and variable renewable energies plus residual fossil-fuel power plants serving as supplements. In line with this, variable renewable producers must be encouraged to invest in technologies that aim to mitigate their intermittency. With an equivalent firm power auction, all technologies would bid in equal conditions. For instance, a pure variable renewable producer would bid less and receive remuneration in line with its bid. For this renewable producer to be able to bid more and receive a higher payment, it would have to join other suppliers able to guarantee greater flexibility, for instance, energy storage systems.

The benefits of this proposal for renewable energy producers would be the creation of incentives for variable generators to increase their capacity margin, a greater focus on the market in terms of flexibility, with variable generators contracting for a flexible generation or investing directly in storage and demand-side flexibility (such as conventional storage and new battery technologies) and radically reduced policy interventions and regulations for their remuneration.

More information can be found at:

- https://doi.org/10.1016/j.jup.2021.101312
- https://doi.org/10.1093/oxrep/grz003
- https://doi.org/10.1016/j.ijepes.2019.105631

8.2.3 Co-optimisation

The secure operation of the power system requires the provision of certain services to guarantee security, quality, reliability, and flexibility to the network. The idea behind this proposal is that all the required services would be procured together. In this way, generators and storage operators would compete based on a bundle of services (energy, capacity, balancing, ancillary services) rather than competing in separate markets for each service. Combined bids could be submitted for energy, capacity, balancing, and system services, considering their risk profiles and thus maximising profits.

The benefits of the combined bids are that they would reward suppliers' capacity and the strengths of generation and storage and facilitate a move towards co-optimising energy. This allows to enlarge the opportunities and economic viability and, furthermore, increases the system's flexibility.

Details can be found at:

- https://doi.org/10.1016/j.ijepes.2019.105631
- https://doi.org/10.1109/EEM.2014.6861288

8.3 New Business Models for Variable Renewable Energy Sources

As previously stated, with the increase of vRES, periods with excess generation and periods with generation deficit are expected. Besides the already mentioned storage bet, it is necessary to investigate the opportunity of renewable producers investing in other business models to have additional revenue streams, so that they can be less dependent on the wholesale electricity market outcomes. New renewable business models (RBM) should be developed to prevent variable renewable energy producers from dropping out of the market. An overview of business opportunities for vRES owners in a post-FiT era follows.

8.3.1 On-Site Power-to-X

One of the emerging business models for renewable producers is power-to-X (PtX). The X in the terminology can refer to one of the following: power-to-ammonia, power-to-chemicals, power-to-fuel, power-to-gas (power-to-hydrogen, power-to-methane) power-to-liquid (synthetic fuel), power to food, power-to-syngas and power-to-power.

One of the most promising X is hydrogen. Assuming that an excess supply of electricity implies the energy market price to be lower, renewable producers can use the surplus of energy to produce hydrogen (power-to-gas) through water electrolysis. The hydrogen produced can be supplied immediately to end-users, or stored for a future reverse conversion using fuel cells, for instance, to generate electricity when the price is high. Generally speaking, the current characteristics of any generic energy system allow five main end-use configurations: power-to-gas (P2G), power-to-mobility (P2M), power-to-industry (P2I), power-to-synfuel (P2FUEL), and power-to-power (P2P).

Power-to-X has tremendous potential, and governments are encouraging research, development, and deployment of pilot installations. Renewable energy generated can be converted into another form (e.g., heat and solid, liquid and gaseous fuels), or its excess can be stored in another form for later shipment. Chemicals and fuels produced using PtX technology are still in the demonstration and pilot project phase. The profitability of the process depends on factors such as investment costs, input electricity costs, the number of total load hours, and revenues in appropriate markets. Current demonstration projects are still small-scale and do not take advantage of economies of scale. The situation is expected to change in the future when the cost associated with production and hence the product's unit price reduces.

Interested readers can find more information in:

- https://doi.org/10.1002/er.5215
- https://doi.org/10.1016/j.erss.2019.01.017
- https://doi.org/10.1016/j.ijhydene.2021.12.248
- https://doi.org/10.1016/j.egypro.2017.03.1833
- https://doi.org/10.1016/B978-0-12-820560-0.00008-4
- https://doi.org/10.1016/j.apenergy.2021.117863
- https://doi.org/10.1021/acsenergylett.0c02249
- https://doi.org/10.3389/fenrg.2020.00191
- https://doi.org/10.1021/acssuschemeng.1c03212

8.3.2 Physical Power Purchase Agreement (PPPA)

The physical power purchase agreements (PPA) directly between producers and consumers could be a solution for market design if the consumption and generation are geographically close. Two entities (producers and consumers) agree at a specific price (which may or may not be indexed and may depend on factors such as technology costs and grid connection costs), over a particular time, on the supply of electricity, therefore creating long-term regulated tariffs with a fixed price. Physical PPA can take two forms:

- Local physical power purchase agreement—The renewable power plant is physically connected to the consumer in this PPA design. Therefore, the contract is between the seller and the buyer. Depending on the regulations in force or if the seller uses storage, the vRES producer may sell the surplus of energy to the grid operator. If the seller does not use storage, the buyer will not have firm power available. This will push the buyer to pay, in parallel, an electricity supplier for energy outside the agreed PPA.

- Sleeved physical power purchase agreement—In this configuration, there is no direct connection between the consumption and the generation sites. The customer buys power generated by the seller at a fixed price under the PPA, for a given period. The power could flow from the producer to the grid and from the grid to the corporate buyer. An intermediary (which may be an aggregator, for instance) may be necessary, whose function is to ensure the transport of power through the network of the quantities contracted under the PPA. Through the payment of a sleeving fee, the intermediary may provide balancing services and supply extra electricity if necessary.

Physical PPA can be advantageous both for producers and consumers. The producers appreciate creating a new business model and an additional demand segment, long-term predictability and remuneration, and derisking, given the diversification of revenue streams. On the demand side, companies do not like the volatility of electricity market prices and may prefer PPA, where there is more transparency and price stability.

Details of PPPA are available in:

- https://doi.org/10.1016/j.jup.2021.101312
- https://doi.org/10.1016/j.apenergy.2019.113577
- https://doi.org/10.54648/eelr2020043
- https://doi.org/10.1111/anti.12847

8.3.3 Virtual Power Purchase Agreement (VPPA)

This is a long-term bilateral contract, usually between ten and twenty years, to supply electricity. Compared to a physical PPA, there is no physical link between the generation and consumption of electricity. In other words, it is a purely financial contract for the implicit value of the energy and not the energy itself.

The key players in a VPPA are the buyer (often referred to as the off-taker) and the renewable energy generator (such as a wind farm or solar facility). The VPPA sets out the terms and conditions of the agreement, including the contract duration, the quantity of electricity to be purchased (usually expressed in megawatt-hours), the price structure, and the location of the renewable energy project. The off-taker does not directly receive renewable energy. Instead, the electricity generated by the renewable project is typically delivered to the power grid. The off-taker continues to source electricity from their regular utility provider or the grid and pays the agreed-upon price per megawatt-hour of renewable electricity to the generator. The difference between the market price of electricity and the fixed price in the VPPA is settled through a financial instrument, such as a Contract-for-Differences (CfD).

In the CfD, the company developing the project (the renewable power plant) can sell the energy in the market at the end of the project through an intermediary third party (the corporate buyer). The corporate buyer guarantees the owner of the renewable PPA project a specific fixed price, called the *strike price*, for the electricity it sells to the grid: if the electricity is sold for less than the guaranteed amount, i.e., the difference between the strike price and the market price is positive, the corporate buyer pays the difference; if the electricity is sold for more than the guaranteed amount, i.e., the difference between the strike price and the market price is positive, the corporate buyer profits from the difference.

More information on VPPA is in:

- https://doi.org/10.1016/j.jup.2021.101312
- https://doi.org/10.1016/j.apenergy.2019.113577
- https://doi.org/10.54648/eelr2020043
- https://doi.org/10.1016/j.apenergy.2019.01.224
- https://doi.org/10.1016/j.rser.2021.111393

8.4 Peer-to-Peer (P2P) Markets

The Peer-to-Peer (P2P) market model allows the automatic sale of excess or provides energy injection to consumers. As with other sharing economies, P2P energy sharing introduces greater efficiencies in the energy market and financial benefits for individual prosumers (consumers who are also producers) and medium and large renewable energy producers. These innovative platforms empower consumers to directly trade energy with each other, creating a localised, community-driven energy exchange.

Peer-to-peer energy markets are digital platforms that facilitate direct transactions of electricity between energy producers (often households with solar panels or small-scale generators) and consumers (households or businesses). Unlike the traditional centralised energy system, P2P energy markets leverage blockchain technology and smart contracts to enable seamless, transparent, and secure trading of energy within a local community.

Peers must be informed at all times, so an information system must transmit market signals, monitor market operations, and provide a virtual channel to the acquiring market. All peers must access the same information, for example, through blockchain-based smart contracts. This means a virtual layer must be put in place, which refers to the technology behind the transactions and data privacy, being implemented on top of the existing physical grid infrastructure, based on blockchain architecture, and the physical layer which uses the traditional grid to provide the physical flow of energy.

The following are features of P2P markets:

- Distributed Energy Production: Participants in P2P energy markets generate their electricity locally through solar panels, wind turbines, or other renewable sources. Surplus energy can be sent back to the grid or shared directly with nearby consumers.
- Blockchain and Smart Contracts: Transactions on P2P energy platforms are recorded on a decentralised blockchain, ensuring transparency and security. Smart contracts automatically execute energy trades based on predefined conditions, such as price, availability, and location.
- Peer-to-Peer Trading: Producers and consumers can set their own energy prices and negotiate directly with each other. This autonomy fosters a competitive and flexible energy market.

Several advantages can be identified regarding P2P energy markets.

- Decentralisation and Energy Independence: P2P energy markets reduce reliance on centralised energy providers, granting consumers greater control over their energy choices and encouraging sustainable practices.
- Enhanced Efficiency: By eliminating intermediaries, P2P energy markets streamline the energy trading process, resulting in reduced transaction costs and more efficient use of renewable resources.
- Sustainability and Environmental Benefits: Facilitating local renewable energy trade encourages the adoption of green energy sources, leading to a reduction in greenhouse gas emissions and a greener, cleaner environment.
- Resilience and Reliability: P2P energy markets can enhance the overall grid resilience by decentralising power generation and reducing vulnerability to single-point failures.

Nevertheless, P2P markets face a few challenges regarding their full deployment.

- Grid Integration: Integrating decentralised energy sources into the existing grid can be challenging due to technical constraints and regulatory barriers.
- Market Design and Regulation: Developing suitable market structures and regulatory frameworks that accommodate P2P energy trading while ensuring fairness, security, and consumer protection poses a significant challenge.
- Data Privacy and Security: As P2P energy markets rely on digital platforms, maintaining data privacy and ensuring cybersecurity becomes crucial to safeguard sensitive consumer information.
- Scalability: While P2P energy markets have proven successful in smaller communities, scaling up to serve larger populations may require addressing technical and operational complexities.

The literature presents three topologies of P2P markets: full P2P, community-based P2P, and hybrid P2P. These names can change according to the authors. For instance, full P2P can be called decentralised market, whereas community-based P2P may be known as coordinated market or centralised market. The degree of centralisation in the trading process and information sharing distinguish the models from each other.

8.4.1 Full P2P

The method of communication among peers and the trading process is considered decentralised in this market structure. Peers directly negotiate a transaction for a certain amount of energy and a price without the intervention of a third entity. Peers directly and freely communicate with each other to sell and buy energy, which is one of the main advantages of the structure. The lack of supervision may make this market not fully efficient.

In a full P2P energy market, energy trading occurs directly between individual producers and consumers without any intermediary involvement. Each participant is a prosumer (producer-consumer) of energy. This topology leverages blockchain or distributed ledger technology to record and verify transactions securely and transparently.

Decentralisation, transparency, and lower costs portray the advantages of full P2P models. The use of blockchain technology ensures transparent and tamper-proof records of transactions, fostering trust among participants, as well as the elimination of intermediaries reduces transaction costs, benefiting both producers and consumers.

As the number of participants increases, the computational and energy demands on the blockchain can become significant, impacting scalability. On the other side, adapting existing energy regulations to accommodate full P2P models can be challenging due to their decentralised nature.

8.4.2 Community-Based P2P

The method of communication among peers and the trading process is considered centralised in this market structure. Unlike full P2P, community-based P2P relies on a third entity named a community manager/coordinator, communicating with each peer inside and between the community and the rest of the system. The coordinator directly controls the energy export and import limits to be shared between peers. After this step, the revenue of the whole community is distributed among participants according to predefined principles. This design is more structured and can be applied to microgrids, a group of neighbouring prosumers, and members inclined to share green energy, although they are geographically distant.

Community-based P2P energy markets focus on energy exchange within specific communities or neighbourhoods. Participants are typically geographically close to one another, enabling more localised energy trading. These models foster a sense of community and cooperation among prosumers.

By operating within a community, this model can enhance grid resilience and reduce vulnerability to external disruptions. Community-based P2P models strengthen social ties and encourage collective responsibility for sustainable energy practices. Due to the local nature of transactions, grid integration of renewable energy sources can be more straightforward.

On the other side, the scope of trading might be limited to the community's energy production and consumption capacities, and in cases where there is an energy surplus in one area and a deficit in another, a balanced energy exchange might be challenging to achieve.

8.4.3 Hybrid P2P

It results from combining the two previous market designs with different layers for energy trading. A centralised system operator influences indirectly by sending price signals rather than directly instructing the import/export of peer energy or the operational status of peer devices. P2P transactions may run between communities, between individual/collective peers, and may also occur within existing markets. Individual/collective energy pairs are involved in P2P transactions with each other, e.g., buildings and their inhabitants are considered a collective energy pair, and they can transact energy to the neighbours.

In short, hybrid P2P energy markets combine elements of both traditional centralised energy systems and P2P models. These models integrate peer-to-peer transactions with interactions with the central grid or utilities, providing flexibility and efficiency.

Hybrid models can stabilise the grid by balancing intermittent renewable energy sources with the centralised grid's reliability. Participants have the option to trade directly with each other or interact with the central grid, providing more options for energy management. Hybrid models offer a transitional approach from traditional energy systems to fully decentralised P2P markets.

On the other side, the integration of centralised and decentralised components can introduce complexity in the market's design and operations and regulators must develop suitable frameworks to accommodate the hybrid nature of these models.

Further literature on the subject of P2P energy markets is available in:

- https://doi.org/10.1109/MSP.2018.2818327
- https://doi.org/10.1016/B978-0-12-820074-2.00012-5
- https://doi.org/10.1016/j.rser.2019.01.036
- https://doi.org/10.1109/EEM.2019.8916543
- https://doi.org/10.1016/j.apenergy.2020.116131
- https://doi.org/10.1016/j.eng.2020.06.002

8.5 Flexibility Markets

Flexibility markets are a fairly new concept in academia and the energy industry. In recent years, a number of projects regarding flexibility markets have been initiated in Europe. This follows EU directives and policies, including the Clean Energy Package, which enforces TSOs and especially DSOs to procure market-based flexibility services, such as distributed electricity generation, demand flexibility, and energy storage options, when these services are cheaper than investing in electricity network expansion.

8.5.1 Flexibility Services Models: TSO-DSO Coordination

Traditionally, the market structure was centralised. The TSO had a primary role and oversaw the wholesale markets where energy exchange took place. The traditional paradigm is disrupted by Distributed Energy Resources (DERs) with high penetration (most often connected to the distribution grid).

Before we proceed, let us discuss what are DERs. Distributed Energy Resources (DERs) are decentralised and often small-scale energy sources or technologies that generate, store, manage, or reduce the consumption of electricity or thermal energy. These resources are typically located near the point of energy consumption, providing an alternative to traditional centralised power generation and distribution. Some common types of DERs include:

- Solar Photovoltaic (PV) Systems: Solar panels installed on rooftops or in small-scale solar farms convert sunlight into electricity. Solar PV is one of the most widely adopted DERs for residential, commercial, and industrial applications.
- Wind Turbines: Small wind turbines, often used in distributed applications, harness wind energy to generate electricity. They are suitable for both on-grid and off-grid installations.
- Energy Storage Systems (ESS): Battery storage systems, such as lithium-ion batteries, store excess electricity generated by DERs or during periods of low demand. ESS can discharge stored energy when needed, providing grid stability and backup power.
- Combined Heat and Power (CHP) Systems: CHP systems, also known as co-generation systems, simultaneously produce electricity and useful thermal energy (heat) from a single energy source, such as natural gas or biomass. They are highly efficient and suitable for various applications.
- Microgrids: Microgrids are localised energy systems that can operate independently or in conjunction with the main grid. They incorporate DERs, energy storage, and control systems to provide reliable power to specific areas or facilities.
- Distributed Generators: Small diesel generators, natural gas generators, or other backup generators can serve as DERs to provide electricity during grid outages or peak demand periods.

- Demand Response (DR) Technologies: Demand response involves reducing energy consumption during peak periods in response to price signals or grid emergencies. Smart thermostats, lighting controls, and other technologies enable DR.
- Electric Vehicles (EVs): EVs can serve as mobile DERs when connected to the grid. They can store excess energy and feed it back to the grid when needed, supporting grid balancing and renewable energy integration.
- Energy Efficiency Measures: Improvements in energy efficiency, such as building insulation, LED lighting, and HVAC (Heat, Ventilation, and Air Conditioning) system upgrades, are considered DERs because they reduce overall energy consumption.
- Hydroelectric and Biomass Systems: In some regions, small-scale hydroelectric or biomass power generation can be considered distributed energy resources.

The fact is that not only does generation no longer come from centralised producing units, but DERs also add a higher level of unpredictability, increasing the demand for flexibility services. As a result, both TSO and DSO are pressured to deal with issues including unpredictability, bidirectional power flows, voltage limit issues, and distribution-level congestion. In the past, the DSO was a passive body tasked with ensuring a secure and consistent power supply. This was mostly accomplished through network expansion investments. This entails considering the financial viability of such an expenditure as well as planning to increase grid extension based on the worst-case scenario, even if it is not very likely to occur.

The problem arose when the DSO started interfering with the TSO's role, on a lower level. The DSO became a more active player, solving their own problems at the distribution level by managing existing DERs to solve the issue. It is crucial then that a coordination mechanism exists so that when the DSO fixes one issue, it does not bring consequences to the transmission level.

All of the coordination mechanisms between TSO and DSO that will be mentioned in this section share comparable prequalification (verifying potential consequences of flexibility procurement by the TSO, at a distribution level and vice-versa), activation, and settlement of flexibility resources.

Said coordination mechanisms can be divided into five different types:

- Centralised ancillary services market.
- Local ancillary services market.
- Shared responsibility for balancing market.
- Common TSO-DSO ancillary market.
- Integrated flexibility market.

Centralised ancillary services market—The DSO plays a completely passive role in this approach, whereas the TSO plays the dominant role. Only the TSO, which is not required to take distribution grid limits into account, may activate resources connected at the transmission and distribution levels for flexibility. No local markets are envisioned for this approach because the TSO operates both the system and the market. DSOs are not permitted to use DERs to solve their issues. Prequalification is sometimes used in order to solve restrictions at the distribution level. While this centralised market removes the concern for low liquidity, its biggest drawback is that TSO cannot employ DERs to address issues with the distribution grid.

Local ancillary services market—The distribution level receives some control under this concept. There are local markets here as well, which contrasts with the centralised market concept. DSOs run them, whereas TSOs continue to run the central market. Since the DSO is in control of the DERs on the distribution, it can use them to solve local congestion and control voltage levels. This makes it so that local markets are cleared before the centralised market. The central market then provides the remaining services provided in the local market. In this case, cooperation between the DSO and TSO is required to prevent the procurement of flexible services from going in the wrong direction.

Shared responsibility for balancing market—Similarly to the previous model, in this case, there are two markets as well: the local market, managed by the DSO, and the central market, run by the TSO. The main difference from the previous model is that the two markets are cleared at the same time and the DSO can obtain the necessary flexibility from the local market whereas the TSO can only obtain flexibility from the central market. Both system operators are responsible for the balance at the distribution grid and for resolving congestions. For such a paradigm to work, TSO and DSO must communicate in order to agree on predetermined schedules that forbid autonomous system operation by TSO and DSO. Due to the local markets' separation from the central market and its operation by DSOs, this architecture may also experience issues with liquidity, higher prices for ancillary services, and greater operational expenses for DSOs (though lower for TSOs). If the DSO is unable to fulfil its balancing obligations for the relevant area, the entire system risks instability, which is the biggest hazard.

Common TSO-DSO ancillary services market—The TSO and DSO jointly run a single market for ancillary services. Since system stability is the model's ultimate goal, neither system operator has precedence; instead, flexibility is given to the entity that needs it the most. There are two possible approaches to implementing the model. The first option is a single central market run by TSO and DSO, which would be financially beneficial since it would make use of the existing DERs to efficiently manage the system but also the inclusion of grid constraints in the market clearing. The con in this option is the fact that since both transmission and distribution are observed at the same time and the solution has to work for both at the same time, the optimisation problem might be that much more difficult, especially for large systems. The second option would be more numerous and varied local markets which would result in increased operational costs and potential financial issues. The mathematical cost of large systems is lessened, however, because initially just local distribution grid restrictions are taken into account, shared with TSO, and dealt with in a second optimisation that also takes into account TSO limitations. Some contend that in order to maintain neutrality, both options require an impartial market referee.

Integrated flexibility market—In this approach, alongside system operators (TSO and DSO), deregulated participants may also procure flexibility. They may both purchase and sell supplementary services. To ensure neutrality and an even playing field for all market players, an independent market operator is a must, and the market clearing procedure should take distribution-level restrictions into account. High market liquidity is ensured by this approach. Due to the fact that all services are offered on a single market, there will be more competition and cheaper pricing, but this could affect trade volumes in the day-ahead and intraday energy markets. Additionally, gained flexibility may be sold, and the cost of maintaining the market should not worry anyone because it may be distributed among a large number of players. The likelihood that TSO might purchase ancillary services to preserve system stability outside of the market if deregulated participants induce so with their market actions, is possibly the biggest drawback of this paradigm. In some cases, the rivalry may even force the activation of services in the reverse direction at the transmission and distribution levels, resulting in high costs for the end customers. This option will be the focus here forward. This is the type of market, at the distribution level, that will be explored in this section.

8.5.2 Local Flexibility Markets

A Local Flexibility Market (LFM) can be described as an electricity flexibility trading platform to trade flexibility in geographically restricted areas such as neighbourhoods, communities, towns, and small cities by analogy with the definitions of a market and a local market. As a result, in an LFM there are the goods or services that are exchanged there, as well as a market operator, participants, and a clearing system.

8.5.2.1 Flexibility Trading: Congestion Management and Voltage Control

Congestion occurs when the demand for electricity exceeds the capacity of the grid infrastructure to deliver it efficiently. This can lead to increased costs, inefficient energy distribution, and even power outages. Voltage is an electrical quantity that must be kept under tight limits ($\pm 10\%$ of the nominal voltage) to allow the secure operation of the grid.

One of the key functions of the DSO is congestion management and voltage control. The DSO can use LFM to solve congestion and voltage violations. Actually, local flexibility markets address both these issues through the following mechanisms:

- Resource optimisation: LFMs allow for the integration and coordination of diverse energy resources at the local level. These resources include distributed energy generation, energy storage systems, electric vehicles, and demand response programs, as seen before. By aggregating these resources and offering them in the market, LFMs enable grid operators to optimise the utilisation of these assets to alleviate congestion and manage voltage deviations. For example, during periods of congestion, grid operators can procure additional energy from locally distributed generators or activate demand response measures to reduce load and stabilise voltages.

- Market-based coordination: LFMs provide a platform for market participants, including energy producers, consumers, and aggregators, to trade flexibility services. Through real-time pricing signals, participants can respond to congestion and voltage conditions by adjusting their energy consumption or generation patterns. Higher prices during congested periods incentivise energy users to reduce their consumption, while participants with excess capacity can offer their flexibility services for compensation.
- Demand response integration: LFMs facilitate the integration of demand response programs into grid operations. Demand response enables energy consumers to adjust their electricity usage in response to grid conditions. During periods of congestion or voltage deviations, grid operators can activate demand response signals, prompting participating consumers to curtail their energy consumption or shift it to off-peak hours.
- Grid visibility and control: LFMs leverage advanced monitoring and control technologies to provide real-time visibility into the grid's operating conditions. This allows grid operators to identify congestion hotspots and voltage deviations more accurately. With this information, operators can make informed decisions regarding resource dispatch, load balancing, and voltage control.
- Grid resilience and local autonomy: LFMs promote grid resilience by allowing localised decision-making and reducing dependence on centralised control. By enabling local resources and demand response measures to participate in the market, the grid becomes more decentralised and responsive to local conditions. This autonomy reduces vulnerability to congestion caused by transmission limitations or disruptions in centralised power plants.

In summary, local flexibility markets provide an effective means to tackle congestion issues and voltage deviations within the energy grid. However, other methods can be used for similar purposes. Existing methods can be categorised into market-based methods, where LFMs are included, and control-based methods.

Market-based methods mainly include other than LFMs, Local Energy Markets (LEM), price-based control and transactive energy. Control-based methods, in turn, include virtual power plants (VPP) and active network management (ANM).

Local Energy Markets

Local Energy Markets (LEM) were created and put into place to manage the high penetration of DERs by promoting regional energy trading. They allow for the integration of various energy resources, such as solar panels, wind turbines, energy storage systems, and even electric vehicles, within a specific locality. Trading can be carried out either peer-to-peer (P2P) or under the supervision of a mediator or even as a combination of the two. LEM allows prosumers to trade among themselves, thereby reducing the load on the overall grid and preventing congestions. However, it may result in them having less negotiating power when exchanging with significant players.

Price-based control

For price-based control, the DSO forecasts potential congestion during specific times and announces the congestion pricing in advance to the flexible demands. Most of the methods consist of the employment of different types of dynamic tariffs, either day-ahead or closer to real-time, to motivate players to shift flexible demands from peak hours, in order to solve congestion problems.

Transactive energy

The transactive energy method may be defined "a set of economic and control mechanisms that allows the dynamic balance of supply and demand across the entire electrical infrastructure using value as a key operational parameter". It uses decentralised decision-making frameworks in order to manage the operation of DERs, allowing for efficient coordination and optimisation of energy transactions in real-time. Transactive energy relies on market mechanisms to coordinate the flow of energy between different participants in the grid. Through real-time pricing signals, prosumers can adjust their energy consumption and generation patterns, including storage devices, to respond to congestion and voltage conditions. Higher prices during congestion periods incentivise energy conservation and load shifting, reducing strain on the grid and alleviating congestion.

Virtual Power Plant

A virtual power plant (VPP) is a cloud-based distributed power plant that aggregates the capacities of DERs for the purpose of enhancing power generation, as well as trading or selling power on the electricity market. The aggregation of DERs will have system visibility, controllability, and impact when they are combined in a VPP, much like a traditional generator would. Additionally, the enhanced utilisation of existing assets made possible by VPP technologies might result in advantages for the DSO, like improved network efficiency and cost savings in the distribution system.

Active Network Management

Active network management (ANM) is nothing more than control systems that manage generation and load for specific purposes, in this case, solving congestion and voltage violations of distribution systems with high DER penetration. It involves real-time monitoring, control, and optimisation of DERs to ensure the reliable and efficient operation of the grid. These resources can be actively controlled and dispatched to alleviate congestion and stabilise voltages. ANM employs advanced monitoring and control systems that provide real-time visibility into the grid's operating conditions.

Finally, we will take a look at how congestion is managed at the interconnections between two countries. The case of Portugal will serve as an example. In Portugal, congestion management at the interconnection level is a responsibility allocated to the TSO. It can be either internal congestion, which is when the congestion happens solely in one area inside the country's borders, or inter-area congestion, which is when the congestion happens on the border between Portugal and Spain. In this case, there has to be coordination between the Portuguese and Spanish TSOs.

The mechanism for interconnection congestion management takes place after the day-ahead market. The Portuguese TSO starts by assessing the interconnection capacity and verifies if it exceeds its limit. It, then, can manage purchasing or selling bids to secure the interconnection. Even if after this action, congestion is verified, the system operator can resort to Coordinated Remedial Actions, like counter-trading, which are energy transactions provoked by the TSO in real-time and that overwrites the existing interconnection program. The resulting counter-trading measures lead to speculative actions (of the same order of magnitude and in the opposite direction) in each system, as symmetric reserves are mobilised upwards or downwards, reserves provided by the respective ancillary services markets.

8.5.2.2 Main Players

It is interesting to understand who are the primary stakeholders and their objectives in markets for local trading flexibility. Generally, the key participants are:

Prosumers

Prosumers can be defined as "individuals who consume and produce value, either for self-consumption or consumption by others, and can receive implicit or explicit incentives from organisations involved in the exchange". They might have energy storage devices, other controlled and unpredictable units, and renewable energy sources like solar panels and wind turbines. They may act independently and even participate in peer-to-peer trade, but given their often tiny size (capacity and power), it makes sense to cluster them and then provide the flexibility of a clustered system.

Aggregators

Aggregators are the newest players in this type of market structure. Aggregators have been defined as "a company which acts as an intermediary between electricity end-users and DER owners and the power system participants who wish to serve these end-users or exploit the services provided by these DERs". Due to their small quantities of flexibility, the individual DER or prosumer has little negotiating power in LFMs; therefore, in order to maximise their flexibility potential, the aggregators collect the individual flexibility to create multiple flexibility services and exchange it in LFMs.

Other than the limited negotiating position that the individual owner of DER or prosumer has when participating in an LFM, the other main reason for not participating individually is if every prosumer or owner of DERs decided to participate in these circumstances, they would end up overloading the market due to the large number of individuals involved, through communication and computation burden.

Aggregators are starting to be considered as the component needed in order for small-sized consumers to start exploiting their own assets as flexibility. There are however some challenges for aggregators to overcome. Consumers and prosumers, most of the time, are unaware of the problems that high loads can cause on networks. So, aggregators should be able to inform their customers in a simple way about their role in fixing the problem of peak loads. Customers should also need smart appliances and control systems in order for aggregators to gather the load data and better manage their flexibility. However, these technologies should require only minimal investment so as to not alienate possible customers.

Distribution System Operator

Unlike aggregators, the DSOs have been mentioned before and are the most well-known of the participants in LFMs. The DSO's major responsibility is to provide customers with power in a way that is affordable, sustainable, and effective. It is also responsible for the distribution system's safe operation as well as the calibre of power delivery services. Traditionally, this has involved planning, maintaining, and managing the distribution level grid, as well as supply interruptions and connecting and disconnecting DERs, which is where the DSO gets its flexibility in order to operate congestion management, loss control, and voltage violations, to name a few.

Balance Responsible Parties

Balance Responsible Parties (BRPs) trade in the energy markets on behalf of their clients, being exclusively responsible for balancing their portfolios. It is their duty to maintain a balance between energy supply and consumption over a specific period of time and incur imbalance fees if they fail to keep this balance. As a result, BRPs look for the flexibility they need to maximise their portfolio and fulfil their energy commitments. Since BRPs are many times simultaneously generators, retailers, aggregators, DSOs or TSOs, they sometimes juggle the responsibilities of being a balance responsible party with other functions.

Local Flexibility Market Operator

The Local Flexibility Market Operator (LFMO) is the entity in charge of running the LFM's most crucial functions. It provides the flexibility trading platform for bidding, clearing, and settlement of the market. Market clearing is the process that gathers flexibility offers and requests and determines trading outcomes (i.e., price and quantity of flexibility to be traded).

Many players can take over the responsibility of LFMO, such as the DSO, the aggregator or a BRP. It can also be a third-party entity. Usually, a third-party entity, preferably neutral, is preferred to take over the role.

8.5.2.3 Models

When it comes to possible models for the proper functioning of LFMs, there are many options under consideration and none are in a clear position for the best solution. In deciding which model is better, it is preferable to declare that there are different market models for different needs or purposes. Either through full peer-to-peer (P2P) trading, trading through a mediator, or a combination of both, as seen before, no model should take precedence over the other without considering the characteristics of the specific market. Different formulations can be classified into four categories and each category is divided into small subcategories according to each model format.

Centralised optimisation methods

In centralised optimisation models, there is usually an optimisation problem when it comes to clearing the market. The optimisation problem is typically expressed as an objective function that needs to be optimised while complying with a number of technical and economic restrictions. The two main types of objective functions are:

- Social Welfare Optimisation—The term originates from economic standpoints and refers to the satisfaction or benefit of numerous market participants, including consumers and producers. Put in a simple way, it is calculated as the summation of the utility of all buyers minus the cost of all sellers.
- Operational Cost Minimisation—From each participant's perspective, the objective is to keep the operational cost of acquiring flexibility in the LFM as low as possible. This objective function seeks exactly that, i.e., to minimise the operational costs for a specific and different stakeholder (DSO, aggregator, etc.).

The most important feature of this model is that it permits the optimisation of an objective function in order to find the best market clearing option. As easy as it is to implement, the market clearing solution is also easy, solving headaches for the operator. However, the big cons of this system are the existent barriers for large-scale systems since it is a model difficult to scale due to limits of communication and computation burden.

Game theory-based models

These models are based on game theory, the study of how interdependent decisions made by economic agents result in outcomes that are consistent with their preferences (or utilities) even when none of the agents may have intended the particular outcomes in question. The focus is usually the competition of all market participants in order to try to maximise profits:

There are two main types of game theory-based models:

- Noncooperative game theory—These models enter into action expecting players to decide their own strategies without consulting other players. It is better suited for situations where participants may only have some of the information. It describes competition among participants who have partially or completely conflicting interests. This is one of the most common models, being used commonly even at the distribution level.
- Cooperative game theory—These models focus on rational cooperative players that exhibit cooperative behaviour. The method can be used in circumstances when it is possible, acceptable, and desirable for individuals to provide information. The DSO is typically not willing (or even permitted) to disclose network information to other players in flexibility markets.

As the name indicates, game theory-based models look at the LFM as a game, wherein a player's ability to make decisions depends on those of the other players, and it mathematically records the game's players' behaviours. It is a scalable model and can be applied to markets with a huge number of players since it is able to represent all the participants in the LFM with competence. However, the model also assumes every player will act rationally, which sometimes does not happen in reality and it might generate situations where all influences cancel each other.

Auction theory-based models

An auction, the base mechanism of this model, functions by allocating resources in such a way that seeks to balance supply and demand through a competitive bidding procedure. In an auction market, multiple sellers and buyers can concurrently submit a number of orders for certain commodities to the mediator. This mediator in the process, the auctioneer, collects the bids and determines the clearing price.

In a *pay-as-cleared* system, the nominated buyers, those who offered a price higher or equal to the cleared price, are required to pay the same amount, and all nominated sellers, whose offered price is lower or equal to the cleared price, are required to pay the same commission. In a *pay-as-bid* system, the nominated sellers receive the price they requested.

Auction theory-based models can be then classified into single-side auction and double-side auction models. In the former, numerous buyers bid for an item being sold, while multiple sellers offer for the item that the buyer requested. The latter, the double-side auction, works with multiple buyers bidding to buy goods from various sellers. The double-sided auction model overcomes the single-sided one, being the most general type, and most appropriate in the vast majority of occasions.

This model, inspired by stock exchange markets, which sees the LFM as an auction market, where the auctioneer determines flexibility allocation as well as the clearing price of the market based on the participants' bidding strategies, has some pros and cons. This model, for one, is also scalable and can be applied even with the participation of many players and permits buyers and sellers to achieve market equilibrium quickly. There can be, at times, unviable auction price spikes, unreasonable and undesirable, due to excessive competition at the auctions.

Simulation models

Last but not least, there are multiagent-based simulation models. It is a pretty flexible modelling framework, built to mimic the behaviour of multiple participants. It captures the dynamics of the power market and is a good option for scenarios with several parties. Each party can be a single variable in a computer program or a smart item with an endless number of actions and decisions.

Even though the flexibility of the model in representing all the participants of the LFM regardless of the strategy, resources, private or public information, permits a really solid representation of the LFM, it requires heavy computational power in order to consider every variable.

Discussion

Every model has its advantages and disadvantages. In a practical situation, a case-by-case study should be carried out to decide and choose the model that best suits the situation at hand.

The centralised optimisation model is ideal when just one player is the subject of interest. This streamlines the model but ignores other elements, such as the utility goals of the rest of the participants who are not being observed. This model also struggles to scale when there are many participants involved. On the other hand, alternative models can imitate market behaviours more accurately because they can include all market participants. Centralised optimisation models are, therefore, simple to implement, however, because of communication and computing issues, they have scalability restrictions and face obstacles for large-scale systems. Market designs based on simulation models, auction theory models, and game theory models are a better option in situations with larger systems since they are definitely more scalable.

Game theory models capture the strategic interactions among market participants in local flexibility markets. They enable the analysis of different decision-making scenarios, taking into account participants' objectives, incentives, and responses to other players' actions. On the other side, game theory models typically assume rational behaviour, assuming that participants always act in their self-interest. However, in practice, participants may exhibit bounded rationality, incomplete information, or other behavioural traits that affect their decision-making, challenging the validity of the model's assumptions. Furthermore, game theory models can be computationally intensive, especially when dealing with a large number of market participants and complex interactions.

The primary benefit of auction-theory-based models is their capacity to swiftly establish equilibrium between the supply and demand curves with little computational work. Auctions provide a transparent process for price discovery in local flexibility markets. They enable market participants to assess the value of flexibility and determine appropriate pricing, allowing for a more efficient and competitive market. The biggest disadvantage is that excessive competition may result in unintended auction price spikes. Moreover, auction theory models are susceptible to market power and collusion risks. Dominant market players or collusion among participants can manipulate bidding strategies to gain an unfair advantage, compromising the competitiveness and integrity of the market.

Since multiagent-based modelling is a simulation, its primary distinguishing feature is that it attempts to mimic potential human behaviour and display it using mathematical models. Simulation models provide a detailed and realistic representation of local flexibility markets. They can capture complex interactions among market participants, considering various factors such as demand, supply, pricing mechanisms, and grid constraints. However, building accurate simulation models requires extensive and reliable data on market participants, infrastructure, demand patterns, and other relevant parameters. Obtaining and maintaining such data can be challenging, especially in emerging local flexibility markets where data availability might be at its early stages.

8.5.2.4 Clearing Methods

Before we begin, it is important to notice that the clearing methods and the market models are intrinsically connected so when opting for a model it is important to also choose a clearing method that matches the model and vice-versa. The clearing methods for local flexibility markets may be divided as follows:

Centralised optimisation

As the name implies, this is the market clearing method used in the explored earlier *centralised optimisation model*. It consists of an optimisation program that searches for the ideal value to minimise or maximise the objective function using a specific algorithm over a given time horizon, and with a set of various constraints. Depending on the constraints, the optimisation problem to solve may be considered by using linear programming (LP), mixed integer linear and non-linear programming (MILP; MINLP), quadratic programming (QP) and mixed integer quadratic programming (MIQP).

The centralised optimisation methods can be solved by direct algorithms and indirect algorithms in accordance with the mathematical characteristics of the optimisation programs. *Direct Algorithms* gather the optimisation problems that can be solved directly with commercial solvers like CPLEX and GUROBI. In contrast, *Indirect Algorithms* need to be slightly altered and converted to a format that can play well with the existing solvers. In general, linear convex centralised optimisation problems and those transformed into that format are solved by direct algorithms. On the other hand, when network limits are taken into consideration, the indirect approach is typically utilised. This rule should be followed so that congestion and voltage issues are taken into account.

Decomposition methods

When dealing with a large number of market participants escalating from small models to huge ones may be hard to accomplish. To bring the computational burden down, a common solution is to break down these large models into small sub-problems. This is how Decomposition Methods work. Dividing and solving sub-problems brings down the burden in computational work, and decentralises the efforts to each respective sub-problem.

One decomposition method is the *Decomposition based on augmented Lagrangian relaxation* which works perfectly with scalability regardless of the number of constraints. However, when the problem is non-convex and has a dual gap, it may struggle to be efficient. When the problem has these features, an *augmented penalty function* is usually applied. Based on this technique, four decomposition methods were developed, namely the *Alternating Direction Method of Multipliers* (ADMM), the *Analytical Target Cascading* (ATC), *Proximal Message Passing* (PMP) and *Auxiliary Problem Principle* (APP). Another decomposition method is the *Decomposition based on Karush-Kuhn-Tucker conditions* (KKT) which looks at first-order KKT optimality conditions and solves them in a decentralised fashion.

Bi-level optimisation

Bi-level optimisation is a suitable solution for situations directed towards processes that require large-scale optimisation and decision-making. They address hierarchical organisations with a leader at the top level and a follower at the lower level, whose actions directly affect the upper-level leader. It is defined as a mathematical program, where an optimisation problem is constrained by another optimisation problem, in addition to the conventional constraints.

Bi-level problems are tackled either using single-level reduction techniques (with KKT conditions or KKT conditions and duality theory) or nested methods. KKT conditions are sufficient to reduce the problem to a single-level equivalent and solve it using readily accessible commercial solvers when both the upper- and lower-level problems are linear. However, they cannot solve bi-level issues by themselves when the upper level is non-linear. In this case, dual theory can help to linearise the upper level before solving the problem. In addition, there are nested methods. They are better applied in bi-level problems when the lower-level problem is non-linear. Depending on the mathematical properties, the nested technique works with each level using the suitable optimisation procedure. It comes, however, at a high burden in mathematical work.

8.5.3 Market Platforms

A decent amount of literature refers to local flexibility markets initiatives pioneered in other European countries, their respective platforms, and examples of other structures and models that can enhance network flexibility. Below are some of these initiatives and other analyses.

8.5.3.1 NODES Marketplace

The NODES marketplace platform is one of four pioneering flexibility marketplace platforms. It is currently being implemented in a number of locations including Northern Europe and the Baltic region and emerged as a joint venture between energy market Nord Pool and Norwegian energy group Agder Energi. Its main function is to make flexibility trading easier while taking into account the location of available resources.

To achieve this goal, NODES, which acts as an independent market operator, locates the resource and introduces a tag with that location, and then assigns another with a price determined according to the potential flexibility value of the analysed resource. Thus, it manages to provide conditions for safe trading at transparent prices.

Flexibility providers—mostly BRPs, but aggregators, microgrids, and other individuals can also be considered—bid for sale on NODES, instead of the owners of flexibility resources, with a model defined by them and by the owners themselves and with the technology that allows the activation of flexibility on the part of the buyers.

Bids can be differentiated, according to how they will be sold, either centrally or locally. Local sales are restricted to specific places in the network and, thus, more risky, as only a few hundred hours are needed each year. Therefore, central selling is usually more advantageous, as it allows it to be used for necessary re-balancing on a larger scale. For this, each bid is included in a catalogue that provides location, availability, time, profile, and order parameters.

The last market entity is the flexibility buyer. This entity can range from system operators like TSOs, DSOs to also BRPs. They present bids through the NODES marketplace platform to buy flexibility in a given location. They can also filter within the flexibility catalogue available on the platform in order to choose those that best suit their needs. For example, DSOs can hire local flexibility to solve grid problems in the form of increased consumption rather than causing renewables to be curtailed. TSOs, on the other hand, gain the ability to contract flexibility that would otherwise be out of their reach in traditional markets, with preference given to activation times and ramping.

The way NODES works makes flexibility products non-standard. This means that, while there may be a better response to more specific flexibility needs, they may not be liquid enough for subsequent activities, due to their lack of standardisation. As such, they make it more challenging to build an order for competitive organisation between products, price transparency and competition between providers.

8.5.3.2 ENERA and the EPEX SPOT Local Flexibility Market

ENERA first appeared as part of SINTEG, a publicly funded program in northern Germany. The main objective was to develop and prove that there could be a solution for an energy supply that not only was scalable but was also aware of ecological and environmental issues, safe and economical with a focus on the role and responsibility of the energy sector, regulatory framework, grid tariffs, and local flexibility markets. Its structure was integrated in 2018 as part of the cooperation agreement between the European Power Exchange (EPEX SPOT) and an energy group, EWE AG, with the aim of addressing the problem of grid congestion, establishing a demonstrable and scalable practical case (in the northeast of Germany).

The EPEX SPOT local market platform is an open, voluntary, market-based network congestion control platform for flexibility providers that centralises local flexibility offerings so that they can easily be used by TSOs and DSOs for the purpose of alleviating congestion. EPEX SPOT acts neutrally as an intermediary between suppliers and buyers, overseeing the establishment of prices and ensuring transparency.

The platform is active during the operation period of the intraday market, separate from other existing markets, and the access interface is universal, i.e., suppliers (aggregators, asset managers and exchanges) and buyers (TSOs and DSOs) use the same medium.

Suppliers can continuously submit bids in their respective order books and buyers can also continuously submit demand orders in the order book that correspond to the market area where flexibility is needed. The buying and selling process is then carried out in a similar way to that practised in several intraday markets across Europe: offers in the same order book are automatically and continuously matched and remunerated according to the pay-as-bid pricing scheme. Flexibility products are defined and standardised by EPEX SPOT itself, in collaboration with network operators, taking into account ascending and descending blocks of energy, during a specific period and location.

The objective is to create conditions for coordination between system operators at the various levels of the electricity grid, depending on a high level of digitalization and automation. Thus, participants are able to add value to their assets, avoiding a potentially more expensive increase in the physical network and allowing for greater reliability, security and coordination.

8.5.3.3 PICLO Flex Marketplace

PICLO Flex emerged from a project funded by the UK government but developed by an independent software company. The goal of the project, aimed at buyers and suppliers of flexibility in the UK, is to provide a smart, flexible, and clean energy system, through the platform that allows flexibility providers to promote their services, as well as encourage new initiatives and models of business.

The pilot phase of the project revealed that both groups of flexibility suppliers and buyers that participated in the platform are quite diverse groups. The suppliers are mainly composed of aggregators, electricity providers, and industrial and commercial users. Also included in this group, although to a lesser extent, are community participants and individual industrialists. But the biggest novelty that happened was the introduction of speculative users. More than 30% of Piclo users were potential investors in flexibility services.

Flexibility buyers included the six British DSOs—UKPN, Scottish and Southern Electricity Networks, Electricity North West Limited, Northern Powergrid, SP Networks, and Western Power Distribution, each using the platform in a particular way. While some used the platform as an improvement of the processes of the existing markets, others used the platform in-depth, making use of the auction's functionalities and seeking to satisfy their requirements in a standardised format. The major causes for the need for flexibility were reinforcement delays (deferral reinforcement), unplanned, pre- and post-failure network interruptions, and also, albeit to a lesser extent, planned network maintenance. During the pilot period, PICLO Flex saw demand for more than 456 MW of flexibility spread across the six DSOs.

In short, PICLO Flex was able to demonstrate that an online platform could be used by several DSOs in order to help maintain the electricity grid, using only distributed resources for deferral reinforcement, planned maintenance, and unplanned interruptions. It appears, however, that PICLO Flex is used only by DSOs so far, with cooperation with the TSO being limited.

In late 2022, the Portuguese DSO partnered with PICLO Flex to launch FIRMe, a local flexibility pilot project. It is being made as a proof of concept to test the approach of local flexibility markets and if the respective market has the due acceptance. While the market is organised by the DSO, which acts as the market operator, the platform for procuring flexibility is the PICLO platform, as a market-based framework.

The operation of the local flexibility market in Portugal is similar to the successful flexibility markets that PICLO has helped implement in the UK and Europe. This market allows companies, consumers, and prosumers to leverage their assets, such as distributed resources, renewables, and electric vehicles, to assist the DSO in managing network stress. Participants will be compensated by the DSO for modifying their generation or demand patterns to alleviate network stress, thereby providing an innovative way to balance supply and demand in the energy market.

8.5.3.4 IREMEL and the Iberian Electricity Market

The last of the four pioneering initiatives is IREMEL. It was launched by the Spanish electricity market operator, OMIE, and IDEA, the Institute for the Diversification and Saving of Energy. Its main objective is to develop a local market model in order to facilitate the efficient integration of DERs such as renewables, prosumers, and storage facilities. It also promotes the participation of DERs as a solution to localised congestion and other needs of DSOs. Other secondary objectives are the development of prototypes of local markets, identifying challenges and opportunities for the role of prosumers and consumers in the market, and demonstrating the feasibility of new technologies to facilitate the management of distributed resources in participating in local markets.

Flexibility providers involved in IREMEL include DSOs, individual DERs, aggregators, proactive consumers, battery producers, etc. Two groups of products are considered, short-term (on order, only when necessary) and long-term for structural problems (long-term contracts, months or years). IREMEL currently has 5 pilot projects in different areas of the Spanish market, with different participants and under different conditions as well.

8.6 Conclusion

In the dynamic landscape of emerging energy markets, one thing is clear: the future of energy is evolving rapidly. As we conclude our exploration of this chapter on "Emerging Energy Markets", we find ourselves at the intersection of innovation, sustainability, and adaptability. In this concluding section, we will delve into some key takeaways and emerging trends that are shaping the energy markets of tomorrow.

We showed that possible alternatives to current marginalist energy markets, which are not adequate for the future energy markets dominated by a high share of renewables, are market segmentation, flexibility compensation, and co-optimisation.

One of the significant shifts in energy markets is the recognition of the diverse energy sources available. While renewables have rightfully gained prominence due to their sustainability benefits, dispatchable energy sources such as natural gas and advanced storage technologies play a crucial role in maintaining grid reliability. The emerging energy markets must acknowledge the importance of both and find a way to split the market effectively, ensuring a balanced and resilient energy ecosystem.

Flexibility is the cornerstone of modern energy systems. To address the intermittency of renewables and maintain system reliability, innovative approaches like Equivalent Firm Power Capacity Auctions and co-optimization have gained traction. These mechanisms incentivise providers to deliver not only energy but also the crucial flexibility required to keep the grid stable, especially during peak demand and unexpected fluctuations.

In the new paradigm of energy markets composed of a predominant share of renewables, new renewable business models, that seek for a proper remuneration of the energy services provided by the renewables, must be found.

Power-to-X (PtX) is a captivating and versatile concept that holds the key to unlocking the full potential of renewable energy sources. The “X” in PtX can represent a multitude of valuable products, including ammonia, chemicals, fuel, gas (hydrogen or methane), liquid synthetic fuels, food, syngas, or even the direct generation of electricity, offering an array of solutions to address the world’s pressing energy and environmental challenges. At its core, PtX is a process that converts surplus electricity, often generated from renewable sources like wind or solar, into various energy carriers or commodities. This transformation not only enhances energy storage capabilities but also facilitates the integration of intermittent renewables into existing energy systems.

Local and sleeved Power Purchase Agreements (PPAs) are revolutionising how renewable energy projects are financed and supported. These agreements enable organisations to directly purchase renewable energy from nearby sources, promoting local sustainability and reducing transmission losses. Such localised arrangements bolster the economic viability of renewable projects, making clean energy more accessible.

Virtual Power Purchase Agreements (VPPAs), particularly those utilising Contract for Difference (CfD) structures, allow companies to invest in renewable energy without the need for physical infrastructure. By committing to buy renewable energy at a predetermined price, businesses can support clean energy projects, thus contributing to both environmental sustainability and their bottom line.

The rise of Peer-to-Peer P2P energy markets represents a shift in energy governance. Full P2P, community-based P2P, and hybrid P2P models enable individuals and communities to trade energy directly with one another. These markets empower consumers to take control of their energy choices, fostering a sense of ownership and autonomy within the energy ecosystem.

Flexibility services models are emerging as a crucial component of modern grid management. These services encompass congestion management, voltage control, and the provision of grid services. As the energy landscape becomes increasingly decentralised, flexibility markets will play a pivotal role in ensuring grid stability.

In this evolving energy landscape, a diverse range of actors has emerged. From traditional utilities to tech startups and renewable energy developers, the ecosystem is expanding. Market platforms have emerged as essential facilitators, connecting buyers and sellers, and streamlining transactions. These platforms enhance market efficiency and transparency.

To ensure fairness and efficiency, emerging energy markets are adopting advanced market clearing methods. These methods consider factors such as real-time supply and demand, pricing dynamics, and the unique characteristics of renewable energy sources. These approaches are critical to achieving optimal market outcomes.

Local flexibility markets are gaining momentum, allowing communities to tap into their local resources effectively. By leveraging distributed energy resources and local demand response, these markets reduce grid congestion, improve reliability, and promote energy resilience at the community level.

Flexibility market platforms like NODES, ENERA, PICLO, and IREMEL are pioneering the transformation of the energy landscape. These platforms serve as pivotal hubs where the supply and demand for flexibility services converge, ushering in a more dynamic and efficient energy ecosystem.

In conclusion, the future of emerging energy markets holds immense promise. By recognising the need to balance renewables with dispatchable sources, compensate for flexibility, and adopt innovative business models, we can create a sustainable, resilient, and equitable energy future. The proliferation of P2P energy markets, flexibility services, and the involvement of diverse market players and platforms are propelling us toward a more decentralised and democratic energy landscape.

MIX
Papier aus verantwortungsvollen Quellen
Paper from responsible sources
FSC® C105338

If you have any concerns about our products,
you can contact us on
ProductSafety@springernature.com

In case Publisher is established outside the EU,
the EU authorized representative is:
Springer Nature Customer Service Center GmbH
Europaplatz 3, 69115 Heidelberg, Germany

Printed by Libri Plureos GmbH
in Hamburg, Germany